P9-DUF-661

BASIC LEGAL RESEARCH
FOR
CRIMINAL JUSTICE AND
THE SOCIAL SCIENCES

James R. Acker, JD, PhD
Associate Professor
School of Criminal Justice
Nelson A. Rockefeller College of Public Affairs and Policy
University at Albany
State University of New York
Albany, New York

Richard Irving, MA, MLS
Bibliographer, Public Affairs, and
Reference Librarian
Thomas E. Dewey Graduate
Library for Public Affairs and Policy
University at Albany
State University of New York
Albany, New York

AN ASPEN PUBLICATION®
Aspen Publishers, Inc.
Gaithersburg, Maryland
1998

This publication is designed to provide accurate and authoritative information in regard to the Subject Matter covered. It is sold with the understanding that the publisher is not engaged in rendering legal, accounting, or other professional service. If legal advice or other expert assistance is required, the service of a competent professional person should be sought. (From a Declaration of Principles jointly adopted by a Committee of the American Bar Association and a Committee of Publishers and Associations.)

Library of Congress Cataloging-in-Publication Data

Acker, James R., 1951-
Basic legal research for criminal justice and the social sciences /
James R. Acker, Richard Irving.
p.   cm.
Includes bibliographical references and index.
ISBN 0-8342-1013-4
1. Criminal justice, Administration of—United States—Legal research.
2. Legal research—United States.   I. Irving, Richard D. II. Title.
KF241.C75A28   1998
340'.07'2073—dc21
97-40978
CIP

Copyright © 1998 by Aspen Publishers, Inc. All rights reserved.

Aspen Publishers, Inc., grants permission for photocopying for limited personal or internal use. This consent does not extend to other kinds of copying, such as copying for general distribution, for advertising or promotional purposes, for creating new collective works, or for resale. For information, address Aspen Publishers, Inc., Permissions Department, 200 Orchard Ridge Drive, Suite 200, Gaithersburg, Maryland 20878.

Orders: (800) 638–8437
Customer Service: (800) 234–1660

About Aspen Publishers • For more than 35 years, Aspen has been a leading professional publisher in a variety of disciplines. Aspen's vast information resources are available in both print and electronic formats. We are committed to providing the highest quality information available in the most appropriate format for our customers. Visit Aspen's Internet site for more information resources, directories, articles, and a searchable version of Aspen's full catalog, including the most recent publications: **http://www.aspenpub.com**
Aspen Publishers, Inc. • The hallmark of quality in publishing
Member of the worldwide Wolters Kluwer group

Editorial Services: Ruth Bloom
Library of Congress Catalog Card Number: 97-40978
ISBN: 0-8342-1013-4

Printed in the United States of America
1   2   3   4   5

To **Jenny**, **Elizabeth**, and **Anna**,
whose understanding and caring are more than I deserve,
and mean more than I can say.

J.R.A.

To **Suzanne**,
for her patience and support.

R.D.I.

Table of Contents

Preface

So why have we written this book, and what's in it for the reader?

One of the book's authors, James Acker, has a law degree and teaches law at the University at Albany School of Criminal Justice. The other author, Richard Irving, has a master of library science degree and is a reference librarian at the University at Albany Graduate Library for Public Affairs and Policy, which includes collections from criminal justice, information science, political science, and other public–policy-related disciplines. We both hold graduate degrees in criminal justice. We have long appreciated that neither one of us knows all there is to know about law, legal research techniques, or criminal justice. By collaborating on this book, we have attempted to combine our strengths and compensate for our respective weaknesses to produce a volume that will help demystify legal research for students and professionals in criminal justice—or virtually any other academic or job-related calling.

The "law" entails a massive storehouse of information. No one person can ever really expect to know more than a small slice of all the concepts and rules embodied in that term. Even relatively familiar legal terrain is likely to be prone to both incremental and rapid change and thus require ongoing study. Becoming familiar with the references that are available to report, describe, analyze, and update different sources of the law, and understanding how to use those references, are the keys to unlocking the law's mysteries. This is what legal research is all about, and why learning how to do it can be so valuable.

Legal research techniques are not difficult to learn. On the contrary, they can be acquired relatively quickly, and they pay rapid dividends. Anyone who is open to a bit of hands-on experimentation and is willing to practice can quickly assimilate and then polish legal research skills.

This book is designed to familiarize readers with printed and computerized legal references and research systems. It explains and illustrates how these sources can be used to help answer questions of law and to assemble other law-related information. In the ensuing pages, we describe many references and offer numerous examples in explanation of the legal research process. Virtually all of these examples involve legal issues related to criminal justice, which allows us to consult references and databases that will be of special interest to those studying or working in this field. However, because legal research strategies truly know no disciplinary boundaries, the authorities and techniques described here should prove useful no matter what specific issues are explored. We cover how to find "black letter" law as well as social-scientific, statistical, and other empirical information about the law.

Although we focus on other resources, we do not mean to slight the Internet, which has many sites valuable to legal research. We appreciate that some researchers will not have access to all of the resources referred to in the book, such as the WESTLAW and LEXIS® systems, which are expensive for private subscribers. Access to U.S. Supreme Court cases and other law-related materials via the Internet can be helpful and undoubtedly is better than no access at all. However, the legal research process emphasized in this book relies extensively on elaborate classification systems and cross-referencing techniques. Legal information on the Internet, even at sites that have attempted some degree of organization, currently is not integrated in a manner that compares to the legal research systems on which we focus. An additional concern is the accuracy and timeliness of information accessed on the Internet, a topic best addressed elsewhere.

Many of the search strategies illustrated in this book exemplify the additional research capabilities made possible by computer technology. The applications we have described are meant to be illustrative rather than exhaustive and will stimulate readers to examine the extensive potential of computer technology. One of the difficulties in describing computer applications is that the software is constantly improving. Both WESTLAW and LEXIS have come out with upgraded software since this text was written. While the newer software offers improvements for the user, the basic search techniques we describe remain essentially unchanged. Of course, the best research tool continues to be the gray matter located between the researcher's ears. We have tried to emphasize throughout this text the importance of critical thinking in the research process.

The book is divided into four chapters. Chapter 1 provides the foundation for the discussion of legal research references and techniques that follows. In this chapter we distinguish secondary and primary legal authorities, identify finding tools, describe the different sources of the law, provide a quick overview of how a case progresses through the state and federal court systems, discuss proper legal citation format, and explain why the art of framing the research question is so important to the overall research process. Chapter 2 introduces three novel and, we hope, interesting hypothetical cases that will serve as research examples in Chapters 2 and 3. We identify and show how to use secondary legal authorities (Chapter 2) and primary legal authorities and corresponding finding tools (Chapter 3), in both print and computer sources, by working through solutions to our hypothetical problems. We encourage readers to go to a library or to sit down at a computer terminal to work through these problems. And through the use of numerous exhibits depicting the various sources, we also do our best to present both the printed and electronic sources to the reader. Chapter 4 invokes 10 research questions in the form of hypothetical problems. We invite readers to devise their own research strategies to produce answers to these issues, and we describe and illustrate our way of trying to solve them.

By the end of the book, those readers who have combined a fair measure of practice time in the library stacks and at the computer with the tips and information provided in these pages should be well on their way to mastering basic legal research techniques. Through using these techniques, you will have almost unlimited capacity to answer questions of law and uncover information about the law as it relates to criminal justice and countless other subject areas. We hope you enjoy the process of legal research and benefit from its products.

James R. Acker
Richard Irving

Acknowledgments

The authors gratefully acknowledge the assistance of **Ms. Jo Anne DeSilva** for her efforts and skill in word processing the several drafts of manuscripts that eventually resulted in this book. We also thank **Peter Vonnegut**, whose illustrations have given the book immensely more character than we could accomplish through words alone, and for all of his assistance with the technical aspects of computer applications that helped this book take shape.

1 The Building Blocks for Legal Research

A WORD ABOUT OBJECTIVES AND METHODS

This book describes legal reference materials, both printed sources and computer databases, and explains how these references can help you discover answers to questions of law and enable you to expand your legal knowledge. Legal research skills are not particularly difficult to acquire. In fact, many techniques for finding the law are quite straightforward. Legal research is made possible by a highly organized, interconnected web of reference materials and indexing systems. After you are introduced to the books, computer options, and finding tools, and have gained a bit of practice working with them, in no time you will be tracking down answers to virtually any legal questions.

Although this book focuses on legal research within the criminal justice system, the same techniques and many of the same references we discuss can be used to research other issues of law, from admiralty to zoning ordinances. The basic research steps are common to questions of law arising in all academic disciplines and professions.

The first encounter with a law library or computer system can be intimidating. Without guidance, finding the answer to a question of law in Harvard Law School's 1.5 million-volume library would be like searching for the proverbial needle in a haystack: It would be impossible to know where to begin. Computer databases can be just as daunting. The sheer quantity of legal information, coupled with its rapidly changing content, means that no one—not even a Supreme Court Justice, law school professor, or Wall Street attorney—can ever know the answers to all legal questions. But the judge, the professor, and the lawyer possess something as valuable as knowledge itself: They have the tools and the know-how for mining information that can produce answers to questions of law. They have mastered the basic research techniques, and so can you.

Understanding the fundamentals of legal research can be invaluable to anyone interested in criminal justice, and students, researchers, and practitioners ignore these issues at their peril; in fact, they can ill-afford *not* to know how to find and use legal reference materials. Crimes and punishments are defined by statute. Po-

lice, prosecutors, courts, corrections personnel, and other criminal justice officials are required to abide by administrative, legislative, and constitutional rules. Appellate courts are continuously engaged in the process of interpreting and determining the constitutionality of laws and assessing the legality of conduct under those laws. Administrators who do not know what the law is, or who abuse it, run the ever-present risk of lawsuits, which can be costly and time-consuming. Empirical research has contributed to our understanding of the formation, evolution, and application of laws, but much research remains to be done on a wealth of legal issues.

Describing legal research techniques in the abstract would be a singularly dull business, as well as unproductive. If you want to learn how to hit a golf ball, or sketch a portrait, or change the oil in your car, there is no doubt that listening to an expert or reading a book or an owner's manual can help you get started. At the same time, if you really want to become proficient at these activities, there is no substitute for driving a bucket of balls, applying pencil to paper, or getting your hands dirty under the hood. Legal research is no different. There will be times when you simply must go to the library or sit down at a computer terminal for some hands-on experience with the sources and techniques described in this book. We subscribe to this philosophy:

> *Tell me, and I will forget.*
>
> *Show me, and I will remember.*
>
> *Involve me, and I will understand.*

Accordingly, as we describe the different law books and databases and tips for using them, we rely heavily on examples, sample problems, and illustrations, which we hope will involve you in the legal research process. Searching for answers to questions of law often resembles detective work. We do not wish to mislead you into believing that bold-print answers will jump out at you from books or that computers will instantly display search results, or even that unambiguous solutions always exist to the legal issues you are investigating. To the contrary, definitive answers may be elusive, and search strategies occasionally run into deadends or take frustrating twists and turns. Creativity, persistence, and hard work are often necessary. The research process can be challenging and, like solving other mysteries, even fun.

Before we jump into a discussion of techniques, we have a few preliminary matters to cover. We begin with a brief overview of the reference materials you will be using in your research. We distinguish between primary and secondary legal authorities and finding tools. We then identify the computerized legal databases with which you should be familiar and discuss social science references that may be of interest for law-related research. We also outline the basic structure of the state and federal judicial systems that coexist in the United States and describe how a typical case progresses through these court systems. The conventions for citing published judicial decisions are explained in this process. When we finish with these introductory matters, we will be ready to explore in much greater detail the available legal reference sources and how these references are utilized.

LEGAL REFERENCE MATERIALS AND SOCIAL SCIENCE RESOURCES

Primary and Secondary Legal Authorities and Finding Tools

In the following chapters we identify an assortment of legal authorities, or sources of information about the law, and rely on examples to help explain how to use these authorities. Some reference materials are classified as **primary** legal authorities, and others are considered **secondary** legal authorities. Still other references are not legal authorities at all but **finding tools** that are used to locate, update, or confirm the continuing validity of different sources of law. These distinctions would not be terribly important if we were concerned only with the process of legal research: No matter what they are called, books are books, and computer commands are computer commands. Yet these classifications are of tremendous significance to the end product of legal research, which is to provide the most authoritative answers to the questions we are examining.

Primary legal authorities are "the law." They form the body of rules that are legally binding on people and that are enforceable in the courts. Of course, not all laws are of equal stature. There are different kinds of primary legal authorities, which have lesser or greater degrees of "clout" or authoritative value. The four kinds of primary legal authorities we consider are: **administrative regulations, statutes, constitutions,** and **judicial decisions** (case law).

In contrast to primary legal authorities, secondary legal authorities have no power to create legally enforceable rights or obligations and thus have no claim to being "the law" as we normally define that term. The secondary legal authorities we consider include **law dictionaries, Words and Phrases, legal encyclopedias, legal treatises** and **casebooks, court briefs, American Law Reports, law reviews,** and other law-related periodicals. Although these references may contain great wisdom and be informative or persuasive, you will not be able to rely on them for authoritative answers to the legal problems you are trying to solve.

Secondary legal authorities serve three principal functions. They typically (1) describe what the law is, (2) explain, analyze, or criticize different aspects of the law, and/or (3) assist us in finding primary authorities and other secondary authorities. But they do not pretend to be, and are not considered to be, binding sources of law. You should never rely on just a law review article—even a very good one—or any other secondary legal authority to establish a point of law. No one has ever been arrested, granted a legal right, or successfully prosecuted solely on the strength of a secondary legal authority, and we are confident that this practice will not soon change.

As their name implies, finding tools help you locate primary and secondary legal authorities. Finding tools are not a form of legal authority, and they should not be cited as such. However, do not underestimate their usefulness for legal research projects. They help you begin your research; they can serve as connecting links between different authorities; and they are necessary to confirm that the authorities you have found remain current and valid. Finding tools include various indexes, digests, and a set of volumes, Shepard's Citations, commonly known as **Shepard's**, which, as we shall see, not only help you find legal references that may be of interest, but allow you to determine whether judicial authority and legislation you have located remain unchanged and effective after the passage of time.

Computer-Assisted Legal Research: WESTLAW and LEXIS

Access to law materials has become increasingly prevalent via electronic media, through commercial online databases, CD-ROMs, and sites available on the Internet. Primary legal documents reside in the public domain and are not subject to copyright law. Hence, many nonprofit groups, mostly affiliated with law schools, have created web sites that make it possible to access the full text of court cases, statutory materials, and even some administrative law documents. One such group is the Legal Information Institute at Cornell University, http://www.law.Cornell.edu. However, in this text we limit our description of electronic legal research techniques, commonly referred to as **computer-assisted legal research** (CALR), to two commercially produced computerized online database systems, **WESTLAW** and **LEXIS®**. We do this because only these two systems have all of the following characteristics:

Reliability. WESTLAW and LEXIS are the products of companies with excellent reputations for publishing American legal documents. WESTLAW is produced by the West Publishing Company, which is this country's premier publisher of federal and state case law and statutory material. LEXIS is based largely on the print publications of Lawyers Cooperative Publishing Company, another highly respected publisher of United States law materials. Indeed, most of the print publications covered in this text are published by one of these two companies. Having accurate and up-to-date information is crucial in legal research. These companies have developed systems for updating the information contained in their publications, and they have earned reputations for accuracy.

Comprehensive Coverage. WESTLAW and LEXIS come close to providing comprehensive coverage of federal and state law. Both provide full-text access to reported cases, both federal and state, as well as to the statutory law for the federal government and all the individual states. They also provide access to federal administrative law documents and some state administrative law sources. In addition, both WESTLAW and LEXIS are constantly building on their collections of full-text secondary sources, such as law reviews.

Indexing and Cross Reference Features. Since these online databases are produced by the same companies that publish most of the printed sources we will be covering, the indexing and cross reference features used in the print sources also are available in the online databases. For example, the **key number** system, which West uses to classify cases by legal subject in its printed sources, can also be used in WESTLAW. This makes it easy for the researcher to switch back and forth between the printed and computer versions of materials.

Additional Search Methods. Both LEXIS and WESTLAW have developed software that provides the researcher with additional access points to documents in the database. For example, the online versions allow the researcher quickly to locate a case in which a specific judge was the primary author of the decision, or in which a particular attorney participated. The software also allows the researcher to search for relevant documents using key words or phrases, in contrast to the subject-based researching regimen required in printed sources.

We should briefly mention some of the advantages and disadvantages of CALR. The most obvious advantage is the physical convenience of being able to access information through a computer. Those who are fortunate enough to have access to WESTLAW or LEXIS through computers in their offices or homes will save many trips to the library. Even if you have to go to the library, the computer saves on constant trips to the shelves to retrieve books and running to the photocopier to make copies of documents. In short, it can be a real time-saver.

Another important advantage is the speed with which new information is accessible through electronic media. For example, WESTLAW and LEXIS make available the full text of U.S. Supreme Court decisions on the same day the decisions are announced.

A third advantage is the greater flexibility afforded the researcher in formulating a research query. The major benefit is the ability to move away from the subject-based access dictated by printed sources to the use of key words or phrases. The subject-based system used in the printed sources is designed to help the researcher find the law as it applies to a particular situation in a specific jurisdiction. Criminal justice practitioners and researchers frequently will rely on legal research designed to find the law. However, they also may wish to compare how different jurisdictions have addressed recurring factual issues or analyze trends in the criminal justice system. Since the indexing in printed finding tools tends to be based on legal principles instead of on fact patterns, the key-word searching capability of computerized databases can facilitate the latter types of searches.

The primary disadvantage of CALR is that some sources are not yet covered. For example, WESTLAW does not provide access to either **Corpus Juris Secundum** (West's encyclopedia of American law) or **Words and Phrases** (a West publication that provides definitions of words and phrases as used in judicial decisions).

A second disadvantage is that CALR can impede the understanding of the legal process, which is reinforced by printed sources. Printed sources are designed in such a systematic and interconnected way that by using them the legal researcher not only will arrive at an answer to the issue being researched, but in doing so will reinforce his or her understanding of the legal process. To the extent that CALR allows the researcher to deviate from the system imposed by printed sources, some knowledge of the legal process itself may be sacrificed. This may be a minor consideration for law students or lawyers, but it deserves some consideration

for criminal justice administrators, field workers, and researchers who are not formally trained in the law.

Another potential pitfall of CALR is the key-word searching capability, which is a double-edged sword. Properly done, key-word searching can be effective, but researchers also must be wary of potential problems. To be effective, a key-word search requires the use of appropriate words and phrases. The researcher must be familiar with the legal terminology used in connection with particular concepts and be aware of the fact that some terms may have many meanings and appear in a wide variety of contexts.

Ideally, researchers will use both printed and computer sources. The relative effectiveness of one format versus the other depends largely on the issues being investigated. Part of the process of formulating a research strategy should include consideration of whether printed sources or computer sources make the best starting point.

Social Science Materials Related to Law and Criminal Justice

Although the primary purpose of this text is to familiarize users with legal research techniques, we recognize that the law does not operate in a vacuum. The law is an instrument for regulating conduct among individuals, groups, and institutions in our society. Criminal justice researchers and practitioners not only must know what the law is when applied to particular fact situations, but they also must be able to determine the intent of the laws, assess how they are implemented, and be able to evaluate the effectiveness of particular laws.

Since much social science research involves the study of human behavior, lawmakers (legislators) and law interpreters (judges) understandably may rely on social science research to help support their actions. Social science research can be classified in many different ways, but for our purposes it is helpful to divide such activity into two categories: descriptive research and applied research. Descriptive research indicates the prevalence of a social phenomenon within society. Applied research attempts to evaluate societal responses to particular social phenomena. Both descriptive and applied research make use of similar theoretical constructs and research methodologies. For example, descriptive research dealing with the topic of spousal abuse might examine the changing patterns of abuse within our society, whereas applied research might explore the effectiveness of criminal sanctions in deterring spousal abuse.

Legislators may be interested in both types of social science research. Descriptive research can help identify social problems that may be alleviated through legislative action. Applied research can serve as a guide for crafting an effective legislative response to a problem and may be particularly helpful in assessing the effectiveness of existing laws.

For example, social science research is frequently cited in evaluating the efficacy of criminal sanctions to control so-called "victimless" crimes, such as drug use, prostitution, and gambling, or to shed light on whether harsher sentences influence the crime rate. Indeed, when legislative bodies hold hearings on proposed legislation, they often solicit the opinions and/or advice of social scientists regarding the proposals. Judges also may rely on social science research for their decisions. For example, the courts may be interested in research that addresses whether a law is being applied in a discriminatory manner, whether sobriety checkpoints are effective in detecting or deterring drunk driving, or whether expert testimony about eyewitness identification would assist jurors in criminal trials.

There are five principal disciplines in the social sciences: anthropology, economics, political science, psychology, and sociology. Each discipline has its own body of literature, consisting mainly of academic journals and books devoted to the subject. Some disciplines have developed particular research methodologies, although the same basic methods are common to all of the social sciences.

Indexing and abstracting sources have been created to provide access to the literature for each discipline. For example, Psychological Abstracts (with a CD-ROM version called PsycINFO) provides bibliographic citations and abstracts to journal articles and selected books in psychology. Sociological Abstracts (CD-ROM version called Sociofile) does the same for the sociology literature. Other indexes, such as the Social Science Index, provide access to the literature of all the social science disciplines, although they do not include as many references as the discipline-specific indexes.

The social sciences also include "fields of study" that are interdisciplinary in nature. Instead of relying on unique theories or research methods, fields of study borrow from the social science disciplines and occasionally from the physical sciences and humanities. They also develop their own bodies of literature. Criminology is a field of study within the social sciences that focuses on criminally deviant behavior and societal responses to that behavior.

The criminal justice field encompasses criminology as well as the study of the criminal justice process, including law and administrative issues. The criminal justice literature is correspondingly expansive, ranging from criminology to professional and management literature to law. Indexing and abstracting sources provide access unique to the criminal justice literature. Criminal Justice Abstracts, formerly called Crime & Delinquency Literature, provides bibliographic citations and abstracts for selected journal articles, books, and research reports in the field. It is available in paper format and on CD-ROM, and is searchable as an online database through the WESTLAW system.

The **NCJRS Database** is another major indexing and abstracting resource for criminal justice literature. It is based on the holdings of the library at the National Criminal Justice Reference Service (NCJRS), which is affiliated with the National Institute of Justice. It provides references to journal articles, books, book chapters, research reports, and agency documents. It is available in CD-ROM format and also can be searched as an online database. Many of the recent documents are available at the NCJRS Internet site, http://www.ncjrs.org/.

THE COURT SYSTEM AND CASE CITATION FORMAT

The State and Federal Trial Courts

Each state has its own court system. One or more federal courts also are located in each state and in the District of Columbia. The courts in the state and federal judicial systems include trial courts and generally one or two levels of appellate courts. State trial court names differ from state to state. They may be called district courts, superior courts, county courts, circuit courts, or, to make matters particularly confusing, supreme courts (in New York). Misdemeanor charges and more serious felony charges usually are resolved in separate divisions of state trial courts.

The trial courts in the federal system are the U.S. district courts. Like their state counterparts, these courts perform other important functions in addition to holding trials. Some 94 U.S. district courts do business in the 50 states, Washington, D.C., and U.S. territories. Many states house only a single federal district court, while other states have two, three, or four courts, depending on the state's size and population. When two or more U.S. district courts exist within a state, they are identified by region, such as the Eastern District, Western District, Northern District, Southern District, or Middle District Court of the state. For example, four U.S. district courts operate in heavily populated New York, in the Northern, Eastern, Southern, and Western districts of the State. Just one federal district court sits in each of the smaller and less populous states, such as Delaware, Idaho, Massachusetts, and South Carolina.

State and federal trial courts have jurisdiction—that is, the authority to decide cases—over different matters, although sometimes their jurisdiction overlaps. The jurisdiction of the state and federal courts varies depending on the type of crime allegedly committed or the subject matter of a civil dispute. Trials for crimes committed in violation of state law are held in state court, and trials for the violation of federal law—such as criminal fraud for nonpayment of federal income tax—can only take place in a federal court. Occasionally, the same conduct is both a state and federal crime, and the accused offender can be tried in either court system or both. For example, a person who tries to shoot and kill the President of the United States in any one of the 50 states commits the crime of attempted murder and can be brought to trial for that offense in the state where the attempt was made. The shooter simultaneously would have committed the federal crime of attempting to assassinate the President, and could be prosecuted for that offense in a U.S. district court as well. Oftentimes civil suits, which seek monetary damages or injunctive relief for conduct such as unlawful discrimination or the violation of constitutional rights, can be litigated in either a state or federal trial court.

The courts that conduct trials also decide pretrial motions in cases, such as a motion to dismiss an indictment because of irregularities with a grand jury, a motion to suppress evidence because a search or seizure was unlawful, or a motion to dismiss a civil suit because it fails to state grounds that would authorize a ruling in favor of the plaintiff. They also decide motions for new trials that are made after a defendant has been convicted of a crime. These postconviction motions typically involve allegations that could not have been resolved on the original appeal because the underlying circumstances were then unknown, or because they involve issues that are not adequately documented in the record available to the appellate court. Examples include claims of newly discovered evidence inconsistent with an offender's guilt, prosecutorial misconduct, and ineffective assistance of defense counsel. Such claims typically involve the presentation of new evidence that requires the trial court to make findings of fact as well as rule on issues of law. Rulings on postconviction motions generally can be appealed.

One very important type of case decided initially in the U.S. district courts is initiated by a **petition for a writ of habeas corpus**. A federal statute grants state prisoners the right to petition a federal district court for habeas corpus relief when their conviction or sentence allegedly was based on a violation of the U.S. Constitution. Prisoners who succeed on their federal habeas corpus petitions normally must be retried, resentenced, or released, although a federal district court's decision granting or denying habeas relief can be appealed. Recent legislative enactments and Supreme Court decisions have made it more difficult for prisoners confined in state penal institutions to present their constitutional claims via habeas corpus, but the consideration of such petitions remains an important part of the U.S. district courts' business.

Judicial systems obviously could not function without the tasks that trial courts perform, but, for legal research purposes, the decisions of trial courts are usually of significantly less interest than are the decisions of appellate courts. Trial courts typically do not announce their rulings through published opinions, which makes it impractical to consult the rulings as legal authority. State trial court judges especially are unlikely to write opinions that become published in the **case reporters**, a collection of written opinions. It is much less unusual for the U.S. district courts to publish case decisions, which then become accessible to legal researchers, although the great bulk of their holdings are not published in the federal case reporters.

The trial courts' principal function is resolving disputes between parties, a task that largely hinges on how factual issues are decided. Trial courts also rule on questions of law, but few of these rulings break new ground. Instead, they typically require the application of established legal principles to the facts of a particular case. When trial courts do make novel legal rulings or apply established law to unusual fact situations, their decisions may be of more general interest. Even so, their opinions may not be widely read or followed. Since the trial courts are at the bottom of the judicial hierarchy, their opinions have little value as **precedent**; that is, they are not binding on other courts. This is simply a fact of judicial life in both the state and the federal court systems.

The federal district courts' authority extends no farther than the geographic district in which each of the individual courts sits. For example, if the U.S. District Court for the Western District of Kentucky interprets a law in a certain way, its ruling may be of interest to other U.S. district courts, such as those in the Eastern District of Kentucky or the Eastern District of Tennessee, but those other courts are not obliged to follow the ruling. In fact, state courts are not formally required to accept a federal district court's interpretation of the law as general precedent. Of course, the state courts must respect a district court's ruling in specific cases, such as those involving state prisoners' habeas corpus claims, unless those rulings are successfully appealed.

Appellate Courts

Appellate courts serve two principal functions: They review trial court decisions for errors that may have been committed in the lower courts' handling of specific cases, and they clarify and announce rules of law that apply generally throughout a jurisdiction.

Although the intermediate and highest appellate courts both perform "error-correction" and "rule-making" functions, there is a decided difference of emphasis in what they do. The error-correction and rule-making responsibilities are loosely divided between the intermediate and highest appellate courts.

In jurisdictions that have intermediate courts of appeals, the losing party in the trial court ordinarily has an automatic right of appeal to that court. The intermediate appeals court's main job is to review the trial record for errors that may have been committed during the trial. (Of course, a prosecutor has no right to appeal in criminal trials that result in an acquittal, because the defendant is protected against double jeopardy.) In contrast, there generally is no right of appeal to the jurisdiction's highest court. The high court ordinarily picks and chooses the cases it will decide based on its assessment of the significance of the issues. It normally exercises its discretionary review powers by granting or denying a **petition for a writ of certiorari**, or **cert. petition**, which simply is a request made by the losing party for the high court to review the lower court's decision.

The high court's decisions frequently involve its rule-making authority, and may result in significant new interpretations or applications of the law. Courts of highest jurisdiction are too busy to review each and every trial court judgment for case-specific error; if they did they would needlessly duplicate the business of the intermediate appellate courts. Accordingly, when the highest appellate court denies a petition for a writ of certiorari, or a request for discretionary review, this does not signal that the lower court's ruling necessarily is correct; it means that the lower court's decision stands. "Cert. denied" by a high court in no way represents a judgment affirming the lower court's decision.

As in the federal court system, many state court systems rely on a three-part structure involving trial courts, intermediate appeals courts, and highest appeals courts, but there are a number of exceptions to this convention. For example, in several states—generally the smaller or less populated ones, including Maine, Montana, New Hampshire, North and South Dakota, Rhode Island, Vermont, Wyoming, and a few others—there are no intermediate appellate courts; appeals go directly from the trial courts to the state supreme court. In most state systems, the intermediate level appeals court is called the **court of appeals**, and the highest court is called the **supreme court**.

A few states have created an appellate court that decides only criminal cases. In both Oklahoma and Texas, the Court of Criminal Appeals is the court of highest authority in criminal cases, and the state Supreme Court makes final decisions in noncriminal appeals. In Alabama and Tennessee, courts of criminal appeals serve as intermediate appellate courts. They make the initial decisions in appealed criminal cases, but their rulings can be reviewed by the respective state supreme courts.

Nor are the names of the different levels of courts always predictable. For example, the highest court in Maryland and New York, as well as in the District of Columbia, is called the Court of Appeals instead of the Supreme Court. The intermediate appellate court in Maryland is known as the Court of Special Appeals, and in New York as the Appellate Division of the Supreme Court. There is no intermediate appellate court in the District of Columbia's municipal court system. The highest court in Maine and Massachusetts is called the Supreme Judicial Court.

The federal judicial system conforms to the normal court structure and uses the standard nomenclature. Appeals from the U.S. district courts are taken in almost all cases to the appropriate circuit court of appeals. The country is divided geographically into 12 federal judicial circuits, consisting of the First through the Eleventh Circuits, and the District of Columbia, or D.C. Circuit. (See Figure 1–1.) As illustrated on the accompanying maps, the numbered federal circuits encompass states in different regions of the United States and its territorial possessions. The U.S. Court of Appeals for the Federal Circuit decides appeals in patent, copyright, and trademark cases, and also decides certain other specialized appeals, including breach of contract claims against the federal government. The other federal courts of appeals handle the appeals taken from the federal district courts within their geographic region. For example, the Fourth Circuit Court of Appeals accepts appeals taken from the U.S. district courts in Maryland, West Virginia, Virginia, North Carolina, and South Carolina.

All district courts are bound by the decisions of the court of appeals that presides over their circuit, but district courts outside of that circuit are not. For instance, the federal district courts in Illinois, Indiana, and Wisconsin must follow the general rules of law announced by the Seventh Circuit Court of Appeals, but district courts in Ohio, Georgia, and elsewhere in the country are not bound by the Seventh Circuit's rulings. They must abide by the decisions of the court of appeals only in their particular circuit. The different courts of appeals occasionally disagree about proper interpretations of the law, thus creating conflicts that may not be resolved until the U.S. Supreme Court decides a case involving the controversial issue.

The U.S. Supreme Court is the court of highest authority in the federal judicial system, and its rulings must be followed by all of the federal courts. In addition, all state courts are bound by the Supreme Court's interpretation of the U.S. Constitution and other federal law. Thus, for some purposes, it is useful to conceive of the U.S. Supreme Court as being at the top of the ladder in both the state and federal court systems. However, there are many state law issues over which the U.S. Supreme Court has no jurisdiction. State court judgments will not be dis-

Figure 1–1

The Thirteen Federal Judicial Circuits. *See* 28 U.S.C.A. § 41.

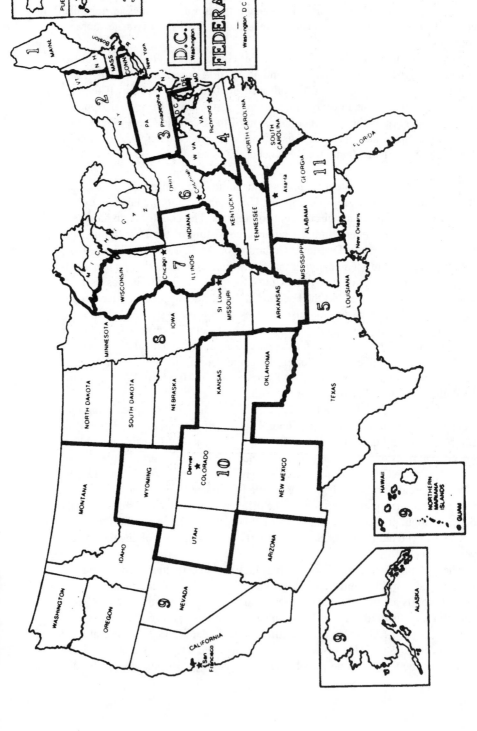

Source: The Thirteen Federal Judicial Circuits—Map of U.S., © West Publishing, used with permission.

turbed by the U.S. Supreme Court when the proper interpretation of a *state* constitution is at issue (as long as federal constitutional rights are not diminished by a state court ruling), or when a state law question—such as the meaning of a state statute or the applicability of a state rule of evidence—involves no federal constitutional issues and does not otherwise implicate federal law.

For example, the Iowa Supreme Court will have final say over whether a state trial court erred when it instructed a jury about the definition of "manslaughter" under Iowa law, or whether a trial court properly admitted a witness's testimony under an exception to the state's hearsay rule. It is the final arbiter of all Iowa's state law issues that do not involve federal constitutional principles or other federal law. Nor can the U.S. Supreme Court disturb a ruling by the Iowa Supreme Court that the Iowa Constitution protects individual rights to a greater extent than the federal Constitution. Of course, state constitutions cannot be interpreted to restrict federal rights. Thus, the Iowa Supreme Court could decide that the Iowa Constitution guarantees a suspect the right to counsel at a lineup when the U.S. Constitution does not so require, but it would not be at liberty to rule that a suspect can be denied counsel when the federal Constitution requires an attorney to be appointed.

Tracing the Path of a Hypothetical Criminal Case through the Courts: Case Reporters and Citation Format

We now are in a position to chart the course of a hypothetical criminal case through the state and federal courts. As we trace this path, we will describe the references in which court decisions are published, and explain the standard format for citing them. Figure 1–2 illustrates one route a criminal case could take through the state courts to the U.S. Supreme Court, and how it then might enter the federal court system through a petition for a writ of habeas corpus and proceed once again to the U.S. Supreme Court. This is not the only path that such a case might follow. We are assuming that the state system includes an intermediate court of appeals and that the case we have chosen cannot be directly appealed to the state supreme court (as might occur, for example, when a murder conviction results in a death sentence). In our hypothetical we have omitted a state postconviction motion and appeal. We further assume that the case involves one or more federal constitutional issues that can be raised through a petition for a writ of habeas corpus.

Let us begin with a case that is being prosecuted in state court. The defendant, James Acker, has been accused of armed robbery. (Depending on the jurisdiction, the defendant has been charged through either an indictment or a prosecutor's information.) When the trial commences, the case is called *State v. Acker* (or, in some jurisdictions, *People v. Acker* or *Commonwealth v. Acker*), and the case is assigned a number on the trial court docket. Acker has entered a plea of not guilty and has exercised his right to a trial by jury. The crime victim has testified that she got a good look at the man who robbed her at gunpoint, and she is certain that Acker is the man. Her

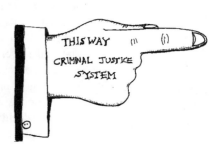

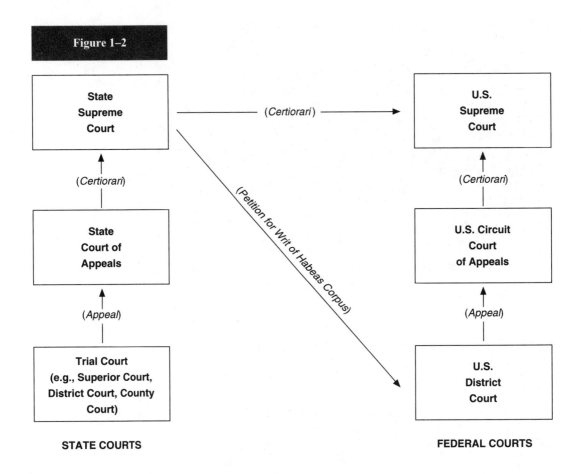

Figure 1–2

State Supreme Court — (*Certiorari*) → U.S. Supreme Court

(*Certiorari*)

(*Petition for Writ of Habeas Corpus*)

State Court of Appeals

U.S. Circuit Court of Appeals

(*Certiorari*)

(*Appeal*)

(*Appeal*)

Trial Court (e.g., Superior Court, District Court, County Court)

U.S. District Court

STATE COURTS   **FEDERAL COURTS**

confidence in this identification has been bolstered because she saw Acker when, just 30 minutes after the robbery, the police brought him by in handcuffs to confirm that they had arrested the right suspect. She further testified that when the police brought Acker by after the crime, her friend, who had been with her at the time of the robbery but who made no appearance at the trial, had exclaimed, "That's the man!" and fainted dead away.

Defense counsel objected that the identification procedure violated Acker's constitutional rights to the assistance of counsel and due process of law, and that the victim's testimony regarding her friend's statement was inadmissible hearsay. The trial judge overruled these objections, and the jury found Acker guilty as charged. Citing Acker's lack of remorse and apparent incorrigibility, the judge imposed a lengthy prison sentence.

State trial courts frequently make written findings of fact and conclusions of law when they decide issues such as the admissibility of eyewitness identification testimony. (Issues of this nature usually are resolved through a hearing conducted outside of the presence of the jury.) Trial judges make written findings and decisions much less frequently when issues such as the admissibility of hearsay evidence arise unexpectedly during a trial, however. Instead, they may simply announce that an objection is sustained or overruled, without further explanation. In neither instance is there much likelihood that the trial judge's ruling will be published in a case reporter. In all likelihood, the ruling will become grist for the appeal to which defendant Acker is entitled.

All states guarantee criminal defendants the right to appeal a conviction. Normally, the party who has the burden of proving an issue in order to prevail in court, or who seeks to upset the status quo after a court has issued a ruling, is the first name listed in the title of a case. This convention dictates that the party who has lost in a lower court is listed first when a case is appealed. While this rule normally is followed in civil cases, many states continue to adhere to the tradition of listing the state first in criminal cases even when the defendant has lost in the trial court and is appealing a conviction. Thus, when Acker appeals his conviction to the state intermediate court of appeals, the case may still be called *State v. Acker*.

As we already have discussed, appellate courts, unlike trial courts, commonly issue written opinions explaining their decisions. When these opinions are published, they are indexed in ways that make them accessible to researchers. Before we describe legal research techniques, we must identify the books in which judicial opinions are published and explain how these books are cited.

In many states, court opinions are printed in two different sets of case reporters. One set of volumes contains what state law has designated as the "official" report of cases, and includes opinions only from that particular state's courts. The opinions in these reporters typically are supplemented by few indexing aids to facilitate legal research efforts. We will spend little time on these official reporters, because far more useful case reporters are available for conducting legal research. You should know about the existence of these state reporters, and you may have to cite them, but we can confidently predict that you will not often be working with these official case reporters. Just so you will be familiar with how a judicial decision is displayed in an official case reporter, the first page of *Blair v. State*, which was decided by the Georgia Court of Appeals, is reproduced in Exhibit 1–1.

The proper citation for the case shown in Exhibit 1–1 is *Blair v. State*, 216 Ga. App. 545 (1995). Almost all legal reference materials conform to the format reflected in this citation. The first number following the case name refers to the **volume** of the case reporter in which the case is found—in this example, volume 216. Then comes the accepted abbreviation for the name of the case reporter. In this example, "Ga. App." stands for Georgia Appeals Reports. The next number indicates the **page** at which the reported opinion begins. Thus, to find *Blair v. State*, we would have to locate the Georgia Appeals Reports in the library, find volume 216 of those reporters, and open the book to page 545. The year in which the Georgia Court of Appeals made its decision in this case, 1995, is identified in the parentheses.

The same cases found in official case reporters are published in West Publishing Company's **regional reporter** system, which is infinitely more useful for

Exhibit 1–1

216 Ga. App.   JANUARY TERM, 1995   545

A94A2749. BLAIR v. THE STATE.
(455 SE2d 97)

BEASLEY, Chief Judge.

At a bench trial, Blair was convicted of loitering and prowling, OCGA § 16-11-36, and possession of less than one ounce of marijuana. OCGA § 16-13-2 (b).

Cobb County Police Sergeant Carter testified that she was doing a security check of an industrial park at night when no businesses in the park were open. The area is patrolled frequently, because stolen cars are found there on occasion and juveniles gather there.

On the night in question, Carter observed a car with the lights out in the parking lot, where authorized vehicles do not remain at night. It was occupied by Carnes in the driver's seat and Blair in the passenger seat. When the officer approached the car, she smelled the odor of burning marijuana. She asked Carnes what they were doing there, and he stated they had been to a movie and had parked to try to decide what to do next.

Carter returned to her patrol car to request another police car. After Officer Sullivan arrived, both officers approached the car in which Carnes and Blair were seated. Sergeant Carter asked Carnes to step out of the car and asked what he and Blair were doing. . . .

Source: Blair v. State, 216 Ga. App. 545 (1995), © Georgia Appeals Reports, used with permission.

legal research purposes. The state court opinions published in the regional reporters are a part of West's comprehensive National Reporter System. West Publishing Company's indexing and referencing system makes case research possible. (We discuss West's "key number" system in much greater detail in Chapter 3. For now, we will simply introduce the case reporters in the National Reporter System.) So popular are West's case reporters that nearly half of the states have discontinued the practice of publishing their own official volumes of judicial opinions and now rely exclusively on the West reporters. These states include Alabama, Alaska, Colorado, Delaware, Florida, Indiana, Iowa, Kentucky, Louisiana, Maine, Minnesota, Mississippi, Missouri, Montana, North Dakota, Oklahoma, Rhode Island, South Dakota, Tennessee, Texas, Utah, and Wyoming.

In West's National Reporter System, a specific state's court opinions are assigned to one of seven regional reporters. The states are grouped geographically, but you may find the regional placement of some states surprising. For example, few people associate Colorado, Kansas, and Oklahoma with the Pacific Ocean, and Kentucky is not generally thought of as being in the Southwest, nor Illinois and Indiana in the Northeast, nor Iowa, Michigan, and Nebraska in the Northwest. The seven regional case reporters, and the states included within each are listed in Table 1–1. The regional system is pictured in Figure 1–3.

The regional reporters have been in existence since the nineteenth century. All are now in their second series, indicated by the "2d" after the abbreviation for the

Table 1–1

Atlantic Reporter, 2d Series (A.2d)

Connecticut	New Hampshire
Delaware	New Jersey
District of Columbia	Pennsylvania
Maine	Rhode Island
Maryland	Vermont

North Eastern Reporter, 2d Series (N.E.2d)

Illinois	New York
Indiana	Ohio
Massachusetts	

North Western Reporter, 2d Series (N.W.2d)

Iowa	North Dakota
Michigan	South Dakota
Minnesota	Wisconsin
Nebraska	

Pacific Reporter, 2d Series (P.2d)

Alaska	Nevada
Arizona	New Mexico
California	Oklahoma
Colorado	Oregon
Hawaii	Utah
Idaho	Washington
Kansas	Wyoming
Montana	

South Eastern Reporter, 2d Series (S.E.2d)

Georgia	Virginia
North Carolina	West Virginia
South Carolina	

South Western Reporter, 2d Series (S.W.2d)

Arkansas	Tennessee
Kentucky	Texas
Missouri	

Southern Reporter, 2d Series (So. 2d)

Alabama	Louisiana
Florida	Mississippi

region. Older cases appear in the first series, such as in the Atlantic Reporter, the North Eastern Reporter, and so forth. We may soon see the emergence of the Pacific Reporter, 3d Series, and the South Western Reporter, 3d Series. The sec-

Figure 1-3   National Reporter System Map—Showing the States in Each Reporter Group.

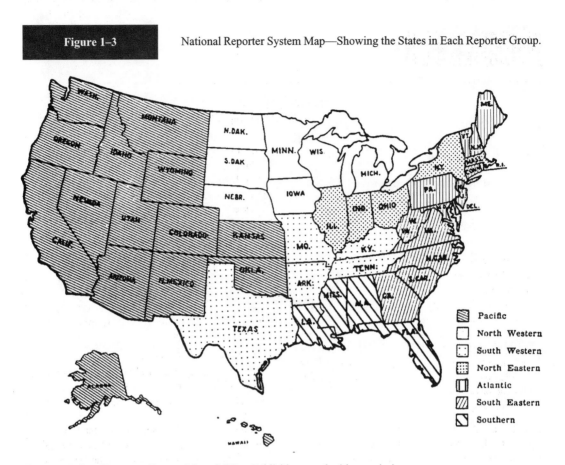

Source: National Reporter System Map, © West Publishing, used with permission.

ond series in each of these reporters now numbers over 900 volumes, and a third series is likely to start after they reach volume 999. Because New York and California are such populous states and have such prolific court systems, West also publishes reporters that contain exclusively those states' judicial decisions. These opinions appear, respectively, in the New York Supplement, 2d Series and the California Reporter, 2d Series.

No legal significance accompanies the groupings of states that West has chosen for its regional case reporters. For example, the fact that Alabama, Florida, Louisiana, and Mississippi court opinions all appear in the Southern Reporter, 2d Series does not imply that the courts in any one of these states must abide by the decisions of the other state courts that are reported in their grouping.

Exhibit 1–1 illustrates how *Blair v. State* is presented in the official Georgia Appeals Reports. You can see how this same case appears in the South Eastern Reporter, 2d Series by referring to Exhibit 1–2. Note the materials that precede the

Exhibit 1–2

BLAIR v. STATE Ga. 97
Cite as 455 S.E.2d 97 (Ga.App. 1995)

216 Ga.App. 545
BLAIR
v.
The STATE.
No. A94A2749.
Court of Appeals of Georgia.
March 8, 1995.

Defendant was convicted of loitering and prowling, and possession of less than one ounce of marijuana, in the State Court, Cobb County, McDuff, J. Defendant appealed. The Court of Appeals, Beasley, C.J., held that: (1) evidence supported conviction for marijuana possession, even though defendant claimed he was merely seated in automobile while his companion consumed drug, and (2) evidence supported conviction for loitering and prowling.

Affirmed.

1. Criminal Law ☞ 1159.6

If totality of circumstantial evidence is sufficient to connect defendant to possession of drugs, conviction will be sustained, even though there is evidence to authorize contrary finding. O.C.G.A. § 24-4-6.

2. Drugs and Narcotics ☞ 117

Evidence supported conviction of defendant for possession of marijuana, even though he claimed that he was simply sitting in automobile while his companion used drug; arresting police officers, who were familiar with symptoms of drug usage, noted that defendant's eyes were glassy and bloodshot, which was consistent with marijuana consumption, and defendant admitted that he placed marijuana pipe next to passenger door after companion had handed it to him. O.C.G.A. § 16-13-2(b).

3. Vagrancy ☞ 3

Evidence supported conviction of defendant for prowling and loitering; police discovered defendant sitting in automobile in location where law-abiding persons would not be at that time, in area where businesses had closed, his presence warranted justifiable and reasonable alarm or concern for safety of property in closed businesses and buildings, and defendant's explanation that he and companion were smoking marijuana did not dispel alarm or concern. O.C.G.A. § 16-11-36(a, b).

Sandy E. Scott, Marietta, for appellant.

Benjamin F. Smith, Jr., Sol., Barry E. Morgan, Chief Asst. Sol., Cindi Yeager, Asst. Sol., Marietta, for appellee.

BEASLEY, Chief Judge.

At a bench trial, Blair was convicted of loitering and prowling, O.C.G.A. § 16-11-36, and possession of less than one ounce of marijuana. O.C.G.A. § 16-13-2(b).

Cobb County Police Sergeant Carter testified that she was doing a security check of an industrial park at night when no businesses in the park were open. The area is patrolled frequently, because stolen cars are found there on occasion and juveniles gather there.

On the night in question, Carter observed a car with the lights out in the parking lot, where authorized vehicles do not remain at night. It was occupied by Carnes in the driver's seat and Blair in the passenger seat. When the officer approached the car, she smelled the odor of burning marijuana. She asked Carnes what they were doing there, and he stated they had been to a movie and had parked to try to decide what to do next.

Carter returned to her patrol car to request another police car. After Officer Sullivan arrived, both officers approached the car in which Carnes and Blair were seated. Sergeant Carter asked Carnes to step out of the car and asked what he and Blair were doing. He admitted they were smoking marijuana and both were arrested.

Source: Blair v. State, 455 S.E.2d 97 (1995), © West Publishing, used with permission.

court's opinion: The opinion itself does not start until the words "BEASLEY, Chief Judge" identify its author.

The information above the beginning of the court's opinion consists of the research devices created by West Publishing Company, which we will consider in Chapter 3. The West editors do not change a single word of the opinions from the official case reports, so both versions are true to what the judges on the court wrote. The case report in the West publication appears in volume 455 of the South Eastern Reporter, 2d Series, and it begins on page 97 of that volume. Thus it is cited: *Blair v. State*, 455 S.E.2d 97 (Ga. App. 1995).

It once was the convention to include both the official and West case reporters when citing a case, such as: *Blair v. State*, 216 Ga. App. 545, 455 S.E.2d 97 (1995). This citation style may still be appropriate if you are preparing a legal document for use in the courts of the relevant state. However, the standard citation format for state court cases generally has been simplified. Now it is conventional to cite only the West regional reporter. The specific state court that decided the case is identified in parentheses, with the appropriate abbreviations, and the year of decision is also provided. Thus, the correct citation of the case to which we have been referring is: *Blair v. State*, 455 S.E.2d 97 (Ga. App. 1995). Examples of case citations involving other regional reporters and intermediate courts of appeals are:

- State v. Morales, 667 A.2d 68 (Conn. App. 1995)
- Smith v. State, 655 N.E.2d 532 (Ind. App. 1995)
- State v. Puffinbarger, 540 N.W.2d 452 (Iowa App. 1995)
- State v. Carlson, 906 P.2d 999 (Wash. App. 1995)
- State v. Looney, 911 S.W.2d 642 (Mo. App. 1995)
- State v. Bailey, 664 So. 2d 665 (La. App. 1995)

Returning to the discussion of our hypothetical case, *State v. Acker*, assume that the intermediate court of appeals has rejected the arguments Acker made in his appeal and has upheld his armed robbery conviction. Acker now has exhausted the one state court appeal to which he is legally entitled. Under most circumstances, he has no right to require the state supreme court to consider his case. A state's high court would not make good use of its scarce resources if it simply reconsidered the issues addressed by the intermediate court of appeals in each and every case appealed. As we discussed earlier, state supreme courts by and large have discretion to choose which cases they will consider. They make their choices by taking into account such factors as the novelty and importance of the issues raised and the extent to which the lower courts have been inconsistent in their resolution of those issues.

Acker can petition the state supreme court to issue a writ of certiorari, or to exercise its discretion to decide the questions presented in his case (see Figure 1–2). Note that the state also can petition for a writ of certiorari and ask the state supreme court to overturn a decision by the intermediate court of appeals had that been in Acker's favor. This situation is unlike a state appealing a jury's not-guilty verdict, an appeal that is precluded by double jeopardy principles. Like the defendant, however, the state is permitted to ask a higher court to review questions of law that may have been erroneously decided by an intermediate appeals court.

Let us assume that the state supreme court grants certiorari, and decides to hear the *Acker* case. Remember that when a court denies certiorari, this does not necessarily mean that the lower court's decision was right, nor does it mean that the high court agrees with or is affirming the lower court's decision. It simply means that the high court did not take the opportunity to consider the issues presented in the case. Conversely, when a supreme court does grant certiorari, it does not necessarily signify that the court intends to reverse the lower court's decision.

Even though Acker was the losing party in the court below, we adhere to the convention used in many jurisdictions and continue to call the case *State v. Acker*. The state supreme court will request the parties to submit new briefs in the case and present oral argument. After the justices on the supreme court have had some time to deliberate, they will issue an opinion explaining their decision. Most state supreme courts have between five and nine judges. A state supreme court opinion almost certainly will be published, since the court carefully selects the cases it decides.

We have been assuming that this case has arisen in a state that has an official case reporter in addition to West's regional reporters. Accordingly, the state supreme court opinion will be published in two separate case reporters, just like the *State v. Acker* opinion of the intermediate court of appeals. State supreme court opinions typically appear in official state reports reserved exclusively for the high court, but the same West's regional reporter series as for the earlier appeal will be involved. Thus, to assume we are in the same jurisdiction as *Blair v. State*, a case opinion might appear in the following volumes: *Hammond v. State*, 264 Ga. 879, 452 S.E.2d 745 (1995). We can simplify the case citation by adopting the conventional format, which requires only identification of the West reporter: *Hammond v. State*, 452 S.E.2d 745 (Ga. 1995). Examples of citations to state supreme court decisions from other jurisdictions include:

- Commonwealth v. Wilson, 672 A.2d 293 (Pa. 1996)
- People v. Morgan, 662 N.E.2d 260 (N.Y. 1995)
- State v. Conklin, 545 N.W.2d 101 (Neb. 1996)
- Simmons v. State, 912 P.2d 217 (Nev. 1996)
- Jones v. State, 916 S.W.2d 736 (Ark. 1996)
- Terry v. State, 668 So. 2d 954 (Fla. 1996)

Now assume that the state supreme court has rejected both of Acker's claims and has ruled that the eyewitness identification procedures violated none of his constitutional rights and the lower courts correctly ruled that the contested portion of the robbery victim's testimony was not inadmissible hearsay. Acker can press his claims no further in the state courts; he must accept the state supreme court's judgment or else seek further review in the federal court system. If he wants to take his case to the federal courts, he has two choices. Referring once again to Figure 1–2, you see either that he can petition the U.S. Supreme Court to decide his case on a writ of certiorari, or he can petition the appropriate federal district court for a writ of habeas corpus.

Acker knows that his chances of getting the U.S. Supreme Court to hear his case are slim. The Supreme Court normally grants certiorari in just 2 to 3 percent of the approximately 6,000 to 7,000 cases in which certiorari petitions are filed each year. It also takes time for the Supreme Court to consider his cert. petition. Although the Court's denial of certiorari signifies nothing about the merits of his claims, he may wish to bypass the Supreme Court for now and file his habeas corpus petition in the U.S. District Court. Let us assume that he opts for this latter route.

Neither the U.S. Supreme Court nor a federal district court has the authority to second-guess a state court's decision on an issue that involves only the proper interpretation of state law. Thus, for all intents and purposes, the hearsay issue, which only concerns the admissibility of testimony under the state's evidence law, has been definitively resolved against Acker. However, the eyewitness identification issue is a different matter. Acker has claimed that the identification procedures violated his right to counsel and to due process of law under the U.S. Constitution. The U.S. Supreme Court and the federal district court retain the authority to decide federal constitutional issues that arise in state criminal cases. (As we discussed earlier, there are some limitations on the federal courts' ability to consider state prisoners' constitutional claims on habeas corpus, but these exceptions need not concern us for the purposes of this hypothetical.)

A petition for a writ of habeas corpus is treated as a civil proceeding even though it involves a challenge to the constitutionality of a criminal conviction or sentence. This fact helps to explain why the case no longer is known as *State v. Acker*. Instead, Acker, who is asking the federal court on the habeas petition to upset his conviction, will become the first-listed party in the newly named case. Since he is filing the habeas petition, he is now known as the **petitioner**.

The other party in a federal habeas corpus proceeding is the person who is alleged to be confining the habeas petitioner unlawfully. This usually is the warden of the prison in which the petitioner is incarcerated or the director or commissioner of the state's entire prison system. Because the other party must respond to the allegations made in the habeas petition, that person is called the **respondent**. Let us assume that the warden of the prison in which Acker is confined is Richard Irving. Thus, when Acker files his petition for writ of habeas corpus in the federal district court, the case will be called *Acker v. Irving*.

Roughly 650 active federal court judges serve in the 94 U.S. district courts. The number of judges assigned to each district varies, depending on the volume of business conducted in the district. These judges almost always preside over and decide cases individually, rather than as a panel of judges. Accordingly, Acker's habeas corpus petition will be considered by a single federal district court judge. The decision will be announced in an opinion that may or may not be published in the case reporter for U.S. district court decisions, the Federal Supplement. District court judges submit only opinions they consider to be of special significance for publication in the Federal Supplement; their other opinions remain unpublished. (Unpublished opinions are largely inaccessible through legal research techniques that depend on printed sources, and they have correspondingly limited authoritative value.)

Some unpublished decisions are available through computerized database systems such as WESTLAW and LEXIS. There is some controversy about whether

WESTLAW and LEXIS should make the text of these decisions available since they have such limited precedential value. Nevertheless, these unpublished cases can provide useful information to criminal justice researchers even when they cannot be relied on as legal authority.

WESTLAW and LEXIS have their own case citation formats. The WESTLAW and LEXIS cites are considered acceptable by the two authoritative citation guides to law materials: **The Bluebook, A Uniform System of Citation** and the **University of Chicago Manual of Legal Citation**. The WESTLAW and LEXIS cites are helpful not only for referencing unpublished cases but also for citing new cases not yet available in one of the printed sets of reporters. As soon as a case is published in one of the printed reporters, WESTLAW and LEXIS provide the reference to the printed version. We illustrate below the proper format for citing cases on WESTLAW and LEXIS by assuming our hypothetical case concerning James Acker has been decided by the District Court for the Middle District of North Carolina. The WESTLAW case citation would appear as:

Acker v. Irving, 1997 WL 55316 (M.D.N.C.).

The cite consists of the name of the case (*Acker v. Irving*), the year of the decision (1997), the abbreviation for the database (WL for WESTLAW), an accession number for the document (55316), and, in parentheses, the court abbreviation (M.D.N.C.).

The citation on LEXIS would be:

Acker v. Irving, 1997 U.S. Dist. LEXIS 6122 (M.D.N.C.).

The LEXIS cite is slightly different in that it explicitly identifies the type of court making the decision (U.S. Dist., for U.S. District Court), and combines that with an accession number.

The **Federal Supplement**, which is abbreviated "F. Supp.", has been the exclusive case reporter for printed U.S. district court opinions since it came into existence in 1932. It is published by West Publishing Company and thus uses the West indexing and research aids that are so important to the process of legal research. Whenever you are referred to an opinion published in the Federal Supplement, you may assume that, with an occasional exception pertaining to a few specialized courts, the case was decided by a U.S. district court. Case citations to the Federal Supplement follow a standard format: case name, volume number of the Federal Supplement, followed by the abbreviation "F. Supp.", the page at which the decision begins and, in parentheses, the abbreviation for the particular district court that issued the opinion and the year of the decision. Thus, depending on the court in which Acker's federal habeas corpus petition was filed, and the year of its decision, the citation to the case should resemble the following examples:

- *Acker v. Irving*, 754 F. Supp. 1350 (M.D.N.C. 1991).
- *Acker v. Irving*, 786 F. Supp. 501 (S.D. Miss. 1992).
- *Acker v. Irving*, 823 F. Supp. 419 (N.D. Cal. 1993).
- *Acker v. Irving*, 861 F. Supp. 1178 (D. Me. 1994).

Federal habeas corpus petitions are filed in the federal district court that presides over the district containing the state trial court in which the petitioner has been convicted. The example above indicates that the federal district court's opinion is reported in volume 754 of the Federal Supplement beginning at page 1350 and that the opinion was written by a judge assigned to the U.S. District Court for the Middle District of North Carolina in 1991. We may thus infer that Acker's conviction occurred in Greensboro, Durham, or another city or county in the jurisdiction of the U.S. District Court for the Middle District of North Carolina. The other hypothetical citations indicate that the opinions were written by, respectively, the judges of the U.S. district courts in the Southern District of Mississippi, the Northern District of California, and the lone federal district in Maine.

Assume that the federal district court in North Carolina holds that the identification procedures used by the police violated Acker's federal constitutional rights. The court ordinarily would order Irving, the respondent prison warden, to release Acker from custody unless the state retries him within a specified number of days. An order of this nature usually is stayed or does not take effect, pending the outcome of an appeal. The state, acting through its agent, prison warden Irving, would not be required to appeal, although it would have the right to do so.

Referring once again to Figure 1–2, you will see that an appeal from a U.S. district court decision is taken to a U.S. court of appeals. The court of appeals in the federal circuit containing the district court that issued the contested ruling will decide the appeal. The federal circuits are identified in Figure 1–1, which illustrates that an appeal from a judgment of the U.S. District Court for the Middle District of North Carolina would be taken to the Fourth Circuit Court of Appeals. Cases appealed from the Southern District of Mississippi, the Northern District of California, and the District of Maine would be resolved, respectively, in the Fifth, Ninth, and First Circuit Courts of Appeals.

The published opinions of the U.S. courts of appeals appear in the **Federal Reporter**, which now is in its third series, and is abbreviated as F.3d. The transition was made from the Federal Reporter, 2d Series (F.2d) to F.3d in 1993, and the first transition, from the Federal Reporter (F.) to F.2d, occurred in 1924. The Federal Reporter first appeared in 1880. The entire Federal Reporter series also is published by West Publishing Company. The F.3d reporters, and F.2d reporters after 1932, contain exclusively the opinions of the U.S. courts of appeals. Prior to 1932 both the district court and court of appeals opinions were published in the Federal Reporter and F.2d series. Cases decided in the nineteenth century, which you will rarely encounter, were published in several different reporters.

Referring to our hypothetical case, *Acker v. Irving*, James Acker has prevailed in the U.S. district court, and the state, through its agent, Warden Irving, now is in the position of having to appeal. The party who is asking a court to overturn the judgment of another court ordinarily is listed first in a citation, so in our case Irving, as the **appellant**, is named first when the case is appealed. Acker, the **appellee**, is listed second, and the case is called *Irving v. Acker* in the U.S. court of appeals. When citing the decision of a U.S. court of appeals, we must identify the court that issues the opinion by its circuit and give the year of the decision. Following our earlier examples of Acker's appeals, the appropriate citation style is as follows:

- *Irving v. Acker*, 953 F.2d 209 (4th Cir. 1992).
- *Irving v. Acker*, 997 F.2d 1196 (5th Cir. 1993).
- *Irving v. Acker*, 14 F.3d 622 (9th Cir. 1994).
- *Irving v. Acker*, 66 F.3d 972 (1st Cir. 1995).

This citation format follows the familiar pattern of case name, volume, abbreviated name of case reporter, page, identification of the circuit, and year of decision.

Let us assume that the Fourth Circuit Court of Appeals rules that the district court erred when it decided that Acker's federal constitutional rights were violated, and that the circuit court reverses the lower court's decision and reinstates Acker's conviction and sentence. Although decisions of the U.S. courts of appeals normally are made by three-judge panels, Acker is entitled to request a hearing before the entire membership of a court of appeals (for example, before all thirteen of the non-senior judges in the Fourth Circuit) to reconsider the decision of the three-judge panel. A proceeding in which all of the judges in a circuit jointly consider a case is called an *en banc* hearing. These hearings are granted only in exceptional cases. If Acker decides not to request an *en banc* hearing, or if his request is denied, there is only one more court that can consider the issues in his case. That court, of course, is the U.S. Supreme Court. (See Figure 1–2.)

Acker can file a petition for writ of certiorari to request the Supreme Court to review the Fourth Circuit Court of Appeals ruling. If at least four of the nine Justices on the Supreme Court vote to accept the case for review, then certiorari (cert.) is granted. Lawyers representing Acker and Warden Irving then will file written briefs and later will present oral arguments before the Court decides the case. On occasion, the Justices will decide a case summarily, without requesting either briefs or oral argument. Acker again is the petitioner under these circumstances and Irving the respondent, and the case reverts to being called *Acker v. Irving*.

Supreme Court decisions are published in three different case reporters. The official case reporter, the **United States Reports**, is published by the U.S. government. Like many official state case reporters, the United States Reports lack indexing or reference aids that are useful for legal research. The other two reporters for Supreme Court decisions, discussed below, are much more useful for legal research. Because they print Supreme Court opinions verbatim, including the pagination used in the official United States Reports, there seldom is a need to make direct use of the latter. Nevertheless, convention demands that the United States Reports (abbreviated as U.S.) be cited when reference is made to Supreme Court decisions. In Chapter 3 we will reproduce a portion of a Supreme Court opinion as it appears in the official reporter.

The other case reporters in which Supreme Court opinions are found are the **Supreme Court Reporter** (abbreviated as S. Ct.), and the United States Supreme Court Reports, Lawyers' Edition or **"Lawyers' Edition,"** which now is in its second series (and is abbreviated "L.Ed.2d"). The Supreme Court Reporter is published by West and thus is indexed consistent with the state regional reporters, the Federal Supplement, and the Federal Reporter. The Lawyers' Edition reporters are published by Lawyers Co-operative Publishing Company (Lawyers Co-op.), which also publishes a number of other significant legal reference materials that we will be describing later. Lawyers Coop. publications make use of their own finding aids and indexing system. They do not conform to the West system, although each company organizes its materials similarly and relies on similar search strategies.

The Supreme Court can be expected to issue a decision a few months after hearing oral arguments in a case. There is considerable delay, often exceeding two years, between the time the Supreme Court decides a case and when the Court's opinion is printed in final form in the official United States Reports. Opinions are published much more rapidly in both the Supreme Court Reporter and in the Lawyers' Edition, with research aids incorporated into each set. The issues of the Supreme Court Reporter and the Lawyers' Edition are not reprinted in hardcover volumes until the Supreme Court opinions become available in the United States Reports, so they can then incorporate the official reporter's pagination. It is a nice convenience for users of the Supreme Court Reporter and the Lawyers' Edition to be able to refer to specific pages of the official report of a decision without having to consult the United States Reports directly.

When reference is made to a U.S. Supreme Court decision, it is permissible, and sometimes the preferred style, to cite only the United States Reports. Thus, the case being used in our example would be cited as *Acker v. Irving*, 501 U.S. 340 (1991). It often is helpful, however, to cite all three reporters in which Supreme Court cases are published. When this practice is followed, the proper format is to cite the official United States Reports first, then the Supreme Court Reporter, and finally Lawyers' Edition 2d. Our case would thus be cited as follows: *Acker v. Irving*, 501 U.S. 340, 111 S. Ct. 2349, 115 L. Ed. 2d 306 (1991).

As discussed above, the judgment reviewed by the Supreme Court in the *Acker* case is that of the U.S. Court of Appeals for the Fourth Circuit. However, recall that a petition for writ of certiorari can also be filed with the U.S. Supreme Court after the state court of highest jurisdiction decides the case. The involved state is cited by name when cert. is granted from a state court. Thus, if Acker's claims had been rejected by the North Carolina Supreme Court and he petitioned the U.S. Supreme Court for a writ of certiorari, the case would no longer be known as *State v. Acker* but as *Acker v. North Carolina*. Many familiar cases conform to this citation format, including *Miranda v. Arizona*, 384 U.S. 436, 86 S. Ct. 1602, 16 L. Ed. 2d 694 (1966), and *Mapp v. Ohio*, 367 U.S. 643, 81 S. Ct. 1684, 6 L. Ed. 2d 1081 (1961).

When a citation uses this three-citation format, you should instantly recognize that the U.S. Supreme Court has decided the case. You immediately know the year of decision and how to locate the case not only in the official United States Reports but in the Supreme Court Reporter and Lawyers' Edition 2d as well. You can also infer from the case name that the petitioner was the losing party in the court below and that a state court judgment from the identified jurisdiction is being reviewed.

Considerable time may pass before a case is prepared for publication and is bound in the official (U.S.) and unofficial (S. Ct. and L. Ed. 2d) reporters. It may be important for you to gain access to a case decision long before the case reporters are available. As we have discussed, a computer system such as WESTLAW or LEXIS can make court decisions widely available much sooner than printed sources. It is important to remember that court decisions were distributed long before the advent of computers, and they continue to be disseminated in printed form notwithstanding computerization.

When courts announce their decisions, the decisions initially are printed and distributed in what are called **slip opinions** or **advance sheets**, which simply are individual case opinions that have not yet been bound as books. (We describe slip opinions, advance sheets, and looseleaf services in greater detail in Chapter 3.) Slip opinions generally are available the same day that decisions are announced, and copies are distributed to libraries and other addresses within a few days of the decision date.

Looseleaf services collect specific types of court opinions and reproduce them in full or provide excerpts or summaries of the court decisions. The opinions are published on sheets of paper, which can then be organized and maintained in three-ring looseleaf notebooks. The cases reported in looseleaf services are selected after an editorial staff assesses their significance to readers interested in a particular subfield of the law. For example, the Criminal Law Reporter, which is published by the Bureau of National Affairs (BNA), reports summaries and excerpts of state and lower federal court criminal opinions that the BNA editors choose from recent case decisions. It also reports the complete text of all decisions in U.S. Supreme Court criminal cases. Issues are mailed to subscribers weekly, and the full text of Supreme Court opinions usually arrive within days of the decision. United States Law Week, which also is published by BNA, and the Supreme Court Bulletin, published by Commerce Clearing House (CCH), are looseleaf services that promptly disseminate all U.S. Supreme Court opinions and excerpts or summaries of select lower court decisions covering a wide range of subjects.

SUMMARY ... AND THEN ON TO LEGAL RESEARCH

We have now laid the groundwork necessary for the upcoming chapters, where we examine the specific materials and techniques of basic legal research. We have briefly described primary and secondary legal authorities, the WESTLAW and LEXIS computer databases, and a few of the social science references that will be useful for your research purposes. All of these materials will be examined in greater detail in the following chapters. We also have outlined the typical structure of state court systems and the federal court system and have traced how a hypothetical case moves through the courts. We have identified the case reporters and explained the standard citation format associated with published judicial opinions. By now, when you see a case citation, you should know how to find the opinion and how to recognize what kind of court decided the case. With this much established by way of background, we are ready to begin the legal research process.

2

Secondary Legal Authorities and Social Science References

INTRODUCTION

Where and how you seek answers to a question are at least partially determined by how much you already know about the problem you are investigating. If your car's engine fails, you may have to consult the owner's manual simply to figure out how to raise the hood. On the other hand, you may be inclined to roll up your sleeves and take apart the exhaust manifold so you can confirm the suspected difficulty and fix it. So it is with legal research. How much you already know about the issue you are examining makes a big difference in where and how you go about conducting your research. If you know very little about a topic, it makes sense to start with different references and to adopt a different research strategy than if you already know quite a bit.

As we begin our description of legal references and the techniques for using them, we will assume that you are at a level equivalent to having to consult the owner's manual to be able to open the hood of your automobile. We have considerably more confidence in being able to demystify the process of legal research than in ever being able to understand how and why the car engine starts when the ignition key is turned. Our assumption that you know very little about the issues you are researching explains why we consider **secondary legal authorities** in this chapter and reserve consideration of **primary legal authorities** until Chapter 3. Secondary legal authorities will help you get started.

Secondary legal authorities generally serve one or more of the following purposes: (l) to provide **descriptive information** pertaining to an issue of law; (2) to offer critical **analysis** of a legal issue; and (3) to serve as a **finding tool** for other sources of legal authority. Through reading secondary legal authorities, you gain background information and other insights about the subject you are exploring.

In this chapter we also briefly consider extralegal references that relate closely to legal issues in criminal justice. We identify a few social science references and statistical sources that may prove useful as you conduct research. Extralegal references may not be relevant to all issues, but the need for the information available in these sources will arise often enough that it is important not to ignore potentially useful social science and statistical materials.

THREE HYPOTHETICAL CASES

If you want to learn about the mountain, you must go to the man who lives on the mountain.

There is much wisdom in this Zen expression. The learning process is greatly enriched by drawing upon the knowledge and experience of others; doubtlessly, the man who lives on the mountain could reveal much about his domain that  would be missed by visitors. But what the mountain dweller has to offer is only a part of the learning experience. The other part requires going to the mountain.

You will never become an accomplished tennis player, or pianist, or ballerina just by reading books. If you are serious about improving your skills, you must take part in the activity. The same is true about learning how to do legal research. Just reading a book is no substitute for actually conducting research. Ideally, you should take this book to your library, and read it as you prowl the stacks or huddle over a computer terminal, and practice the steps you learn as you go. Because this exercise may not always be practical, we will do our best to bring a bit of the library to you. Remember, however, that reading this volume is no substitute for getting into a library and actually going through the process of finding the books, booting up the computer, and researching questions of law on your own. You must go to the mountain.

We introduce available legal and extralegal reference materials, and the techniques for using them, with the assistance of three hypothetical cases. These cases raise issues of law that can be researched through printed references, the use of computer databases, or a combination thereof. The cases we have selected involve issues of criminal procedure and substantive criminal law, and a civil suit alleging the denial of federal constitutional rights. As we attempt to find answers to the questions raised by these cases, we will point out the different references that can be consulted, and give examples of how they can be used.

Case 1. John Winston has been charged with the first-degree sexual abuse of his eight-year-old stepdaughter, Janice. Winston allegedly subjected Janice to sexual contact (defined as "any touching of the sexual parts of another person for the purpose of gratifying sexual desires") on numerous occasions over a several-month period. Winston has steadfastly maintained his innocence, and blames Janice's mother, from whom he recently separated, for pressuring Janice into making a false accusation. He has confidence that Janice will tell the truth when she testifies at his trial. However, the prosecutor is concerned that if Janice is forced to testify in the presence of Winston, she will be intimidated, psychologically traumatized, and unable to describe what happened. The prosecutor has asked the trial judge to allow Janice to give her testimony outside the courtroom and to have the testimony broadcast to the jury via closed-circuit television. Under the requested procedures, Janice would testify under oath, and Winston's lawyer would be allowed to cross-examine her, but only the prosecutor and the defense lawyer would be in the room with Janice. Winston would remain in the courtroom, where he would be allowed to view and hear Janice's testimony on the television screen, just as the jury would. He also would be able to communicate with his attorney through a telephone hookup while Janice testifies. Through counsel, Winston has vigorously objected to this proposed procedure. Should the trial judge approve the prosecutor's request? Are any of Winston's constitutional rights violated if the request is granted? Are additional facts necessary before the judge can issue a ruling?

Case 2. Andrew Adams has been infected by the human immunodeficiency virus (HIV), which is responsible for causing acquired immune deficiency syndrome (AIDS). Adams, who was wanted for failing to appear in court on a charge of driving while intoxicated, knew that he had tested positive for HIV. Officer Fiegel recognized Adams as Adams crossed the street in front of Fiegel's patrol car. The officer got out of his car and informed Adams that he was placing him under arrest on the authority of the outstanding warrant. Adams thereupon told Officer Fiegel, "I'm a dying man. If you try to arrest me, I'm taking you with me." As the officer attempted to handcuff Adams, a struggle ensued, and Adams bit Officer Fiegel on the hand, drawing blood. "I warned you," said Adams, who then spit in Officer Fiegel's face. "Now you have AIDS, too." When the prosecutor was informed of these facts, she brought a two-count indictment charging Adams with attempted murder and assault with a deadly weapon. The indictment alleged that Adams attempted to murder Officer Fiegel by biting him and spitting on him, and that he intended to cause the officer's death by transmitting HIV. The indictment further alleged that Adams' teeth and spit constituted deadly weapons because Adams is HIV-positive. Adams' attorney has moved to dismiss the indictment. Should the judge grant the motion or instead allow Andrew to be tried on the charges alleged in the indictment?

Case 3. Deborah Miller was incarcerated in state prison, where she was serving a five-year sentence for robbery. She and her cellmate, Felicia Liggett, were locked in their common cell approximately 14 hours each day. Liggett smoked during much of this time, typically going through at least two packs of cigarettes a day. Miller hated the tobacco smoke and feared that inhaling the second-hand smoke would result in serious health problems. She repeatedly had asked Warden Myers to transfer either Liggett or herself to a different cell so she could be housed with a nonsmoker. Warden Myers ignored Miller's requests. Finally, Miller filed a lawsuit in federal court alleging that Warden Myers' deliberate and willful refusal to arrange a transfer so that she was not exposed on a daily basis to the second-hand cigarette smoke caused by her cellmate constituted cruel and unusual punishment, in violation of her federal constitutional rights. She sought both money damages and an injunction forbidding the warden from continuing to confine her in the same cell as a smoker. Citing her ample discretion to manage the prison, and the speculative nature of adverse health effects associated with second-hand cigarette smoke, Warden Myers has asked the federal court to

dismiss Miller's lawsuit. Should Warden Myers' motion to dismiss the lawsuit be granted, or should the federal court deny the motion and allow Miller's complaint to proceed to trial?

DEFINING THE RESEARCH ISSUES

Now that we have presented these case scenarios, our first order of business is to arrive at a succinct statement of the legal issues arising from the three fact patterns. Being able to identify the issue or issues you are investigating correctly and precisely is an important part of the legal research process. It also presents something of a catch-22 situation: It may be impossible for you to define the issue adequately until you know more about the subject you will be researching, but you may not be able to learn more about the subject until you complete some research about it. The only way out of this dilemma is to take a stab at stating the issue as best you can at the outset, with the understanding that you may have to modify your statement of the issue as you continue to do the research. There is nothing wrong with this approach. You frequently will have to make at least minor adjustments in your search strategy as you refine your understanding of the issues suggested by the problem you are researching. Nevertheless, we must try to frame the legal issues arising from our fact presentations. The questions we presently pose are:

1. Are Winston's constitutional rights violated if the judge grants the prosecutor's request for the televised presentation of Janice's testimony?
2. Can Adams be charged with attempted murder or assault with a deadly weapon?
3. Should Warden Myers' motion to dismiss Miller's complaint be granted?

When you define an issue of law suggested by a set of facts, you should do your best to frame it in such a way that someone who is unfamiliar with the fact pattern can understand the issue simply by listening to your statement of the question. Notice how the questions above leave an observer who has not read the case scenarios totally in the dark about the true nature of the issues. On the other hand, you must avoid describing a page-long statement of the issues that essentially restates all of the case facts, with a question mark tacked on at the end.

Below is a series of questions that attempt to identify the issues of law arising from the facts of the three cases. These questions certainly are not the only way the issues could be stated. Because we have some familiarity with the topics, we also have the benefit of knowing enough about the issues to avoid some of the false starts occasioned by the catch-22 situation we mentioned before.

Case 1: Is a defendant's right to confront accusing witnesses at a criminal trial violated when the testimony of an eight-year-old witness, who allegedly was sexually assaulted by the defendant, is presented via closed-circuit television with the child testifying outside of the physical presence of the defendant, if the child would suffer emotional trauma and be impeded from testifying if required to be in the defendant's presence?

Case 2: Can a charge of attempted murder or assault with a deadly weapon be based on the actions of a defendant who bites and spits on his victim, knowing he is HIV-positive and intending to expose his victim to the risk of AIDS?

Case 3: Is the constitutional right of a prisoner not to be subjected to cruel and unusual punishment violated by the actions of a prison warden who deliberately refuses to transfer the prisoner or her cellmate from the cell they jointly occupy, so that the prisoner is not involuntarily exposed to the second-hand smoke caused by her cellmate's practice of smoking two or more packs of cigarettes daily?

Admittedly, each of these statements of the issues arising from our three cases is a mouthful. They may be overly ambitious in their attempt to include important facts and thus be unwieldy. Still, they should reinforce the point that the statement of a question should be sufficiently comprehensive to inform someone who is unfamiliar with the facts of a dispute about the essence of the controversy. If our cases were based on real facts, each would arise in a specific jurisdiction, which would allow us to focus our legal research strategies accordingly. We assume here, however, that Case 1 and Case 2 are being prosecuted under the authority of an unspecified state, which will thus allow us to consult multiple jurisdictions to demonstrate research techniques. Case 3 involves a lawsuit filed in federal court, which allows us to concentrate our research strategy on federal authorities.

THE DIFFERENT SOURCES OF AUTHORITY AND HOW THEY ARE USED

We begin exploring the issues raised by our three case scenarios by consulting the most basic, elementary forms of secondary legal authority. We then review progressively more comprehensive authorities. We conclude by describing a few extralegal authorities—social science periodicals and statistical sources—that may prove to be useful as you pursue research on law-related issues.

Law Dictionaries

The most basic form of secondary legal authority is a **law dictionary**, which is used just like a **Webster's Dictionary** is, and for similar purposes. The law has no monopoly on obscure language, but neither does it lack for words that are not in common usage and thus require definition. From time to time you may run into words in the legal lexicon—in English, Latin, or another language—with which you are wholly unfamiliar; words such as *bot, usufruct, weregild, Witan,* and *writ of error coram nobis.* Other words may spark a glimmer of recognition, but you may wish to get a more precise definition or a better understanding of their meaning; words such as *corpus delecti, malum in se, quid pro quo, tortfeasor,* and *writ of certiorari.* Law dictionaries come in especially handy when a topic and its vocabulary are both new to you. Some law dictionaries not only define words, but provide the bonus of citing one or more cases or other authorities from which the definitions of terms were extracted. Law dictionaries also are excellent sources to explain legal abbreviations. As we noted in Chapter 1, case reporters and other legal references usually are cited by abbreviation, so a listing of commonly used abbreviations is helpful for legal research.

The most widely used of these references is **Black's Law Dictionary**, which is published by West. Black's Law Dictionary comes in a bulky hardcover volume of over 1,500 pages and also in a slimmer abridged paperback edition. Among the numerous other law dictionaries that may be purchased or consulted in a library are **Ballentine's Law Dictionary**, which is made available by Lawyers Cooperative Publishing Company, and **A Dictionary of Modern Legal Usage**, published by Oxford University Press.

We have defined the issue in Case 1 as whether John Winston's right to confront accusing witnesses would be violated if the child he is charged with sexually assaulting is allowed to testify outside of his presence via closed-circuit television. Exhibit 2–1 shows what we find if we look up the word "**confrontation**" in Black's Law Dictionary (6th ed., 1990). If we have started our research knowing little about the concept of confronting accusing witnesses, you will see that we gain helpful information by reading the definition of "confrontation." For instance, we learn that confrontation rights are guaranteed under the Sixth Amendment to the U.S. Constitution. The definition of "**confrontation**," which of course does not conclusively answer the question we are researching, seems to pull us in different directions concerning the appropriate resolution of our issue: It first suggests that the right of confrontation consists of "setting a witness face to face with the accused," which did not happen in Case 1. However, note that the definition

---

Exhibit 2–1

Confreres /kónfrerz/kənfrérz/. Brethren in a religious house; fellows of one and the same society.

Confrontation. In criminal proceedings, the accused has a right to be "confronted with the witnesses against him." This Sixth Amendment right consists of the act of setting a witness face to face with the accused, in order that the latter may make any objection he has to the witness, or that the witness may identify the accused; and, does not mean merely that witnesses are to be made visible to the accused, but imports the constitutional privilege to cross-examine them.

In fact, the essence of the right of confrontation is the right to cross-examination. Davis v. Alaska, 415 U.S. 308, 94 S.Ct. 1105, 39 L.Ed.2d 347. A disruptive defendant may, however, lose his right to be present in the courtroom, and, as a result, lose his right to confront witnesses. Illinois v. Allen, 397 U.S. 337, 90 S.Ct. 1057, 25 L.Ed.2d 353.

Confrontation clause. *See* Confrontation.

Confusio /kənfyúwz(h)(i)yow/. In the civil law, the inseparable intermixture of property belonging to different owners; it is properly confined to the pouring together...

Source: Black's Law Dictionary, "Confrontation," p. 300 (6th ed. 1990), © West Publishing, used with permission.

goes on to say that "the essence of the right of confrontation is the right to cross-examination." Under this definition, then, Winston is clearly permitted to cross-examine young Janice. This example should remind you that a law dictionary is a very elementary legal research tool. It may be of help in the early stages of research but should not be relied on to provide final answers.

Note also that the definition of "**confrontation**" supplied by Black's Law Dictionary includes two case citations, *Davis v. Alaska*, 415 U.S. 308, 94 S. Ct. 1105, 39 L. Ed. 2d 347 [(1974)], and *Illinois v. Allen*, 397 U.S. 337, 90 S. Ct. 1057, 25 L. Ed. 2d 353 [(1970)]. The ground we covered in Chapter 1 should trigger your memory as to the court that issued these decisions. If you have any hesitation about concluding that these cases were decided by the U.S. Supreme Court, you should review the citation conventions we discussed in Chapter 1. These citations provide our first example of how secondary legal authorities can serve as a finding tool: They cite other authorities you may wish to consult.

We round out our discussion of law dictionaries by considering the issues presented in our other two cases. The indictment in Case 2 has charged Andrew Adams with attempted murder and assault with a deadly weapon, based on Adams' (1) biting and spitting on Officer Fiegel, and (2) being infected with HIV. To gain a bit more insight into the definition of "**deadly weapon**," we consult Ballentine's Law Dictionary (3d ed. 1969) (see Exhibit 2–2). We learn that one meaning of a "deadly weapon" is "[a]n instrument which is likely to or which will cause or produce death or great bodily harm when used in the manner contemplated by its design and construction." Once again, this definition does not, and should not, be expected to resolve the question presented in Case 2. However, in addition to this definition, we are provided with a citation to a case and to two other references we discuss in greater detail later in this chapter: the annotation in "**ALR**" (**American Law Reports**) and two articles in the first edition of the legal encyclopedia **American Jurisprudence** (abbreviated in the Exhibit as "**Am J**

Exhibit 2–2

---

deadly. See mortal.

deadly weapon. An instrument which is likely to or which will cause or produce death or great bodily harm when used in the manner contemplated by its design and construction. Barboursville ex rel. Bates v Taylor, 115 W Va 4, 174 SE 485, 92 ALR 1093; 26 Am J1st Homi § 7; 56 Am J1st Weap § 2.

Whether an unloaded firearm is to be considered a deadly weapon depends ordinarily upon whether the manner in which the instrument is used or attempted to be used, whether as a firearm or a bludgeon. 56 Am J1st Weap § 4.

dead man's part (ded´ manz part). That portion of the personal property of a married man of which
...

Source: Ballentine's Law Dictionary, "Deadly Weapon," p. 308 (3d ed. 1969). Reprinted with permission of LEXIS-NEXIS, a division of Reed Elsevier Inc. LEXIS and NEXIS are registered trademarks of Reed Elsevier Properties Inc. FREESTYLE, KWIC, SuperKWIC and MEGA are trademarks of Reed Elsevier Properties Inc. SHEPARD'S and SHEPARDIZE are registered trademarks of Shepard's Company, a Partnership.

1st"). (Note that for purposes of brevity a publication may adopt its own citing conventions—omitting dates and periods and spacing around abbreviations. We are not bound by these original formats, but should maintain a consistent citation style.) You should be able to locate the cited case, "Barboursville ex rel. Bates v. Taylor, 115 W Va 4, 174 SE 485 [(1934)]," by consulting volume 115 of the official **West Virginia Reports**, at page 4, or by referring to West's regional case reporter, volume 174 of the **South Eastern Reporter** at page 485. Note that these references are now quite dated, and we should be most reluctant to assume that a definition from a 1934 West Virginia case applies in other jurisdictions.

In Case 3 we are investigating whether a prisoner exposed involuntarily to second-hand tobacco smoke may have a valid claim that she is being subjected to cruel and unusual punishment. When we look up "**cruel and unusual punishment**" in A Dictionary of Modern Legal Usage (2d ed. 1995) (Exhibit 2–3), we learn that the right to be free from such punishment is guaranteed by the Eighth Amendment to the U.S. Constitution. We get a general definition of "cruel and unusual punishment," and the citation of a U.S. Supreme Court case, *Coker v. Georgia*, that may help us gain a better understanding of the meaning of that phrase.

Words and Phrases

One step above law dictionaries in the hierarchy of secondary legal authorities is a West publication called **Words and Phrases**. As the title suggests, by using this series the researcher can gain information about words and phrases that are commonly used in the law. Unlike law dictionaries, the series of volumes called Words and Phrases typically provides page after page of definitions of the same term. All definitions are taken directly from judicial decisions in which the word or phrase has been considered. Citations to these decisions are always provided,

---

Exhibit 2–3

cruel and unusual punishment. The Eighth Amendment states: "Excessive bail shall not be required, nor excessive fines imposed, nor cruel and unusual punishments inflicted." U.S. Const. amend. VIII. The U.S. Supreme Court has construed the phrase *cruel and unusual punishment* to include not just barbarities such as torture but also punishment that is excessive for the crime committed. See *Coker v. Georgia*, 433 U.S. 584, 598 (1977) (stating that a death sentence was a disproportionate punishment for rape because "rape . . . in terms of moral depravity and of the injury to the person and to the public . . . does not compare with murder, which does involve the unjustified taking of human life.").

Source: From A DICTIONARY OF MODERN LEGAL USAGE, SECOND EDITION by Bryan A. Garner. Copyright © 1995 by Bryan A. Garner. Used by permission of Oxford University Press, Inc.

making for a convenient and direct link between this secondary authority and case law. In contrast to law dictionaries, which can be purchased and kept available for home use, you will have to go to a library to consult Words and Phrases. This collection consists of some 90 volumes and spreads out over several good-size library shelves. Entries are arranged alphabetically by word or phrase. To give you an idea of the expansive nature of Words and Phrases, the first volume covers just "A to Accident." Volume 1A includes "Accidental to Across," and so on.

Because law is a very rapidly changing discipline, to keep up with these changes, the published sources of the law must undergo continuous supplementation and revision. Book series such as Words and Phrases consist of voluminous, bulky, and rather elegant-looking bound volumes, which can be expensive—much too expensive to discard, modify, and replace on an annual basis. Law book publishers need a mechanism for updating reference materials without reprinting new editions each year. The common solution is to insert a paper supplement, or "**pocket part**," placed in a special slit in the back cover of a book. *It is absolutely imperative that you develop the practice of checking hardbound legal reference books for these paper supplements,* which incorporate developments in the law occurring after the publication of the original volume. This practice is not optional. You must do it, as a matter of course. If you do not, you run the risk that the law as stated in the bound volume has changed, or is incomplete, or is no longer even worth the paper on which it is printed. When you consult Words and Phrases or many of the other printed legal authorities, you must refer not only to the material in the original hardbound volume, but to the pocket part as well.

Any of a variety of words and phrases that may be of interest to you can be looked up in Words and Phrases. Thus, if you want to know how "statutory rape" is defined in different jurisdictions, or what "probable cause" means, Words and Phrases can supply definitions derived from the discussion of those phrases in judicial decisions. We illustrate what Words and Phrases has to offer by investigating our hypothetical case issues with the assistance of this reference.

Let us pursue the meaning of "deadly weapon," since that phrase is relevant to the biting and spitting by the HIV-positive defendant, Andrew Adams, in Case 2. To look up "deadly weapon," we simply scan the spines of the Words and Phrases volumes until we come to the one containing the alphabetical sequence that would include "deadly weapon." We see that volume 11 includes the entries from "Dacion" through "Deciduous Plant," so we thumb through the pages until we locate "**Deadly Weapon**." The result is pictured in Exhibit 2–4. Note the convenient subclassifications under this entry, which unfortunately do not include "teeth" or "spit," or "AIDS," or another topic directly implicated by our case facts. Still, we may be able to learn by analogy something interesting from the subtopic, "**Hands and feet**." When we consult "Hands and feet" (see Exhibit 2–5), we find a half dozen excerpts from different jurisdictions that generally relate to whether hands and feet are considered deadly weapons when used to injure another person. We then look up the cases cited—for example, *State v. Smith*, 83 P.2d 749, 754, 196 Wash. 534

CORPUS JURIS SECUNDUM V.26 1996-97

POCKET PART IN BACK COVER SLOT—THE LATEST!

Exhibit 2–4

DEADLY WEAPON
  In general—p. 206
  Automobile—p. 209
  Ax—p. 210
  Ax handle—p. 210
  Baseball bat—p. 210
  Billiard cue—p. 210
  Blackjack—p. 210
  Bludgeon, see Pitchfork used as bludgeon—
    p. 226
  Bottles—p. 210
  Bow and arrow—p. 210
  Brass knuckles—p. 210
  Brick or brickbat—p. 211
  Buggy trace—p. 211
  Chair—p. 211
  Chisel—p. 211
  Club or stick—p. 212
  Crowbar—p. 213
  Dangerous weapon included—p. 213
  Eye hoe—p. 213
  Fence pole—p. 214
  Fingernail file—p. 214
  Fist—p. 214
  Gun or rifle—p. 215
  Hammer—p. 215
  Hands and feet—p. 216
  Hoe—p. 216
  Iron weight—p. 216
  Knife—p. 216
  Likely to produce death—p. 219
  Piece of metal—p. 222
  Piece of pipe—p. 222
  Pin—p. 222
  Pistol—p. 222
    In general—p. 222

  Toy pistol—p. 225
  Unloaded or defective—p. 225
    Used as club or bludgeon—p. 225
  Pitchfork used as bludgeon—p. 226
  Razor—p. 226
  Rifle, see Gun or rifle—p. 215
  Rock—p. 226
  Shoes—p. 226
  Shotgun—p. 227
    In general—p. 227
    Used as club—p. 228
  Size and manner of use—p. 228
  Sledge hammer—p. 229
  Sling shot—p. 229
  Steel screw-driver—p. 229
  Steel wedge—p. 229
  Steel-yard—p. 229
  Stick, see Club or stick—p. 212
  Stone—p. 229
  Surgical instruments and drugs—p. 229
  Telephone—p. 230
  Tire iron—p. 230
  Whip—p. 230

Cross References
Armed With a Deadly Weapon
Assault and Battery with a Dangerous and
  Deadly Weapon
Assault with Dangerous or Deadly Weapon
Assault with a Deadly Weapon
Assault With Intent to Kill By Means of a
  Deadly Weapon
Dangerous And Deadly Weapon
Or any Dangerous or Deadly Weapon
Other

Source: Words and Phrases, "Deadly Weapon" (Index), vol. 11, p. 206, © West Publishing, used with permission.

[(1938)]. We prefer to locate this case in West's regional reporter, the **Pacific Reporter**, 2d Series, rather than in the official **Washington Reports**. We would expect to find a discussion resembling the passage presented in Words and Phrases at page 754 in volume 83 of the Pacific 2d case reports. (The *Smith* case actually begins at page 749, as the citation indicates.)

We are not yet finished with the Words and Phrases volume, however. It is hoped that our recent admonition that you must check the pocket part will occur to

Exhibit 2–5

Hands and feet

Hands and feet, which defendant, according to indictment, used in killing deceased, were not "deadly weapons," per se, and they could become such only in manner used. Ray v. State, 266 S.W.2d 124, 128, 160 Tex.Cr.R.12.

Hands and feet are not "deadly weapons" within statute condemning willful and malicious cutting, stabbing, or striking another with knife or other deadly weapon with intent to kill. Reed v. Com., Ky., 248 S.W.2d 911, 914.

Hands are not "deadly weapons" within meaning of statute relating to the offense of maliciously cutting and wounding another with a deadly weapon with intent to kill. Bradley v. Com., 236 S.W.2d 266, 267, 314 Ky. 457.

Hands and feet are not "deadly weapons" within meaning of statute relative to assault with deadly weapons with intent to kill. McIntosh v. Com., 120 S.W.2d 1031, 1033, 275 Ky. 126.

The fists, though not generally a "deadly weapon," may become deadly by blows often repeated, long continued, and applied to vital and delicate parts of the body of a defenseless, unresisting man. State v. Smith, 83 P.2d 749, 754, 196 Wash. 534.

A person's hands and feet are not "deadly weapons," within the meaning of the law, and, where death results unintentionally from their use in an assault, the result is not murder but involuntary manslaughter. Thomas v. Com., 86 S.W. 694, 695, 27 Ky. Law.Rep. 794.

Source: Words and Phrases, "Deadly Weapon: Hands and Feet," vol. 11, p. 216, © West Publishing, used with permission.

you and that you will look for information that became available after the hardbound edition of Words and Phrases was published, which in this case was 1971. We thus locate the pocket part placed in the back of volume 11, which collects relevant definitions from cases decided between the publication dates of that bound volume and the most recent paper supplementation. We consult the 1995 Supplement and find cases from three jurisdictions that address whether hands and feet are considered deadly weapons (see Exhibit 2–6).

By employing these same techniques, we could use Words and Phrases to look up "**Attempted murder**," which is another possible charge in Case 2, or, more generally, "**Attempt to commit crime**." For Case 1 and Case 3, respectively, we may wish to find "**Confrontation**" and "**Cruel and unusual punishment**" in the appropriate volumes of Words and Phrases. You would benefit by visiting a library, finding the Words and Phrases series on the shelves, and going through this exercise in order to practice using these volumes.

Legal Encyclopedias

There are two legal encyclopedias of national scope and several encyclopedia series that focus on legal issues in individual states. The two national legal encyclopedias are **Corpus Juris Secundum**, which is published by West, and **American Jurisprudence 2d**, published by Lawyers Cooperative Publishing Company.

Exhibit 2–6

...highly capable of causing death or serious bodily harm and was, therefore, "deadly weapon" for purposes of first-degree robbery. Goolsby v. State, Ala.Cr.App., 492 So.2d 635, 637.

Hands and feet

Defendant's use of his hands to attack his wife did not constitute use of "deadly weapon" for purpose of conviction for aggravated battery committed through use of deadly weapon. State v. Townsend, 865 P.2d 972, 977, 124 Idaho 881.

Hands may be considered "deadly weapons,"

depending upon manner in which they were used and relative size and condition of parties involved. State v. Grumbles, 411 S.E.2d 407, 409, 104 N.C.App. 766.

Although neither hand nor pair of underpants is deadly weapon per se, hands and underpants were "deadly weapons" where defendant used his hands and a pair of underpants to apply pressure to victim's nose and mouth to cut off her air supply. Morales v. State, Tex.App.–Hous. [1 Dist.], 792 S.W.2d 789, 790.

Source: Words and Phrases, "Deadly Weapon: Hands and Feet," vol. 11, 1995 Supp., p.78, © West Publishing, used with permission.

As their names suggest, each of these sets is in its second edition. Corpus Juris Secundum is universally referred to simply as "C.J.S.," and American Jurisprudence 2d is commonly known as "Am. Jur. 2d."

Legal encyclopedias in many respects resemble the **World Book Encyclopedia**, the **Encyclopedia Britannica**, and the other general encyclopedias with which you are familiar. They provide descriptive articles about countless topics relevant to the law. However, they differ from the World Book type of encyclopedia in some important attributes. In order to find articles about particular topics in a legal encyclopedia you must first consult a **General Index**. The process of locating relevant articles is not difficult, but it is not quite as straightforward as opening a World Book at the logical alphabetical entry point. Another difference is that legal encyclopedias make extensive use of footnotes, which typically cite and sometimes describe judicial decisions related to the points made in the text of the encyclopedia article. The authorities cited in the footnotes can provide very useful leads for further research.

We can get a better understanding about what information legal encyclopedias have to offer, and how they are used, by returning to our case scenarios. We initially focus on the confrontation issue presented in Case 1, and the smoking-in-prison issue raised in Case 3 by consulting Am. Jur. 2d in the library. We first examine the alphabetically arranged General Index in Am. Jur. 2d with the hope that words and concepts related to the issues we are researching are listed and refer to helpful encyclopedia articles.

We might begin our search through the General Index with any of a number of terms. For Case 1, we could try "**confrontation**," "**witness**," "**child**," "**television**," or several other logical starting places. For Case 3, terms and phrases such as "**cruel and unusual punishment**," "**smoke**," "**prisons**," "**civil rights**," and others come to mind. The General Index should be cross-referenced in such a way that no matter where we start, we eventually end up being pointed to the same articles in Am. Jur. 2d. If we start with "confrontation," we quickly discover the topic, "**Confrontation of witnesses**." As we scan the subtopics under the heading, as illustrated in Exhibit 2–7, we note with particular interest "**Television testimony, Witn § 721**," and "**Videotape, Crim L § 960; Trial § 345; Witn § 721**." The double-squiggle symbol § indicates a section of the encyclopedia. The abbreviations "Witn" and "Crim L," and the word "Trial" refer us to other subjects arranged alphabetically in the Am. Jur. 2d volumes. Thus we find that the general subject we are interested in apparently is discussed in different contexts in the encyclopedia.

To explore the issues in Case 3, we could begin by looking up "**Prisons**" in the General Index, and we would be referred to "**Penal and correctional institutions**." Among the interesting subtopics under this heading are "**Conditions of incarceration**, generally, Penal Inst § 26–110," and as a subheading of "**Constitutional law**" we find "**Cruel and Unusual Punishment** (this index)." (See Exhibit 2–8.) When we look up "**Cruel and unusual punishment**" in the General

---

Exhibit 2–7

AMERICAN JURISPRUDENCE 2d

CONFRONTATION OF WITNESSES
—Cont'd
Sentence and punishment, Crim L § 723
State court decisions, generally, Crim L § 956–966
Stenographers, Crim L § 728
Telecommunications, Crim L § 960
Television testimony, Witn § 721
Understanding of testimony by accused, Crim L § 724
Unemployment compensation, hearings on entitlement to benefits, Unempl C § 210
Verdict, Crim L § 722
Videotape, Crim L § 960; Trial § 345; Witn § 721

CONFRONTATION OF WITNESSES
—Cont'd
View by jury, applicability of rights, Crim L § 723, 915
Waiver and estoppel, Crim L § 473, 724, 730, 965, 966; Estop § 163
Welfare laws, constitutional requirement of confrontation of witnesses in administrative hearing, Welf § 107
When right prevails, Crim L § 723, 958
CONFUCIUS
Constitutional law, definition of religious freedom, Const L § 465

Source: Am. Jur.2d, General Index (C-E), "Confrontation of Witnesses: Television Testimony; Videotape," p. 292. Permission has been granted by the current copyright holder, West Group. Further reproduction of any kind is strictly prohibited. For additional information, please contact West Group Customer Services representative at 1-800-328-4880.

Exhibit 2–8

GENERAL INDEX

PENAL AND CORRECTIONAL INSTITUTIONS—Cont'd

Commissioners of county. County and county commissioners, infra

Commitment

– defined, Penal Inst § 7

– **Imprisonment** (this index)

– mental incompetents, Penal Inst § 153, 156

– warrant of, Crim L § 587

Communicable diseases, Penal Inst § 201

Communication, Mail or correspondence of inmates, infra

Commuted life sentences, Penal Inst § 230

Compacts, Crim L § 404–407, 622, 650; Penal Inst § 151

Compensation

– costs and expenses, infra

– garnishment of compensation of jailer, Attach § 180

– injuries to inmates, Penal Inst § 186, 187, 202, 206, 209

– labor by inmates, infra

– police power, Const L § 408

– prison compensation act, Penal Inst § 186, 187

– sheriff and jailer, Sheriff § 68

– **Workers' Compensation** (this index)

Competency. Insane or incompetent persons, infra

Complaint. Pleadings, infra

Compulsory process to procure attendance of convicts, Crim L § 954

Concurrent or consecutive sentences, Penal Inst § 228, 231, 233

Condemnation of property, Em Dom § 40; Penal Inst § 9

Condition precedent

– habeas corpus, prisoner's violation of conditions of pardon or parole, Hab Corp § 75

– **Pardon and Parole** (this index)

Conditions of incarceration, generally, Penal Inst § 26–110

Conduct. Behavior, supra

Confessions (this index)

PENAL AND CORRECTIONAL INSTITUTIONS—Cont'd

Confidential information, Penal Inst § 109, 129, 132

Confinement. Imprisonment, infra

Confiscation of property. Search and seizure, infra

Conflict of interest, Penal Inst § 135

Confrontation of witnesses, Penal Inst § 132, 235

Conjugal visits, Penal Inst § 84

Consecutive or concurrent sentences, Penal Inst § 228, 231, 233

Consent

– access to files, Penal Inst § 109

– marriage of inmates, Penal Inst § 102

– medical treatment, Penal Inst § 97

– place of incarceration, Penal Inst § 15

– pretrial detainees, Penal Inst § 113

– search, Penal Inst § 80, 214; Search § 84, 89

– transfer of inmates, Penal Inst § 153, 158

– wrongful conviction and incarceration, Penal Inst § 210

Conspiracy, Penal Inst § 18

Constitutional law

generally, Penal Inst § 14, 120–122

– assault, constitutional prohibition against imprisonment for debt, Const L § 621

– assembly or demonstrations, Const L § 520, 532; Penal Inst § 127

– association, Const L § 538, 550; Penal Inst § 189

– civil rights and discrimination, supra

– **Cruel and Unusual Punishment** (this index)

– demonstrations or assemblies, Const L § 520, 532; Penal Inst § 127

– discrimination. Civil rights and discrimination, supra

– due process, infra

– equal protection, infra

– escape from prison, Const L § 801; Escape § 9, 11, 17, 18

– ex post facto, Penal Inst § 151, 227

– Fifth Amendment, Penal Inst § 133

Source: Am. Jur.2d, General Index (M–Q), "Penal and Correctional Institutions: Conditions of Incarceration; Constitutional Law—Cruel and Unusual Punishment," p. 639. Permission has been granted by the current copyright holder, West Group. Further reproduction of any kind is strictly prohibited. For additional information, please contact West Group Customer Services representative at 1-800-328-4880.

Index, the most interesting subtopics appear to be "**Cell occupancy and conditions; Penal Inst § 87-89**." (See Exhibit 2–9.)

We now can proceed from the Am. Jur. 2d General Index to the encyclopedia articles corresponding to the topics we have uncovered. Let us first consult "Witnesses." We locate "**Witnesses**" in volume 81 of Am. Jur. 2d., according to the alphabetical arrangement of topics appearing on the books' spines. We could turn directly to section 721 under "Witnesses," but instead we decide to get an overview of the organization of the article addressing this topic, and the numerous subtopics covered. We do so by turning to the very front of the "Witnesses" article and examining the general outline there (Exhibit 2–10). We also inspect the more detailed outline, including the section to which we have been referred, § 721 (Exhibit 2–11). Note in Exhibit 2–11 that Witnesses § 721 corresponds to "**Constitutionality of televised or videotaped testimony**," a subtopic that appears interesting for our purposes. But also note § 720: "**Television or videotape testimony; child witness in sex offense case**," a subtopic that also seems to be on point. You justifiably might wonder why we were not referred by the General Index to Witnesses § 720. This is a fair question; the answer is that neither the indexing systems nor our ability to use them are perfect. Perhaps we would have been directed to this section with a bit more digging. The lesson to take from this example is that it pays to consult both the General Index and the outlines at the beginning of an encyclopedia article to make sure you have covered all the bases. Another lesson is that you should smile when fate unexpectedly rewards you with a serendipitous discovery in the course of your research.

Exhibit 2–9

CRUEL AND UNUSUAL PUNISHMENT
Generally, Crim L § 625–631; Penal Inst § 26, 189
Abatement of nuisance, Crim L § 627
Aliens and citizens, Aliens § 1863; Crim L § 627
Appeal and review, Penal Inst § 121
Assault and battery, Asslt & B § 108; Penal Inst § 190, 193, 197
Banishment, Crim L § 624, 627
Bread and water diets, Crim L § 631
Burning alive, Crim L § 627
Capital offenses and punishment, Crim L § 627, 628, 631
Castration, Crim L § 627
Cell occupancy and conditions, Penal Inst § 87–89

Civil death statute prohibiting marriage, Crim L § 1035
Civil rights and discrimination, Civ R § 19; Crim L § 628
Construction and interpretation, Const L § 132, 147; Crim L § 626
Contempt, Contempt § 228, 232, 235
Coram nobis and post-conviction procedure, Coram N § 47
Corporal punishment, Penal Inst § 141, 142
Costs, Costs § 110; Crim L § 1049
Death and death actions, Death § 665
Death penalty, Crim L § 627, 628, 631
Definition, Crim L § 626

Source: Am. Jur.2d, General Index (C-E), "Cruel and Unusual Punishment: Cell Occupancy and Conditions," p. 618. Permission has been granted by the current copyright holder, West Group. Further reproduction of any kind is strictly prohibited. For additional information, please contact West Group Customer Services representative at 1-800-328-4880.

Exhibit 2–10

WITNESSES   81 Am Jur 2d

Outline

I. ATTENDANCE OF WITNESSES [§§ 1–49]
   A. DUTY OF WITNESS TO ATTEND COURT [§ 1]
   B. POWER OF COURT TO COMPEL ATTENDANCE [§§ 2–6]
   C. SUBPOENA [§§ 7–17]
   D. SUBPOENA DUCES TECUM [§§ 18–33]
   E. SECURING ATTENDANCE OF WITNESSES FROM WITHOUT STATE IN CRIMINAL PROCEEDINGS [§§ 34–49]
II. CALLING OF WITNESSES [§§ 50–67]
   A. IN GENERAL [§§ 50–59]
   B. NOTICE OR DISCLOSURE OF WITNESSES [§§ 60–67]
III. COMPENSATION AND FEES FOR ATTENDANCE [§§ 68–74]
IV. DUTY TO TESTIFY [§§ 75–162]
   A. IN GENERAL [§§ 75–79]
   B. PRIVILEGE AGAINST SELF-INCRIMINATION [§§ 80–130]
   C. IMMUNITY [§§ 131–151]
   D. WAIVER OF PRIVILEGE [§§ 152–162]
V. COMPETENCY [§§ 163–284]
   A. GENERAL REQUISITES [§§ 163–168]
   B. DETERMINATION OF COMPETENCY; IN GENERAL [§§ 169–177]
   C. PARTICULAR FACTORS SIGNIFICANT TO DETERMINATION OF COMPETENCY [§§ 178–209]
   D. COMPETENCY OF PARTICULAR PERSONS [§§ 210–284]
VI. PRIVILEGED RELATIONS AND COMMUNICATIONS [§§ 285–556]
   A. IN GENERAL [§§ 285–295]
   B. MARITAL PRIVILEGE [§§ 296–336]
   C. ATTORNEY-CLIENT PRIVILEGE [§§ 337–435]
   D. PHYSICIAN AND PATIENT [§§ 436–512]
   E. PRIVILEGE AS TO COMMUNICATIONS TO OR BY CLERGY OR OTHER SPIRITUAL ADVISERS [§§ 513–523]
   F. PRIVILEGE AS TO GOVERNMENTAL MATTERS AND INFORMATION [§§ 524–536]
   G. OTHER PRIVILEGED COMMUNICATIONS [§§ 537–556]
VII. TRANSACTIONS WITH PERSONS SINCE DECEASED OR INCOMPETENT; DEAD MAN'S STATUTES [§§ 557–707]
   A. IN GENERAL [§§ 557–571]
   B. ACTIONS AND PROCEEDINGS WITHIN STATUTE [§§ 572–595]
   C. SUBJECT MATTER OF TESTIMONY [§§ 596–633]
   D. PRESENCE OF OR TRANSACTION WITH THIRD PERSON [§§ 634–637]
   E. PERSONS SILENCED [§§ 638–678]
   F. PERSONS PROTECTED [§§ 679–685]

continues

| Exhibit 2–10 | continued |

81 Am Jur 2d   WITNESSES

G. STATUTES AS TO SURVIVING PARTY TO CONTRACT OR CAUSE OF ACTION [§§ 686–693]
H. WAIVER OF BENEFIT OF DEAD MAN'S STATUTE [§§ 694–707]
VIII. OATH [§§ 708–712]
IX. EXAMINATION [§§ 713–768]
   A. IN GENERAL [§§ 713–742]
   B. QUESTIONS TO WITNESSES [§§ 743–756]
   C. ANSWERS OF WITNESSES [§§ 757–768]
X. REFRESHING MEMORY [§§ 769–799]
   A. IN GENERAL [§§ 769, 770]
   B. BY WRITINGS OR MEMORANDA [§§ 771–799]
XI. CROSS-EXAMINATION [§§ 800–861]
   A. IN GENERAL [§§ 800, 801]
   B. RIGHT OF CROSS-EXAMINATION [§§ 802–805]
   C. PERSONS SUBJECT TO CROSS-EXAMINATION [§§ 806–810]
   D. SCOPE AND EXTENT OF CROSS-EXAMINATION [§§ 811–848]
   E. CONDUCT AND MODE [§§ 849–857]
   F. APPELLATE REVIEW; EFFECT OF ERROR [§§ 858–861]
XII. IMPEACHMENT [§§ 862–1000]
   A. IN GENERAL [§§ 862–867]
   B. GROUNDS OR BASES, IN GENERAL [§§ 868–963]
   C. IMPEACHMENT OF ACCUSED [§§ 964–977]
   D. IMPEACHMENT OF PARTY'S OWN WITNESS [§§ 978–991]
   E. CONTRADICTION BY EXTRINSIC EVIDENCE [§§ 992–1000]
XIII. CORROBORATION [§§ 1001–1026]
   A. IN GENERAL [§§ 1001–1005]
   B. CHARACTER AND REPUTATION [§§ 1006–1010]
   C. PRIOR CONSISTENT STATEMENTS [§§ 1011–1026]
XIV. CREDIBILITY OF WITNESSES [§§ 1027–1045]
   A. IN GENERAL [§§ 1027–1037]
   B. FACTORS IN DETERMINATION OF CREDIBILITY [§§ 1038–1045]

Source: Am. Jur.2d, vol. 81, pp. 4–5, "Witnesses" (Outline). Permission has been granted by the current copyright holder, West Group. Further reproduction of any kind is strictly prohibited. For additional information, please contact West Group Customer Services representative at 1-800-328-4880.

We now turn to sections 720 and 721 in the Witnesses article in Am. Jur. 2d. (See Exhibit 2–12.) The text of the article describes the general legal principles related to the subtopics we have found. Also, note the wealth of cases and other authorities cited in the accompanying footnotes. Pay special attention to the U.S. Supreme Court case cited in footnote 1 under section 720 and in footnote 11 under section 721: "Maryland v Craig, [497 U.S. 836], 111 L Ed 2d 666, 110 S Ct 3157 (1990)]." The encyclopedia article suggests that the practice of presenting a child

Exhibit 2–11

Source: Am. Jur.2d, vol. 81, p. 27, "Witnesses" (Outline). Permission has been granted by the current copyright holder, West Group. Further reproduction of any kind is strictly prohibited. For additional information, please contact West Group Customer Services representative at 1-800-328-4880.

witness's testimony over closed-circuit television in a sex offense prosecution does not inevitably violate the defendant's right to confront accusing witnesses. *Maryland v. Craig* is cited in support of this proposition. We strongly recommend that you never rely on a legal encyclopedia article—or any other secondary legal authority—as constituting the last word on a question of law. You should read the accompanying authority cited and confirm that the asserted principle is fairly supported by that authority. In this instance, you certainly will want to track down *Maryland v. Craig* and read this case decision itself.

Exhibit 2–12

§ 720. Television or videotape testimony; child witness in sex offense case

A statute may provide that a witness, under specific circumstances, may testify by means of videotaped testimony [99] or closed-circuit television, [1] as where in the case of a child victim-witness in a sex offense prosecution, presenting the testimony in such a manner is in the best interest of the child witness, [2] where an appropriate individualized showing of necessity or compelling need is made [3]—for instance, where by clear and convincing evidence an adequate showing has been made of the child witness's vulnerability to severe mental and emotional harm [4]—or where the court makes a specific finding of a substantial likelihood that the child witness and sex abuse victim would suffer at least moderate emotional or mental harm if required to testify in open court.[5]

> **IIII** *Observation:* It has been observed that if a statute permitting the closed-circuit televised testimony of a child witness in a sex offense prosecution were interpreted as allowing the determination of the vulnerability of the child witness to be based solely upon the trial court's own observations, such a statute would not be facially unconstitutional; such statute must be construed as also requiring the same findings by clear and convincing evidence.[6]

In the absence of a statute, the trial court does not possess the inherent power to substitute examination of witnesses by closed-circuit television for a live, in-court examination.[7]

§ 721. —Constitutionality of televised or videotaped testimony

Generally, the admission of testimony by a witness who is not present in the courtroom for examination, by means of videotape[8] or a live closed-circuit television, does not constitute an inherent violation of the defendant's right to confront witnesses against him, as guaranteed under the Sixth[9] and Fourteenth[10] Amendments of the United States Constitution and the relevant provisions of certain state constitutions.[11] Particularly in a misdemeanor prosecution, the examination of an expert witness for the prosecution, on closed-circuit television does not infringe on the defendant's rights to due process, counsel, or fair trial.[12] If the state makes an adequate showing of compelling reason[13] or necessity, the state's interest in protecting child witnesses from the trauma of testifying in a child abuse case is sufficiently important to justify the use of a special procedure permitting a child witness to testify at trial in the absence of a face-to-face confrontation with the defendant.[14] However, it also has been stated that the videotaping of a minor witness's testimony, in a . . .

94. Re Martin (Minn App) 458 NW2d 700.

95. Carter-Wallace, Inc. v Otte (CA2 NY) 474 F2d 529, 176 USPQ 2, 176 USPQ 452, cert den 412 US 929, 37 L ED 2d 156, 93 S Ct 2753, 178 USPQ 65.

96. Hoffman on behalf of NLRB v Beer Drivers & Salesmen's Local Union, etc. (CA9 Cal) 536 F2d 1268, 92 BNA LRRM 3302, 79 CCH LC ¶ 11489, 21 FR Serv 2d 1442.

97. Lebeck v William A. Jarvis, Inc. (CA3 PA) 250 F2d 285.

98. United States v Aluminum Co. of America (DC NY) 1 FRD 48.

Practice References: 33 Federal Procedure, L Ed, Witness § 80:43.

99. Vigil v Tansy (CA10 NM) 917 F2d 1277, 31 Fed Rules Evid Serv 689, cert den (US) 112 L Ed 2d 1078, 111 S Ct 995 (applying New Mexico law); State v Vincent, 159 Ariz 418, 768 P2d 150, 26 Ariz Adv Rep 33; State v Darby, 19 Conn App 445, 563 A2d 710, app den 213 Conn 801, 567 A2d 833; Leggett v State (Fla) 565 So 2d 315, 15 FLW 375; Casada v State (Ind App) 544 NE2d

continues

Exhibit 2–12   continued

189; State v Lamb, 14 Kan App 2d 664, 798 P2d 506; Commonwealth v Dockham, 405 Mass 618, 542 NE2d 591; State v Taylor (RI) 562 A2d 445; Ochs v Martinez (Tex App San Antonio) 789 SW2d 949.

1. Maryland v Craig (US) 111 L Ed 2d 666, 110 S Ct 3157, 30 Fed Rules Evid Serv 1, on remand 322 Md 418, 588 A2d 328; state v Vess (App) 157 Ariz 236, 756 P2d 333; Spoerri v State (Fla App D3) 561 So 2d 604, 15 FLW 959; People v Schmitt (4th Dist) 204 Ill App 3d 820, 149 Ill Dec 913, 562 NE2d 377, app den 137 Ill 2d 670, 156 Ill Dec 567, 571 NE2d 154; State v Eaton, 244 Kan 370, 769 P2d 1157; Commonwealth v Willis (Ky) 716 SW2d 224; State v Crandall, 120 NJ 649, 577 A2d 483; People v Cintron, 75 NY2d 249, 552 NYS2d 68, 551 NE2d 561.

2. People v Schmitt (4th Dist) 204 Ill App 3d 820, 149 Ill Dec 913, 562 NE2d 377, app den 137 Ill 2d 670, 156 Ill Dec 567, 571 NE2d 154.

IIII *Observation:* 18 USCS § 3509, added to protect the rights of child victims who become witnesses in a prosecution for crimes of child abuse, provides that the government attorney, the child's attorney, or a guardian ad litem may apply for a court order that the child's testimony be taken outside the courtroom and be televised by two-way closed-circuit television, or that a deposition be taken of the child's testimony on videotape.

Law Reviews: T.R. Finn, Child Witness Practice in Child Abuse Proceedings. Suffolk U Law Rev 271–343 (1989).

3. State v Vincent, 159 Ariz 418, 768 P2d 150, 26 Ariz Adv Rep 33; People v Cintron, 75 NY2d 249, 552 NYS2d 68, 551 NE2d 561; Commonwealth v Amirault, 404 Mass 221, 535 NE2d 193, companion case 407 Mass 927, 556 NE2d 83 (compelling need); State v Vess (App) 157 Ariz 236, 756 P2d 333 (compelling need).

4. People v Cintron, 75 NY2d 249, 552 NYS2d 68, 551 NE2d 561; People v Henderson (2d Dept) 156 App Div 2d 92, 554 NYS2d 924, app den 76 NY2d 736, 558 NYS2d 898, 557 NE2d 1194.

5. Leggett v State (Fla) 565 So 2d 315, 15 FLW 375 (videotape); Spoerri v State (Fla App D3) 561 So 2d 604, 15 FLW 959 (closed-circuit television).

6. People v Cintron, 75 NY2d 249, 552 NYS2d 68, 551 NE2d 561.

As to constitutionality of statutes allowing videotaped or televised testimony, see § 721.

7. Hochheiser v Superior Court (2nd Dist) 161 Cal App 3d 777, 208 Cal Rptr 273.

Annotations: Closed-circuit television witness examination, 61 ALR4th 1155 § 3.

8. State v Murrell, 302 SC 77, 393 SE2d 919.

9. US Const Amend 6.

10. US Const Amend 14.

11. Maryland v Craig (US) 111 L Ed 2d 666, 110 S Ct 3157, 30 Fed Rules Evid Serv 1, on remand 322 Md 418, 588 A2d 328; Commonwealth v Willis (Ky) 716 SW2d 224; Kansas City v McCoy (Mo) 525 SW2d 336, 80 ALR3d 1203; State v Warford, 223 Neb 368, 389 NW2d 575, 61 ALR4th 1141; State v Sheppard, 197 NJ Super 411, 484 A2d 1330; State v Crandall, 120 NJ 649, 577 A2d 483; People v Cintron, 75 NY2d 249, 552 NYS2d 68, 551 NE2d 561 (by implication); People v Guce (2d Dept) 164 App Div 2d 946, 560 NYS2d 53, app den 76 NY2d 986, 563 NYS2d 775, 565 NE2d 524; People v Algarin, 129 Misc 2d 1016, 498 NYS2d 977; Re Burchfield (Athens Co) 51 Ohio App 3d 148, 555 NE2d 325.

Annotations: Closed-circuit television witness examination, 61 ALR4th 1155 § 4.

Source: Am. Jur.2d, vol. 81, pp. 589–590, "Witnesses" (sec. 720, 721). Permission has been granted by the current copyright holder, West Group. Further reproduction of any kind is strictly prohibited. For additional information, please contact West Group Customer Services representative at 1-800-328-4880.

We are not finished with Am. Jur. 2d.'s treatment of this issue. Again we remind you that you must check the pocket part of a legal encyclopedia, or of any other hardbound reference of this nature. The pocket supplement to a legal encyclopedia should describe later legal developments related to the issue discussed in the main volume, and cite more recently decided cases and other authorities. You will note the more recent cases cited in the pages of the pocket supplement corresponding to Witnesses § 720 (see Exhibit 2–13), including two particularly interesting cases decided by the supreme courts of Florida (*Myles v. State*, 602 So. 2d 1278 (Fla. 1992)), and South Carolina (*State v. Murrell*, 302 S.C. 77, 393 S.E.2d 919 (1990)).

We cover similar ground in Am. Jur. 2d to discover information about Case 3; that is, whether a prisoner's sustained, involuntary exposure to second-hand tobacco smoke can be a form of cruel and unusual punishment. Recall that we were referred by Am. Jur. 2d's General Index to the topic "**Penal and correctional institutions**" and that sections 87 through 89 of that article pertain to "**cell occupancy and cell conditions**." Because topics are arranged alphabetically, we locate this article easily in volume 60 of Am. Jur. 2d (which includes the topics "Patents" through "Penal and Correctional Institutions"). When we turn to the detailed outline at the beginning of the article, we see that section 89, "Sanitary condition of cell," is the most promising starting place. (See Exhibit 2–14.) On consulting this section of the article, however, we find nothing directly on point concerning the issue of second-hand smoke in prisons. (See Exhibit 2–15.) But we are not finished, and by this time you should know why: When we check the pocket part under "**Penal and Correctional Etc**. § 89," we strike pay dirt. (See Exhibit 2–16.) The first case discussed in the pocket part is a U.S. Supreme Court decision, *Helling v. McKinney*, [509 U.S. 25, 113 S. Ct. 2475, 125 L. Ed. 2d 22 (1993)], which we will be most interested to read. This case seems to suggest that a prisoner's claim along the same lines as Deborah Miller's complaint in Case 3 can state a legitimate basis for a lawsuit alleging cruel and unusual punishment, in violation of the Eighth Amendment to the U.S. Constitution.

When we look up the same issues in C.J.S. that we just examined in Am. Jur. 2d, we would expect to find roughly similar information and case authorities. Since C.J.S. is published by West and Am. Jur. 2d by Lawyers Cooperative, one can expect the secondary authorities cited in the encyclopedias will emphasize other materials produced by the respective publishing companies. For this reason, it usually pays to check both sets of encyclopedias. The coverage given various issues also differs to some extent in these two encyclopedias. C.J.S. is used in the same way Am. Jur. 2d is: By locating the topic of interest in a general index, consulting the topic or a specific section of the topic in the encyclopedia, making note of the cases and authorities cited in the footnotes accompanying the encyclopedia article, and checking the pocket part for developments that have occurred and authorities that have become available after the hardcover volume was published. We can illustrate this procedure by using C.J.S. to research the issues raised in Case 2.

To make use of C.J.S.'s General Index, we identify words and concepts that characterize our issue. For instance, we might try "**Deadly Weapon**," "**Attempted Murder**," "**Attempt**," "**Assault with a Deadly Weapon**," "**HIV**,"

Exhibit 2–13

WITNESSES

§ 720. Television or videotape testimony; child witness in sex offense case

Practice Aids:

The child victim/witness: balancing of defendant/victim rights in the emotional caldron of a criminal trial, 62 J Kans B 1:38 (1993).

The use of videotaped child testimony: Public policy implications, 7 Notre D J Law E&PP 387 (1993)

Children as witnesses after Maryland v. Craig [110 S. Ct. 3157], 65 S. Cal. L. Rev. 1993, (May 1992)

Case Authorities:

Before ordering closed-circuit testimony from a child witness the trial court must : (1) conduct an inquiry in which evidence is received on whether the closed-circuit procedure is necessary to protect the welfare of the particular child; (2) find that the child witness will be traumatized, not by the courtroom generally, but by the presence of the defendant; and (3) find that the emotional distress suffered by the child witness in the presence of the defendant is more than de minimis, i.e., more than mere nervousness or excitement or some reluctance to testify. Myles v State (1992, Fla) 602 So 2d 1278, 17 FLW S 444.

Defendant adjudicated guilty of sexual battery upon a child less than 12 years old failed to preserve his arguments concerning the child victim's testimony via closed circuit television as to whether the judge made sufficient findings as required under FS § 92.54, that there was substantial likelihood that the child would suffer at least moderate emotional or mental harm if required to testify in open court, where only objection made prior to the closed circuit testimony was a very general objection that did not raise the sufficiency of the factual findings of the judge and therefore did not preserve the issue for review. Hopkins v State (1992, Fla App D1) 608 So 2d 33, 17 FLW D 1774.

Where defendant was convicted of capital sexual battery, court did not err in denying defendant's request to have the child victim to testify via closed circuit television, because the child victim was found to be unavailable to testify at all within the meaning of FS § 90.803(23), and the Confrontation Clause does not allow the defendant to direct the State to call particular witnesses. Seaman v State (1992, Fla App D3) 608 So 2d 71, 17 FLW D 2422.

Complainant's videotaped grand jury testimony was not unduly prejudicial merely because she had bandages on her mouth as result of being shot in jaw, since she appeared tastefully dressed in her hospital gown, had no exposed wounds, and did not express visible pain during taping. People v Rafajlovski (1989, 2d Dept) 152 App Div 2d 608, 543 NYS2d 715, habeas corpus dismissed (ED NY) 1992 US Dist LEXIS 8454.

In determining whether a videotape procedure pursuant to § 16-3-1530 is necessary to protect a minor witness, the following procedure must be followed by the trial judge. First, the trial judge must make a case-specific determination of the need for videotaped testimony. In making this determination, the trial court should consider the testimony of an expert witness, parents or other relatives, other concerned or relevant parties, and the child. Second, the court should place the child in as close to a courtroom setting as possible. Third, the defendant should be able to see and hear the child, should have counsel present both in the courtroom and with him or her, and communication should be available between counsel and the defendant. A decision as to whether to utilize a videotape procedure is subject to reversal only if it is shown that the trial judge abused his or her discretion in making such a decision or failed to follow the appropriate procedure upon deciding that a witness was entitled to special protection. State v Murrell (1990) 302 SC 77, 393 SE2d 919.

Source: Am. Jur.2d, vol. 81, 1995 Supp., p. 22, "Witnesses" (sec. 720). Permission has been granted by the current copyright holder, West Group. Further reproduction of any kind is strictly prohibited. For additional information, please contact West Group Customer Services representative at 1-800-328-4880.

Exhibit 2–14

60 Am Jur 2d	PENAL AND CORRECTIONAL ETC.

§ 67.   Access to court files
§ 68.   Access rights of prisoners in disciplinary confinement
§ 69.   Access rights of jail prisoners
　　　　　2. ACCESS TO LEGAL ASSISTANCE
§ 70.   Access to counsel
§ 71.   —Denial of access to particular attorney for security reasons
§ 72.   Access to legal assistance of paraprofessionals and students
§ 73.   Access to legal assistance of other prisoners
§ 74.   Access to law libraries
§ 75.   —Effect of legal representation or assistance
§ 76.   —Limitations on time for use of library
§ 77.   Access to legal research materials
　　　　　H. INMATES' VISITORS AND VISITS
　　　　　　　1. IN GENERAL
§ 78.   Generally
§ 79.   Time, frequency, and duration of visits; number of visitors
§ 80.   Searches of visitors
§ 81.   Exclusion of intruders
§ 82.   Monitoring of visitors
　　　　　　　2. PARTICULAR TYPES OF VISITS OR VISITORS
§ 83.   Contact visitation
§ 84.   Conjugal visits
§ 85.   Visits with minor children
§ 86.   Interviews with media
　　　　　I.   CELL OCCUPANCY AND CELL CONDITIONS
§ 87.   Overcrowding, generally
§ 88.   —Double-bunking in cells
§ 89.   Sanitary condition of cell

Source: Am. Jur.2d, vol. 60, p. 1123, "Penal and Correctional Institutions," (Outline) sec. 87–89. Permission has been granted by the current copyright holder, West Group. Further reproduction of any kind is strictly prohibited. For additional information, please contact West Group Customer Services representative at 1-800-328-4880.

"**AIDS**," "**Bite**," "**Teeth**," "**Spit**," and other items relevant to the case scenario. The C.J.S. General Index is organized alphabetically, so we can easily scan it for the words we have chosen. Had we started with "**Deadly Weapon**," the index would instruct us to see "**Homicide, this index**," and "**Weapons, generally, this index**." When we turn to "Homicide," we find not only "Deadly Weapons," but also other subtopics that may be of interest, including "**Attempts**" and "**Assault with intent to murder or kill**." (See Exhibit 2–17.) Note that one of the subdivisions under "**Attempts**" is "**HIV, biting victim, intent to kill, Homic § 95**." This seems to be right on point, so we will be sure to consult section 95 of the article entitled "**Homicide**." Upon checking "**AIDS**" in the General Index, we are di-

---

Exhibit 2–15

PENAL AND CORRECTIONAL ETC.　　　　　　　60 Am Jur 2d

§ 89. Sanitary condition of cell

The combined impact of conditions in a jail, including the filthy condition of the inmates' mattresses, may constitute cruel and unusual punishment.[92] But the serving and consumption of meals in the same cell where bodily functions were performed does not constitute cruel and unusual punishment.[93]

Relief from unconstitutional living conditions imposed on inmates in penal institutions, judged under the generally applicable standards of the Eighth Amendment's proscription of cruel and un-

usual punishment,[94] should not be denied because the conditions are created in whole or in part by prisoners. The prisoners' conduct should not be a factor. Prison administrators must see that unsanitary conditions do not continue unabated even though the conditions are first caused by the inmates. If constitutional conditions cannot be maintained in one location because of inmates' misconduct, then the inmates who are not responsible must be moved to a location where their rights can be secured or the unruly inmates must be relocated.[95]

Source: Am. Jur.2d, vol. 60, pp. 1184–1185, "Penal and Correctional Institutions," sec. 89. Permission has been granted by the current copyright holder, West Group. Further reproduction of any kind is strictly prohibited. For additional information, please contact West Group Customer Services representative at 1-800-328-4880.

---

rected to "**See Acquired Immune Deficiency Syndrome (AIDS), generally, this index**." When we follow this instruction we observe two subtopics that look interesting: "**Assault and battery, teeth infected with, Asslt & B § 79**" and "**Attempted murder, biting victim, intent to kill, Homic § 95**." (See Exhibit 2–18.) Other points of entry in the General Index should be pursued, using the alternative words we identified, but we certainly will want to take a look at "**Homicide § 95**" and "**Assault and Battery § 79**."

Let us first examine the "Homicide" article in C.J.S. The C.J.S. volumes are arranged alphabetically by topic. Volume 40 of C.J.S. includes "Highways 175" through "Homicide 175," so it will have section 95 of the "Homicide" article. We start by perusing the general topic outline at the beginning of the article (see Exhibit 2–19), and then the more detailed outline (see Exhibit 2–20) to get an idea about the article's scope. Then we can turn to section 95 of the article, which addresses attempts to commit murder. Like the examples we saw in Am. Jur. 2d, the C.J.S. article combines a textual description of attempt to commit murder, with extensive footnoting of case authority. (See Exhibit 2–21.) Nothing appears in the text of the article about attempted murder and biting inflicted by an HIV-positive defendant; nothing, that is, until we turn to the pocket part. There we find specific discussion of this issue and the citation of a case from New Jersey, *State v. Smith*, which apparently would permit a conviction for attempted murder under the circumstances described in Case 2. (See Exhibit 2–22.)

We follow the same procedure to find the section of the other C.J.S. article that appeared to be of interest, "**Assault & Battery § 79**." We present the results of this search as they appear in the pocket supplement corresponding to "Assault &

Exhibit 2–16

PENAL AND CORRECTIONAL ETC.

§ 89. Sanitary condition of cell
Practice Aids:

Singer, "To Be Or Not To Be: What is the Answer?" The Use of Habeas Corpus to Attack Improper Prison Conditions. 13 NEJ Crim & Civ Con, Summer, 1987.

Case authorities:

On remand to a Federal District Court, where the United States Supreme Court—having held that a state prison inmate stated a cause of action under the cruel and unusual punishment clause of the Federal Constitution's Eighth Amendment by alleging that defendant prison personnel had, with deliberate indifference, exposed him to levels of environmental tobacco smoke (ETS) that posed an unreasonable risk of serious damage to the inmate's future health—affirms a Federal Court of Appeals' judgment reversing and remanding a United States Magistrate's decision granting a directed verdict in favor of the defense, the inmate must show (1) that he himself is being exposed to unreasonably high levels of ETS, as to which determination the fact that the inmate has been

moved to another prison and is no longer the cellmate of a five-pack-a-day smoker is relevant; (2) that the risk he complains of is not one that today's society chooses to tolerate, as the determination whether the inmate's conditions of confinement violate the Eighth Amendment requires, in addition to a scientific and statistical inquiry into the seriousness of the potential harm and the likelihood that such an injury to health will actually be caused by exposure to ETS, an assessment whether society considers the risk complained of to be so grave that it violates contemporary standards of decency to expose anyone unwillingly to such a risk; and (3) deliberate indifference on the part of prison personnel, to be determined (a) in light of prison authorities' current attitudes and conduct, which may have changed considerably since the judgment of the Court of Appeals, and (b) with consideration of arguments regarding the realities of prison administration. Helling v McKinney (US) 125 L Ed 2d 22, 93 CDOS 4501, 93 Daily Journal DAR 7681, 7 FLW Fed S 452.

Source: Am. Jur.2d, vol. 60, 1995 Supp., p. 60, "Penal and Correctional Institutions," sec. 89. Permission has been granted by the current copyright holder, West Group. Further reproduction of any kind is strictly prohibited. For additional information, please contact West Group Customer Services representative at 1-800-328-4880.

Exhibit 2–17

HOMICIDE

HOMICIDE—Continued
Assault with intent to murder or kill, **Homic § 98 et seq.**
Ability to kill, evidence as to, **Homic § 319**
Attempts, **Homic § 95**
   Evidence as to, **Homic § 319**
Avert acts, evidence as to, **Homic § 319**
Circumstantial evidence, sufficiency, **Homic § 319**
Corpus delicti, evidence as to. **Homic § 319**
Defenses, **Homic § 100**

Depositions, abuse of discretion, **Depos § 37**
Distinctions, attempts, **Homic § 95**
Elements of offense in general, **Homic § 161**
Evidence,
   Contemporaneous circumstances,
      admissibility, **Homic § 238 et seq.**
   Extent of injury, **Homic § 199**
   Flight, **Homic § 246**
   Identity of accused, **Homic § 213**
   Intent, **Homic §§ 179, 180**
      Admissible evidence, **Homic § 195**

continues

Exhibit 2–17   continued

HOMICIDE

HOMICIDE—Continued
Assault with intent to murder or kill—Continued
    Evidence—Continued
        Justification in general, admissibility,
            Homic § 252 et seq.
        Physical condition of parties, **Homic § 219**
        Physical conditions, **Homic § 242**
        Subsequent circumstances, **Homic § 241**
        Subsequent hostile statements, admissibil-
           ity to show malice, **Homic § 209**
        Sufficiency, **Homic § 319**
        Weapons, possession of, **Homic § 244**
    Excuse, **Homic § 252 et seq.**
    Extent of injury, evidence, **Homic § 199**
    Flight, admissibility of evidence as to, **Homic
        § 246**
    Indictments and informations,
        Conviction,
            Assault and battery, charge, **Ind&Inf
                § 233**
            Indictment for, **Ind&Inf § 229**
        Elements of offense in general,
            Homic § 161
        Manner, **Homic § 162**
    Instructions to jury, **Homic § 342**
        Elements of, **Homic § 342**
        Evidence warranting, **Homic § 358**
        Specific intent, **Homic § 342**
    Intent,
        Evidence as admissible to show, **Homic
            § 195**
        Murder, **Homic § 98**
    Justification, **Homic § 252 et seq.**
    Malice,
        Acts and preparations, admissibility,
            Homic § 210
        Evidence, **Homic § 319**
        Prior difficulties as evidencing,
            Homic § 206
        Subsequent hostile acts and declarations
            by accused, admissibility, **Homic § 209**
    Manslaughter, **Homic § 99**
    Motive, proof of as essential, **Homic § 319**

HOMICIDE—Continued
    Motor vehicles, see **Title Index to Motor
        Vehicles**
    Participation in crime, evidence as to, **Homic §
        319**
    Previous declarations and threats by accused,
        admissibility, **Homic § 204**
    Reasonable doubts, evidence beyond as
        essential, **Homic § 319**
    Surrounding circumstances, antecedent
        circumstances, previous difficulties,
        admissibility, **Homic § 235**
    Weapons, possession of, evidence, **Homic §
        244**
Attachment, availability of writ, **Attach § 17**
Attempts, **Homic § 95** ◀
    Admissibility, committing other offenses,
        Homic § 196
    Evidence, **Homic § 320**
    HIV, biting victim, intent to kill, ◀
        Homic § 95
    Indictment or information, **Homic § 163**
    Intent, instructions as to, **Homic § 337**
    Mental state, **Homic § 95**
    Other offenses, **Homic § 309**
        Instructions to jury, **Homic § 340**
    Previous attempts, **Homic § 233**

HOMICIDE

Deadly force, necessity, **Homic § 105** ◀
Deadly weapons,
    Evidence, intent, **Homic § 180**
    Grade or degree of offense, **Homic §§ 190,
        191**
    Malice, use of as showing, **Homic § 302**
    Premeditation and deliberation,
        Homic § 303
        Presumption from use of,
            Homic § 182
    Self-defense, evidence as to reputation of
        deceased using on issue of,
        Homic § 261

Source: C.J.S., General Index (E-M) (1995), pp. 483, 485, 486, "Homicide: Assault with intent to murder or kill;" "Homicide: Attempts;" "Homicide: Deadly weapons," © West Publishing, used with permission.

Exhibit 2–18

ACQUIESCENCE

Injunction, defenses, **Zon&LP § 342**
Violations, **Zon&LP § 33**
ACQUIRED IMMUNE DEFICIENCY SYN-DROME (AIDS)
Assault and battery, teeth infected with, **Asslt&B § 79**
Attempted murder, biting victim, intent to kill, **Homic § 95**

Civil rights, pretext for discrimination, dentists refusal to treat, **Civil R § 22**
Constitutional law,
Government employee publicly disclosing individuals infected with, **Const L § 640**
Nondisclosure of persons infected with, right to privacy, **Const L § 640**
Prison inmate testing, **Const L § 642**

Source: C.J.S., General Index (A-D) (1995), p. 44, "Acquired Immune Deficiency Syndrome: Assault and battery, teeth infected with; Attempted murder, biting victim, intent to kill," © West Publishing, used with permission.

Exhibit 2–19

HOMICIDE
Analysis

I. **IN GENERAL, §§ 1–28**
    A. GENERAL CONSIDERATIONS, §§ 1–10
    B. CAPACITY AND RESPONSIBILITY, §§ 11–19
    C. PARTIES TO OFFENSES, §§ 20–28
II. **MURDER, §§ 29–68**
    A. IN GENERAL, §§ 29–44
    B. HOMICIDE IN COMMISSION OF OTHER CRIME; FELONY MURDER, §§ 45–55
    C. DEGREES, §§ 56–68
III. **MANSLAUGHTER, §§ 69–92**
    A. IN GENERAL, §§ 69–73
    B. VOLUNTARY MANSLAUGHTER, §§ 74–87
    C. INVOLUNTARY MANSLAUGHTER, §§ 88–92
IV. **CRIMINALLY NEGLIGENT HOMICIDE, §§ 93–94**
V. **ATTEMPTS, SOLICITATIONS, AND THREATS, §§ 95–97**
VI. **ASSAULT WITH INTENT TO MURDER OR KILL, §§ 98–100**
VII. **JUSTIFIABLE OR EXCUSABLE HOMICIDE; OTHER EXTENUATING CIRCUM-STANCES, §§ 101–138**
    A. IN GENERAL, §§ 101–112

Source: C.J.S., vol. 40, p. 340, "Homicide," (Outline), © West Publishing, used with permission.

Exhibit 2–20

HOMICIDE   40 C.J.S.

Source: C.J.S., vol. 40, p. 344, "Homicide," (Outline), © West Publishing, used with permission.

Battery § 79." Note the assertion that "[t]he teeth of one infected with AIDS can be considered a deadly weapon," and the accompanying citation of *United States v. Moore*, 846 F.2d 1163 [(8th Cir. 1988)]. (See Exhibit 2–23.)

Legal encyclopedia articles can be helpful to introduce you to the issues you are researching, and to locate citations for related case law, such as the examples we have used in Am. Jur. 2d and C.J.S. illustrate. You also should recall that several state legal encyclopedias exist that focus almost exclusively on the law of a specific state. Remember, however, that legal encyclopedia articles tend to present general overviews, which may not address the important finer points of the law. Legal encyclopedias are a relatively elementary form of secondary authority. Their articles can be useful for legal research, but you should recognize their limitations and be careful not to rely on them as if they were a definitive authority.

Exhibit 2–21

40 C.J.S.   **HOMICIDE § 95**

V. ATTEMPTS, SOLICITATIONS, AND THREATS
§ 95. ATTEMPTS
An attempt to commit murder consists of a specific intent to kill accompanied by a sufficient overt act tending to effectuate such intent, and there must be at least an apparent ability to commit the crime.

Library References
Homicide ⚷ 25, 31.
LaFave & Scott Substantive Criminal Law Vol. 2 § 6.2.

An attempt to commit murder is a misdemeanor at common law,[82] although, by statute, such an attempt may be raised to the grade of felony.[83] A general attempt statute is applicable to an attempt to commit murder,[84] unless a specific statute is controlling.[85] There are no degrees of attempted murder.[86]

To constitute an attempt to murder, there must be a specific intent to kill[87] and some overt act[88] or substantial step[89] toward the offense of murder. The attempt must have been made with such intent or under such circumstances that, if consummated, the homicide would have been murder, or murder of the particular degree . . .

82. Md.—Hardy v. State, 482 A.2d 474, 301 Md. 124.

Miss.—Saunders v. State, 114 So. 747, 148 Miss. 685.

83. U.S.—U.S. ex rel. Di Stefano v. Moore, D.C.N.Y., 46 F.2d 308, affirmed, C.C.A., 46 F.2d 310, certiorari denied U.S. ex rel. Di Stefano v. Pulver, 51 S. Ct. 364, 283 U.S. 830, 75 L. Ed. 1443.

Miss.—Saunders v. State, 114 So. 747, 148 Miss. 685.

84. U.S.—U.S. ex rel. Di Stefano v. Moore, D.C.N.Y., 46 F.2d 308, affirmed, C.C.A., 46 F.2d 310, certiorari denied U.S. ex rel. Di Stefano v. Pulver, 51 S. Ct. 364, 283 U.S. 830, 75 L. Ed. 1443.

85. Okl.—Minter v. State, Cr., 129 P.2d 210, 75 Okl.Cr. 133.

86. Nev.—Keys v. State, 766 P.2d 270, overruling to the extent that it holds to the contrary Wheby v. Warden, 598 P.2d 1152, 95 Nev. 567.

87. Ala.—Paige v. State, Cr.App., 494 So.2d 795.

Cal.—People v. Croy, 221 Cal.Rptr. 592, 710 P.2d 392, 41 C.3d 1, rehearing denied.

Ill.—People v. Myers, 426 N.E.2d 535, 55 Ill.Dec. 389, 85 Ill.2d 281.

Mass.—Commonwealth v. Maloney, 506 N.E.2d 1147, 399 Mass. 785.

Tex.—Martinez v. State, App. 4 Dist., 705 S.W.2d 772, review refused.

88. Ala.—Chaney v. State, Cr.App., 417 So.2d 625.

Cal.—People v. Koontz, 3 Dist., 208 Cal.Rptr. 519, 162 C.A.3d 491.

La.—State v. Huizar, 414 So.2d 741.

Tex.—Holman v. State, App. 1 Dist., 697 S.W.2d 824.

Va.—Nobles, IV v. Commonwealth, 238 S.E.2d 808, 218 Va. 548.

89. Conn.—State v. Sharpe, 491 A.2d 345, 195 Conn. 651.

Ill.—People v. Migliore, 2 Dist., 525 N.E.2d 182, 121 Ill.Dec. 376, 170 Ill.App.3d 581, appeal denied 530 N.E.2d 257, 125 Ill.Dec. 229, 122 Ill.2d 587.

N.H.—State v. Allen, 514 A.2d 1263, 128 N.H. 390.

Or.—State v. Lavender, 682 P.2d 823, 68 Or.App. 514, review denied 685 P.2d 998, 297 Or. 547.

Pa.—Commonwealth v. Ford, 461 A.2d 1281, 315 Pa.Super. 281.

continues

Exhibit 2–21   continued

40 C.J.S.   **HOMICIDE § 95**

...ever, something more is required than mere menaces,[1] preparation, or planning.[2]

Ability to commit crime.

To constitute an attempt to murder, there must be an apparent ability to commit the intended crime.[3] However, it is not necessary that the contemplated murder be factually possible.[4]

1. Cal.—People v. Miller, 42 P.2d 308, 2 C.2d 527.
2. Miss.—Jackson v. State, 254 So.2d 876.
Mont.—State v. Rains, 164 P. 540, 53 Mont. 424.
Tex.—Fuller v. State, App.-Corpus Christi, 716 S.W.2d 721, review refused.

Attempt and preparation distinguished

Preparation for a crime consists in devising or arranging means or measures necessary for commission of offense, while "attempt" is direct movement towards commission after preparations are made.
Nev.—Moffett v. State, 618 P.2d 1223, 96 Nev. 822.
3. Miss.—Stokes v. State, 46 So. 627, 92 Miss. 415.
Wash.—State v. Gay, 486 P.2d 341, 4 Wash.App. 834.
4. Ariz.—State v. Lenahan, 471 P.2d 748, 12 Ariz.App. 446.

Source: C.J.S., vol. 40, pp. 483–485, "Homicide," sec. 95, © West Publishing, used with permission.

Exhibit 2–22

§ 95 HOMICIDE
Page 484

Any lesser mental state, such as recklessness, does not support the crime of attempted murder.[93.5]
93.5 Wash.—State v. Dunbar, 817 P.2d 1360, 117 Wash.2d 587.
96. N.M.—State v. Gillette, App., 699 P.2d 626, 102 N.M. 695, affd. in part, revd. in part on oth. grds. 17 F.3d 308, reh. den.

Doctrine of transferred intent
Ariz.—State v. Rodriguez-Gonzales, App., 790 P.2d 287, 164 Ariz. 1.

It has been held that knowledge of the status of the victim is not an element of the offense of attempted murder.[96.5]

96.5 Law enforcement officer
Fla.—Carpenter v. State, App. 1 Dist., 587 So.2d 1355, on motion for reh., certification den., review den. 599 So.2d 654.

99. N.M—State v. Gillette, App., 699 P.2d 626, 102 N.M. 695, affd. in part, revd. in part on oth. grds. 17 F.3d 308, reh. den.

page 485

A defendant could be found guilty of attempted murder upon proof that the defendant, who had tested positive for HIV, intended to kill the victim by biting him, regardless of whether it was medically impossible for a bite to transmit HIV, as it was sufficient that the defendant believed he could cause death by biting his victim and intended to do so.[4.5]

4.5 N.J.—State v. Smith, 621 A.2d 493, 262 N.J.Super. 487, certification den. 634 A.2d 523, 134 N.J. 476.

Source: C.J.S., vol. 40 (1995 Supp.), p. 12, "Homicide," sec. 95, © West Publishing, used with permission.

Exhibit 2–23

ASSAULT AND BATTERY §79
Page 460

N.C.—State v. Carson, 249 S.E.2d 417, 296 N.C. 31.

Tex.—Davidson v. State, Cr., 602 S.W.2d 272. Stevens v. State, App. 10 Dist., 636 S.W.2d 857, review ref.

Pocket knife

Fla.—State v. Nixon, App., 295 So.2d 121.

N.C.—State v. McKinnon, 283 S.E.2d 555, 54 N.C.App. 475.

Tex.—Hudson v. State, App. 2 Dist., 629 S.W.2d 227.

Dagger

Cal.—People v. Cabral, 124 Cal.Rptr. 418, 51 C.A.3d 707.

Broken drinking glass

Wash.—State v. Pomeroy, 573 P.2d 805, 18 Wash.App. 837.

Scissors

Ind.—Johnson v. State, App., 409 N.E.2d 699.

54. Mich.—People v. Buford, 244 N.W.2d 351, 69 Mich.App. 27.

Ohio—State v. Orlett, 335 N.E.2d 894, 44 Ohio Misc. 7, 73 O.O.2d 30.

55. Fla.—Dixon v. State, App. 5 Dist., 603 So.2d 570, review den. 613 So.2d 9.

Ga.—Quarles v. State, 204 S.E.2d 467, 130 Ga.App. 756.

Ky.—Roney v. Commonwealth, 695 S.W.2d 863.

Mass.—Com. v. Appleby, 402 N.E.2d 1051, 380 Mass. 296.

Mich.—People v. Van Diver, 263 N.W.2d 370, 80 Mich.App. 352.

Or.—State v. Wier, 540 P.2d 394, 22 Or.App. 549.

Bare hands or teeth

Mass.—Com. v. Davis, 406 N.E.2d 417, 10 Mass.App. 190, 8 A.L.R.4th 1259.

57. La.—State in Interest of Ruschel, App. 4, Cir., 411 So.2d 1216.

Foot used to stomp victim

N.Y.—People v. Carter, 423 N.Y.S.2d 559, 73 A.D.2d 986, affd. 423 N.E.2d 30, 53 N.Y.2d 113, 440 N.Y.S.2d 607.

Pa.—Com. v. Stancil, 334 A.2d 675, 233 Pa.Super. 15.

Feet whether shod or unshod

Alaska—Wettanen v. State, App., 656 P.2d 1213.

58. Ala.—Hollis v. State, Cr.App., 417 So.2d 617.

Ga.—Kirby v. State, 245 S.E.2d 43, 145 Ga.App. 813.

Miss.—Pulliam v. State, 298 So.2d 711.

Mo.—State v. Gardner, App., 522 S.W.2d 323.

R.I.—State v. Zangrilli, 440 A.2d 710.

S.C.—State v. Carpenter, 205 S.E.2d 141, 262 S.C. 401.

Against infant

Iowa—State v. Chatterson, 259 N.W.2d 766.

The teeth of one infected with AIDS can be considered a deadly weapon.[58.5]

58.5 U.S.—U.S. v. Moore, C.A.8 (Minn.), 846 F.2d 1163, reh. den.

59. Ala.—Hopkins v. State, Cr., 286 So.2d 920, 51 Ala.App. 510.

Ariz.—State v. Bustamonte, 593 P.2d 659, 122 Ariz. 105.

Source: C.J.S., vol. 6A (1995 Supp.), p. 99, "Assault and Battery," sec. 79, note 58.5 and accompanying text, © West Publishing, used with permission.

Treatises and Case Books

Legal treatises, which are sometimes called **hornbooks**, are scholarly books that describe and summarize a body of law and often analyze legal doctrine in the process. **Case books** are the texts most commonly assigned in courses taught in law schools. They present edited judicial decisions that students read and are quizzed over in class. The cases typically are preceded or followed by brief comments, questions, and occasionally excerpts from other books and periodicals. Both kinds of books can be helpful secondary legal authorities.

One reason a treatise or case book is useful is that it collects a great amount of information about the topics it covers, distills that information to manageable proportions, and then presents it conveniently between the covers of a single book. These types of secondary authority necessarily draw upon, and cite, a great many references as they present the compiled information and can direct you very quickly to important cases and other authorities related to the issues you are investigating. They also can provide a rapid introduction to subjects that later can be explored in greater depth.

You can purchase treatises and case books at bookstores, which will gladly order them for you if they do not already carry them, or you can locate them in libraries with the assistance of the card catalog or its computer equivalent. Treatises may or may not be updated by pocket supplements; those that are not obviously cannot include judicial decisions and other legal developments occurring after they went to press. Case books frequently are kept up to date with annual cumulative supplements that are maintained separately from the main volume.

Many treatises and case books cover legal issues relevant to criminal justice. We will not attempt an exhaustive listing of these references, but we identify a few of them and give some examples of what you may find in them. One good treatise on criminal procedure law is Charles H. Whitebread & Christopher Slobogin, **Criminal Procedure: An Analysis of Cases and Concepts** (3d ed., Foundation Press, 1993), with its accompanying paper supplement. We can learn more about the confrontation issue presented in Case 1 by looking for that topic in either the book's table of contents or index. Both starting points refer us to a section discussing the right of criminal defendants to have a face-to-face encounter with trial witnesses. On turning to that section of the book, we find a discussion of two Supreme Court cases that appear to be highly relevant to the issue of whether a defendant's right of confrontation is violated when a witness is allowed to testify out of the defendant's presence, over closed-circuit television: *Coy v. Iowa*, 487 U.S. 1012, 108 S. Ct. 2798, 101 L. Ed. 2d 857 (1988), and *Maryland v. Craig*, 497 U.S. 836, 110 S. Ct. 3157, 111 L. Ed. 2d 666 (1990). (See Exhibit 2–24.)

Many other treatises may be of interest to you. Some focus on subtopics of law relevant to criminal justice and give detailed coverage of those topics. For example, Wayne R. LaFave, **Search and Seizure: A Treatise on the Fourth Amendment** (West Publishing Co., 3d ed. 1996) is an excellent five-volume treatise that makes detailed examination of search-and-seizure law. This oft-cited reference is kept current through pocket supplements. Nathan R. Sobel, **Eyewitness Identification: Legal and Practical Problems** (Clark Boardman, 2d ed. 1995, with inserted updates), Thomas W. Hutchinson & David Yellen, **Federal Sentencing Law and Practice** (West 2d ed. 1994, with pocket supplement), the multivolume **Wigmore on Evidence** (Little, Brown & Co., Tillers rev. 1983, with pocket supplements), Wayne R. LaFave & Austin W. Scott, Jr.'s two-volume **Substantive Criminal Law** (West Publishing Co. 1986, with pocket supplements), Rollin M. Perkins & Ronald N. Boyce, **Criminal Law** (Foundation Press,

Exhibit 2–24

28.05 Challenging Witnesses in the Courtroom

When a prosecution witness does appear in court, two issues arise: when, if ever, the defendant may be denied a face-to-face encounter with that witness, and the extent to which cross-examination of the witness may be limited.

(a) The Right to a Face-to-Face Encounter. In lay terms, one does not "confront" another unless there is a face-to-face encounter. Nonetheless, under narrow circumstances, the state may prevent the accused from confronting prosecution witnesses who appear in court. In *Coy v. Iowa*,[88] the Court found unconstitutional a state law which permitted a large screen to be placed between the defendant and two 13 year-old girls who testified that he had sexually assaulted him. Justice Scalia, writing for the Court, emphasized that "the Confrontation Clause guarantees the defendant a face-to-face meeting with witnesses appearing before the trier of fact." But two members of the six-member majority appeared to disagree with this latter statement. Justice O'Connor wrote a concurring opinion, joined by Justice White, which noted that the statute in Coy *presumed* that trauma would occur any time a youthful victim testified in such a case. She suggested that had there been an individualized finding that the child witnesses needed special protection, she might support a different result.

In *Maryland v. Craig*,[89] such a finding was made by the trial court pursuant to a Maryland statute which permits a one-way television procedure if the judge determines that face-to-face testimony "will result in the child suffering serious emotional distress such that the child cannot reasonably communicate." Under the statute, once this finding is made, the witness, prosecutor, and defense counsel withdraw to a separate room; the judge, jury, and defendant remain in the courtroom. The defendant can watch direct and cross-examination of the child over the video hookup and remains in electronic communication with his counsel. In a 5-4 decision, the Court upheld this procedure, in an opinion written by O'Connor.

Although recognizing that requiring a face-to-face encounter between defendant and witness forms the "core" of the Confrontation Clause, the majority concluded that the "central concern" of the Clause "is to ensure the reliability of the evidence against a . . .

88. 487 U.S. 1012, 108 S.Ct. 2798 (1988).
89. 497 U.S. 836, 110 S.Ct. 3157 (1990).

Source: C. Whitebread & C. Slobogin, *Criminal Procedure: An Analysis of Cases and Concepts,* pp. 741–743, sec. 28.05 (a), 3d ed. 1993, © Foundation Press, used with permission.

3d. ed. 1982), and Ira P. Robbins ed., **Prisoners and the Law** (Clark Boardman, Callaghan, 1985 and insert supplements), are among the many hornbooks that are available on a wide variety of subjects. Incidentally, we can put these criminal law treatises to good use for the topics we are researching in Case 2 and Case 3. For example, section 6.3 in LaFave & Scott's Substantive Criminal Law has an enlightening discussion titled "Attempt—The Limits of Liability," and an excellent article is reprinted in Robbins, Prisoners and the Law, beginning at page 7–29, that discusses "The Right to a Smoke-Free Environment in Prison."

We want to call one particular legal subject and a related treatise to your attention, because they involve a domain of law that sometimes is neglected. This subject is state constitutional law. Individuals whose cases are tried in the state courts naturally are entitled to the federal constitutional protections that the U.S. Supreme Court has ruled must be observed. For a period of several years, the Supreme Court's enforcement of federal constitutional rights in state cases so dominated the legal terrain that federal rights often came to be considered citizens' exclusive protections. Yet, in truth, when the U.S. Supreme Court announces what federal constitutional rights must be observed, it only sets the *minimum* standards that the state courts must follow. State constitutions cannot take away from the federal rights that citizens enjoy, but they can give individuals *greater* protections than the U.S. Constitution does. Thus, it will be important for you to check state constitutional law sources in appropriate cases before you consider your legal research complete.

We will address state constitutions at greater length in Chapter 3, when we consider primary legal authorities. For now, we simply alert you to the potential significance of state constitutions as a source of legal rights. These rights may be no more comprehensive than those provided by the federal Constitution, but sometimes they are construed as being more extensive. The only way to know for sure is to investigate the meaning given to state constitutional provisions by the corresponding state courts. One treatise that can be of help in this area is Barry Latzer's **State Constitutional Criminal Law** (Clark Boardman, Callaghan, 1995, with inserts to update).

The table of contents and index of Latzer's treatise provide more information about state constitutional rulings relevant to the confrontation issue we have been investigating in Case 1. We discover that "Face-to-Face Confrontation" is covered in section 6.3. In addition to the U.S. Supreme Court cases we have already found, *Coy v. Iowa* and *Maryland v. Craig*, we are treated to a lengthy discussion about how some state courts have departed from *Craig*. These courts have interpreted their state constitutions to give criminal defendants greater rights than the federal Constitution provides regarding the presentation of televised criminal trial testimony. Reproduced below is a small portion of this discussion, which focuses on the Indiana Supreme Court's ruling in *Brady v. State*, 575 N.E.2d 981 (Ind. 1991). Note how the state supreme court apparently relies on a provision of the Indiana Constitution to justify its decision. (See Exhibit 2–25.) Thus, the Latzer treatise should motivate you to read the *Brady* decision, consult the Indiana Constitution, and take other steps that we describe in Chapter 3 to follow up on the lead found in this treatise.

Case books also can be a handy reference with which to begin research. These books typically collect judicial decisions that are important to different legal issues, present excerpts of those decisions, and discuss and cite related cases and authorities. Scores of case books cover criminal justice topics. The most widely used criminal procedure case book probably is Yale Kamisar, Wayne R. LaFave & Jerold H. Israel eds., **Modern Criminal Procedure: Cases—Comments— Questions** (West Publishing Co., 1994, with paper supplements). This compre-

Exhibit 2–25

Confrontation of Adverse Witnesses   § 6:3

The Indiana Confrontation Clause,[27] which also provides for "face-to-face" meetings, requires that the defendant be visible to the witness, although a closed circuit televised image of the accused will suffice.[28] Brady v State[29] voided the portion of a statute which specified that a child witness who is being videotaped shall not be able to observe or hear the defendant.

Brady reasoned that the statute, which satisfied Sixth Amendment requirements as construed by *Craig,* ran afoul of the state Confrontation Clause, which "has a special concreteness and is more detailed."[30] As construed by the Indiana courts, this provision grants two different rights: the right to cross-examination, and the right to meet witnesses face-to-face. The latter right is not satisfied by granting the former.

> Indiana's confrontation right contains both the right to cross-examination and the right to meet the witnesses face to face. It places a premium upon live testimony of the State's witnesses in the courtroom during the trial, as well as upon the ability of the defendant and his counsel to fully and effectively probe and challenge those witnesses during trial before the trier of fact through cross-examination. The defendant's right to meet the witnesses face to face has not been subsumed by the right to cross-examination. That is to say, merely ensuring that a defendant's right to cross-examine the witness is scrupulously honored does not guarantee that the requirements of Indiana's Confrontation Clause are met. . . .

27. Ind Const Art 1, § 13 says in part: "In all criminal prosecutions, the accused shall have the right . . . to meet the witnesses face to face."

28. Brady v State (1991, Ind) 575 NE2d 981. The Indiana right to meet the witness face-to-face may be waived by the defendant. State v Owings (1993, Ind) 622 NE2d 948 ("Where there is no showing in the record that a defendant is unable to attend a deposition and he makes no objection to it proceeding, the defendant waives his right to confrontation even if the witness is unable to testify at trial." Id. at 952.).

29. Brady, 575 NE2d 981.

30. Id. at 987.

Source: B. Latzer, State Constitutional Law, pp. 6–10, 6–11, © 1995, Clark Boardman Callaghan, used with permission.

hensive book has slimmer paperback offshoots that cover less material, **Basic Criminal Procedure** and **Criminal Procedure and the Constitution**, both of which appear in a new edition annually. We illustrate the contents of Modern Criminal Procedure in Exhibit 2–26. The page reproduced presents an excerpt of

Exhibit 2–26

1472   ADVERSARY SYSTEM AND GUILT OR INNOCENCE   Pt. 4

. . . this cross-examination and was available to assist his counsel as necessary. * * * Any questions asked during the competency hearing, which respondent's counsel attended and in which he participated, could have been repeated during direct examination and cross-examination of the witnesses in respondent's presence. * * *

"Moreover, the type of questions that were asked at the competency hearing in this case were easy to repeat on cross-examination at trial. * * * In Kentucky, as in certain other States, it is the responsibility of the judge, not the jury, to decide whether a witness is competent to testify based on the witness' answers to such questions. * * * [T]hat responsibility usually continues throughout the trial. A motion by defense counsel that the court reconsider its earlier decision that a child is competent may be raised after the child testifies on direct examination. * * * At the close of the children's testimony, respondent's counsel, had he thought it appropriate, was in a position to move that the court reconsider its competency rulings on the ground that the direct and cross-examination had elicited evidence that the young girls lacked the basic requisites for serving as competent witnesses. Thus, the critical tool of cross-examination was available to counsel as a means of establishing that the witnesses were not competent to testify, as well as a means of undermining the credibility of their testimony.

"Because respondent had the opportunity for full and effective cross-examination of the two witnesses during trial, and because of the nature of the competency hearing at issue in this case, we conclude that respondent's rights under the Confrontation Clause were not violated by his exclusion from the competency hearing of the two girls.[b] * * * [But] respondent [also] argues that his rights under the Due Process Clause of the Fourteenth Amendment were violated by his exclusion from the competency hearing. The Court has assumed that, even in situations where the defendant is not actually confronting witnesses or evidence against him, he has a due process right 'to be present in his own person whenever his presence has a relation, reasonably substantial, to the fulness of his opportunity...

b. Compare *Coy v. Iowa,* 487 U.S. 1012, 108 S.Ct. 2798, 101 L.Ed.2d 857 (1988), where a state trial court, acting pursuant to a state statute designed to protect child victims of sexual abuse, allowed two 15-year-old female complainants to testify from behind a screen which blocked defendant from their sight but did allow defendant to dimly perceive them and to hear them. The Court held this violated defendant's Sixth Amendment right of confrontation. That right, the Court noted, extended beyond cross-examination and included the right to a "face-to-face meeting with the witnesses appearing before the trier of fact." The majority added that it would "leave for another day" the questions of whether this element of a face-to-face confrontation, like other aspects of the Sixth Amendment guarantee, "may give way to other important interests." Even if the "confrontation in-

terest at stake here" could be outweighed by "the necessity of protecting victims of sexual abuse," the state may not rely simply upon a "legislatively imposed presumption of trauma," without requiring "individualized findings that these particular witnesses needed special protection."

In *Maryland v. Craig,* 497 U.S. 836, 110 S.Ct. 3157, 111 L.Ed.2d 666 (1990), the Court upheld, 5-4, a statutory procedure that allows the use of one-way closed circuit television to provide the testimony of a child witness who is alleged to be the victim of child abuse (in *Craig,* the testimony of a six-year-old alleged to have been sexually abused by defendant while attending her preschool center). Use of the televised testimony is conditioned on the trial court first determining, after a factfinding hearing, that requiring the child to give courtroom testimony would result, because of

continues

Exhibit 2–26   continued

the presence of the defendant, in the child "suffering serious emotion distress, such that the child cannot reasonably communicate." The Court ruled that the state interest in protecting the child from trauma caused by defendant's physical presence, "at least where such trauma would impair the child's ability to communicate," as supported by a case-specific finding of necessity, justified dispensing with element of face-to-face confrontation. It stressed that all other elements of confrontation—oath, cross-examination by defense counsel present in the room in which the child testified (with defendant in electronic communication with counsel), and observation of demeanor by the judge and jury (who remained in the courtroom)—would be preserved.

Source: Y. Kamisar, W. LaFave & J. Israel, eds., *Modern Criminal Procedure: Cases—Comments—Questions,* p. 1472, © West Publishing, used with permission.

the Supreme Court decision in *Kentucky v. Stincer*, 482 U.S. 730, 107 S. Ct. 2658, 96 L. Ed. 2d 631 (1987), which addresses the confrontation issue. We were referred to this page by looking up "confrontation" in the index, and found the subtopic, "face-to-face confrontation." Note that the editors have summarized the holdings of *Coy v. Iowa* and *Maryland v. Craig* in the footnotes they have added to the edited case book version of *Kentucky v. Stincer*.

Other good casebooks abound in many different areas of the law. A few that are relevant to criminal justice include Stephen A. Saltzburg & Daniel J. Capra eds., **American Criminal Procedure** (West Publishing Co., 5th ed. 1996 and supplement), Ronald J. Allen, Richard B. Kuhns & William J. Stuntz eds., **Constitutional Criminal Procedure: An Examination of the Fourth, Fifth, and Sixth Amendments and Related Areas** (Little, Brown & Co., 3d ed. 1995 and supplement), Sanford H. Kadish & Stephen J. Schulhofer eds., **Criminal Law and Its Processes: Cases and Materials** (Little, Brown & Co., 6th ed. 1995 and supplement), Fred E. Inbau, James R. Thompson, James B. Zagel & James P. Manak eds., **Criminal Law and Its Administration** (Foundation Press, 5th ed. 1990 and supplement), Lloyd B. Weinreb ed., **Criminal Law: Cases, Comment, Questions** (Foundation Press, 5th ed. 1993); Ronald L. Carlson, Edward J. Imwinkelried & Edward J. Kionka, **Evidence in the Nineties: Cases, Materials and Problems for an Age of Science and Statutes** (Michie, 3d ed. 1991 and supplement), Eric D. Green & Charles R. Nesson eds., **Problems, Cases and Materials on Evidence** (Little, Brown & Co., 2d ed. 1995 with supplement), John W. Strong, Kenneth S. Broun & Robert P. Mosteller eds., **Cases and Materials on Evidence** (West Publishing Co., 5th ed. 1995), Francis B. McCarthy & James G. Carr eds., **Juvenile Law and Its Processes: Cases and Materials** (Michie, 2d ed. 1989 and supplement), and John Monahan & Laurens Walker eds., **Social Science in Law: Cases and Materials** (Foundation Press, 3d ed. 1994). Another casebook, which covers an interesting and diverse range of criminal justice topics is Fred Cohen, **The Law of Deprivation of Liberty: Cases and Materials** (Carolina Academic Press, 1991).

Case Briefs

When a case is appealed or a court agrees to hear a case after granting a petition for a writ of certiorari, both sides file **briefs** with the appellate court arguing their respective positions. Brief writers are required to assemble legal and other authorities in the briefs they file with appellate courts, and many do so in exhaustive fashion. Other briefs sometimes are filed by *amici curiae*, or "friends of the court." *Amicus curiae* briefs are submitted by government bodies (for example, the U.S. Solicitor General might file an *amicus curiae* brief on behalf of the United States or the Attorney General of California might submit an *amicus curiae* brief on behalf of the State of California), by public interest groups and other organizations (for example, the American Civil Liberties Union or the NAACP Legal Defense Fund), and sometimes by individuals who are not parties to a case but have an interest in how the case will be decided. Briefs thus can be another helpful reference for legal research purposes, especially when the researcher is focusing on an issue similar to the one argued in the case in which the briefs were filed.

The court should have copies of the pertinent briefs available in printed form or on microfiche, and the attorneys representing the parties or the *amici* in those cases also are likely to have retained copies. Thus, if other avenues for getting access to case briefs fail, you can always contact the clerk of the court in which the briefs were submitted or the offices of the lawyers who prepared the briefs. Briefs also may be available through other sources, including WESTLAW and LEXIS. Many courts require multiple copies of briefs and in turn distribute copies to law schools and/or the state library in the court's jurisdiction. Briefs filed in U.S. Supreme Court cases are more widely available—in printed form, on microfiche, and through computer databases. Because briefs are filed before a court decides a case, they are usually classified by the names of the parties and by the docket number assigned to the case by the appellate court, rather than by citation to the court's eventual decision.

At least two sources reproduce copies in print of the briefs filed in selected U.S. Supreme Court cases. These sources are available in many libraries. A series edited by Philip B. Kurland & Gerhard Casper, **Landmark Briefs and Arguments of the Supreme Court of the United States; Constitutional Law** (University Press of America), chooses a few important cases decided each year by the Supreme Court, and reprints the certiorari petition and all briefs filed in those cases.

This series covers only a few of the Supreme Court cases decided each year. Examine the spines of the books to find the specific cases in which briefs are reproduced. Exhibit 2–27 from Kurland and Casper reproduces the cover sheet, the table of contents, and a portion of the table of authorities filed by the State of Maryland in *Maryland v. Craig*, the Supreme Court case we encountered in our investigation of Case 1. The body of the brief contains the arguments and a discussion of the authorities cited in the table of contents. As the table of contents from volume 198 of the Landmark Briefs and Arguments series indicates, several *amici curiae* filed briefs in *Maryland v. Craig*. (See Exhibit 2–28.)

Law Reprints, a publisher located in Washington, D.C., publishes **Petitions and Briefs of the Supreme Court of the United States, Criminal Law Series**. This series includes the briefs filed in all criminal cases decided each year by the

Exhibit 2–27

No. 89-478

IN THE

Supreme Court of the United States
OCTOBER TERM, 1989

STATE OF MARYLAND,
Petitioner,

v.

SANDRA ANN CRAIG,
Respondent.

ON WRIT OF CERTIORARI TO THE
COURT OF APPEALS OF MARYLAND

BRIEF FOR PETITIONER

J. JOSEPH CURRAN, JR.
Attorney General of Maryland
GARY E. BAIR*
ANN N. BOSSE
Assistant Attorneys General
200 Saint Paul Place
Baltimore, Maryland 21202
(301) 576-6422
WILLIAM R. HYMES
State's Attorney for Howard County,
  Maryland
Counsel for Petitioner

*Counsel of Record

continues

Exhibit 2–27   continued

TABLE OF CONTENTS

continues

Exhibit 2–27   continued

TABLE OF AUTHORITIES

Cases:	Page
Baltimore City Department of Social Services v. Bouknight, 58 U.S.L.W. 4184 (U.S. Feb. 20, 1990)	22
Barber v. Page, 390 U.S. 719 (1968)	18
Barker v. Wingo, 407 U.S. 514 (1972)	20
Batson v. Kentucky, 476 U.S. 79 (1986)	25
Blanton v. City of North Las Vegas, 489 U.S. ___, 109 S.Ct. 1289 (1989)	20
Bourjaily v. United States, 483 U.S. 171 (1987)	19,25
Brady v. State, 540 N.E.2d 59 (Ind. Ct. App. 1989)	23
California v. Green, 399 U.S. 149 (1970)	20
Chambers v. Mississippi, 410 U.S. 284 (1973)	25,31
Coy v. Iowa, 487 U.S. ___, 108 S.Ct. 2798 (1988)	*passim*
Craig v. State, 316 Md. 551, 560 A.2d 1120 (1989)	*passim*
Craig v. State, 76 Md. App. 250, 544 A.2d 784 (1988)	1,13,24
Davis v. Alaska, 415 U.S. 308 (1974)	20
Delaware v. Fensterer, 474 U.S. 15 (1985) (per curiam)	20
Delaware v. Van Arsdall, 475 U.S. 673 (1986)	20
Dowdell v. United States, 221 U.S. 325 (1911)	19
Dutton v. Evans, 400 U.S. 74 (1970)	20,29
FCC v. Pacifica Foundation, 438 U.S. 726 (1978)	22
Ginsberg v. New York, 390 U.S. 629 (1968)	22
Glendening v. State, 536 So.2d 212 (Fla. 1988), *cert. denied,* ___U.S.___, 109 S.Ct. 3219 (1989)	22
Globe Newspaper Co. v. Superior Court, 457 U.S. 596 (1982)	22,27,28,29

Source: Copyright © 1997 Congressional Information Service, Inc. Used with permission. All other reproduction is strictly prohibited without the express written consent of CIS.

U.S. Supreme Court. The certiorari petition and briefs are bound in separate volumes for each case. As in the Landmark Briefs and Arguments series, both the parties' briefs and *amicus curiae* briefs are included for each case. The briefs filed in *Maryland v. Craig* are in volume 21, number 38 of this series, which covers the Supreme Court's 1989–1990 term.

If you have an opportunity to consult case briefs during your legal research, keep in mind that the parties who have written and submitted briefs are advocates whose job it is to convince the court to rule in their favor. Furthermore, notwithstanding their ostensibly neutral posture as friends of the court, *amici curiae* almost always are interested in achieving a particular outcome in the cases in which they submit briefs. Consequently, you should be especially cautious about accept-

Exhibit 2–28

TABLE OF CONTENTS

State of Maryland v. **Craig**, 497 U.S. ___ (1990)

Source: Copyright © 1997 Congressional Information Service, Inc. Used with permission. All other reproduction is strictly prohibited without the express written consent of CIS.

ing at face value the interpretations that brief writers give to cases, statutes, constitutional provisions, and other legal authorities. Case briefs, nevertheless, can be a big help in raising issues that may not have occurred to you, and in citing authorities that are relevant to the questions you are researching. Legal researchers do not tend to consult case briefs routinely, and they are not traditionally classified as a form of legal authority. Still, you should be aware that briefs are a potentially helpful source of information, and that they are available for your consideration.

American Law Reports

Printed References

Economy and efficiency are important concerns in legal research. No special awards are given for conducting legal research as if you were the first person ever to try to answer a particular question. If you are sufficiently stubborn, or just misguided, you could accept the challenge of researching a legal issue by retreating to a library and beginning the long and solitary task of piecing together each and every authority relevant to your topic as if you were a pioneer in unmapped territory. You would soon disappear behind towers of books and reams of computer printouts. Friends and loved ones would wonder what has become of you. Your task would threaten to consume you. Although it is possible to tackle a legal research question as if no one has previously assembled information about it, if you ever slip into this way of thinking, try to find a quiet corner of the library, take several deep breaths, and do your best to come to your senses. Someone probably has already traveled the path you intend to follow. There is no dishonor in relying on the trail markers others have left. Of course, you must give proper credit for their assistance and be prepared to verify and then attempt to improve on their efforts, but do not approach your research task as if it is virtuous to be a trailblazer or a maverick. Your mantra should be "I need not reinvent the wheel."

American Law Reports (A.L.R.s) are splendid references that help you avoid reinventing the wheel as you conduct legal research. Published by Lawyers Cooperative Publishing Company, the editors of the A.L.R. choose a case that presents an issue of law, which is made the subject of an **annotation**. The case is reprinted in full, either immediately before the annotation or in a separate section in the same volume. After parsing the case issue into many subtopics presented in a detailed outline, the annotation discusses other cases relevant to those subjects. Some annotations focus on state law; others consider federal law; and still others examine U.S. Supreme Court decisions. By the time you have finished with an A.L.R. annotation, including its pocket supplement, you will have been referred to a comprehensive collection of case law relating to the issue you are investigating, with an accompanying textual discussion of that issue. A.L.R. annotations typically present a jurisdiction-by-jurisdiction description of how courts have resolved legal issues. They describe the different analyses courts have employed and the different results they have reached. A.L.R.s can be a highly useful starting point for legal research, so you should make it a point during the early stages of your investigation of an issue to determine whether an annotation addressing your topic has been prepared.

A.L.R. annotations are published in a succession of series. The oldest volumes are the American Law Reports, 1st (A.L.R. 1st). In 1948, A.L.R.2d was added, followed by A.L.R.3d in 1965, A.L.R. Federal in 1969, A.L.R.4th in 1980, and A.L.R.5th in 1992. A.L.R. Federal exclusively presents annotations discussing federal case law. State cases are examined in A.L.R.3d, A.L.R.4th, and A.L.R.5th. Before 1969 federal and state cases were intermingled in the initial volumes of A.L.R.3d, and in A.L.R.2d and A.L.R.1st. United States Supreme Court decisions also may be the subject of annotations. Note that these A.L.R. annotations, which discuss related Supreme Court holdings, are tucked away in the back pages of the U.S. Supreme Court Reports, Lawyers' Edition, and Lawyers' Edition 2d, which, as discussed before, are the Lawyers Cooperative case reporters for U.S. Supreme Court decisions.

Although the A.L.R. annotations are dispersed throughout A.L.R. Federal, A.L.R.1st through A.L.R.5th, and Lawyers' Edition and Lawyers' Edition 2d, for the most part these volumes share a common indexing system that makes them convenient to use. For all but the oldest annotations, access is gained to the

A.L.R.s through an alphabetically arranged **descriptive-word index**, which is usually shelved in the vicinity of the massive A.L.R. collection. The index is found in the volumes identified on their spines as A.L.R. Index, ALR 2d-3d-4th-5th, Federal, LEd2d. The first series of A.L.R., which contains the oldest volumes, has its own Quick Index and Word Index. Many of the annotations that appear in A.L.R.1st, A.L.R.2d, and elsewhere in the A.L.R. series have been superseded or supplemented by more recent annotations. The pocket supplement to the annotation indicates whether an annotation has been superseded or supplemented, as will the Annotation History Table in the index.

A.L.R. Digests also can be used to find A.L.R. annotations. Separate digests exist for A.L.R.1st and for A.L.R.2d. However, A.L.R.3d, A.L.R.4th, A.L.R.5th, and A.L.R. Federal share a common Digest. An explanatory how-to-use guide is provided in the first volume of each digest. The digests are not specifically indexed; instead, they are divided into over 400 general topics, which are presented in a list that extends over several pages. Choose one or more topics from this list that are likely to include the issue you are researching and then examine the annotation titles presented under those topics in the Digest to try to find an annotation on point. There is a hit-or-miss quality about the A.L.R. Digests that generally makes the A.L.R. Index easier to use and more reliable. We recommend resorting to the Index, rather than the digests, for most research issues, unless you have been given a cross-reference from another Lawyers Cooperative publication, such as Am. Jur. 2d or an L. Ed. case report, to the appropriate digest topic and number for an annotation.

Once you have been referred to an A.L.R. annotation, you receive tips for locating related annotations and other references. The annotations begin by listing Total Client-Service Library References in which other Lawyers Cooperative publications are identified, including Am. Jur. 2d encyclopedia articles, that pertain to the subject of the annotation. The more recently published annotations cite other Research Sources, including encyclopedia articles from both Am. Jur. 2d and C.J.S., law review articles, and West digest key numbers (which we will discuss in Chapter 3). They also describe the computer search queries that were used as a basis for the annotation. These features can be very useful in helping you to identify other references related to your topic.

We return to the issues presented in our hypothetical cases to give directives on how to find A.L.R. annotations and to illustrate their contents. Case 2 raises the issue of whether a charge of attempted murder or assault with a deadly weapon can be supported against Andrew Adams, who, knowing that he had tested positive for HIV, bit and spat on Officer Fiegel. We have posited that this case has arisen in state court, in an unspecified jurisdiction. Accordingly, we would be especially pleased if we could uncover an annotation in one of the more recent A.L.R. series that focuses on state cases—that is, A.L.R.3d, A.L.R.4th, or A.L.R.5th. We begin our hunt for an A.L.R. annotation much as we have begun our other searches—by identifying words and concepts that relate to our issue and then trying to locate these terms in the appropriate index. We thus consult the A.L.R. Index for A.L.R.2d-3d-4th-5th, Federal, and L. Ed. 2d. We want to look for entries under the following terms and/or phrases: **"Attempt," "Murder," "Assault," "Deadly Weapon," "AIDS," "HIV,"** and perhaps others.

If we were to begin this search by looking for "**AIDS**" in the A.L.R. Index, we would be referred to "**Acquired Immune Deficiency Syndrome**." On consulting the latter phrase in the bound volume of the Index, we would be tempted to conclude that we had struck out, because none of the annotations addresses an issue analogous to the one we are researching. Undaunted, we refer to the pocket supplement to see if other annotations have been added after the bound volume was published (in 1992). Our efforts are rewarded. We spot an annotation that looks like it may be of interest: "Criminal offense, transmission or risk of transmission of human immunodeficiency virus (HIV) or acquired immunodeficiency syndrome (AIDS) as basis for prosecuting or sentencing defendant for criminal offense, 13 ALR5th 628." (See Exhibit 2–29.)

We turn from the Index supplement to find volume 13 of A.L.R.5th and the annotation introductory notes, which begin on page 628. (See Exhibit 2–30.) The table of contents alerts us to the fact that "**Aggravated assault**—with deadly weapon . . . " is discussed in section 5, and that "**Attempted Murder**" is addressed in section 9. We then examine the other annotations, encyclopedia articles, law review articles, and other references that are listed under the Total Client-Service Library References and the Research Sources at the beginning of the annotation. The query used for a computer-based search and the West Digest key numbers related to our topic are also provided (see Exhibit 2–31). Note the misspelling of homicide, "homocide," in the Electronic Search Query. The misspelling should be corrected before the search is entered. Otherwise the misspelling would distort the search results.

Exhibit 2–29

ALR INDEX

ACCOUNTS RECEIVABLE—Cont'd
. . .ceivable as core proceeding in bankruptcy under 28 USCS § 157(b), 124 ALR Fed 531

ACQUIRED IMMUNE DEFICIENCY SYNDROME
Criminal offense, transmission or risk of transmission of human immunodeficiency virus (HIV) or acquired immunodeficiency syndrome (AIDS) as basis for prosecuting or sentencing defendant for criminal offense, 13 ALR5th 628

Disclosure, state statutes or regulations expressly governing disclosure of fact that person has tested positive for human immunodeficiency virus or acquired immunodeficiency syndrome (AIDS), 12 ALR5th 149

Insurance, rescission or cancellation of insurance policy for insured's misrepresentation or concealment of information concerning acquired immunodeficiency syndrome (AIDS), human immunodeficiency virus (HIV), or related health problems, 15 ALR5th 92

Source: A.L.R. Index, ALR 2d-3d-4th-5th, Federal, LEd2d (A-B) p. 3 (1996 Supp.), "Acquired Immune Deficiency Syndrome." Permission has been granted by the current copyright holder, West Group. Further reproduction of any kind is strictly prohibited. For additional information, please contact West Group Customer Services representative at 1-800-328-4880.

Exhibit 2–30

---

13 ALR5th 628

**TRANSMISSION OR RISK OF TRANSMISSION OF HUMAN IMMUNO-
DEFICIENCY VIRUS (HIV) OR ACQUIRED IMMUNODEFICIENCY SYN-
DROME (AIDS) AS BASIS FOR PROSECUTION OR SENTENCING IN
CRIMINAL OR MILITARY DISCIPLINE CASE**

by
Alan Stephens, J.D.

Table of Contents

> **Research References**
> **Index**
> **Jurisdictional Table of Cited Statutes and Cases**

ARTICLE OUTLINE

   Source: A.L.R.5th, vol. 13, pp. 628–629 (Title, and Article Outline). Permission has been
granted by the current copyright holder, West Group. Further reproduction of any kind is
strictly prohibited. For additional information, please contact West Group Customer Services
representative at 1-800-328-4880.

Exhibit 2–31

Research References

TOTAL CLIENT-SERVICE LIBRARY® REFERENCES

The following references may be of related or collateral interest to a user of this annotation:

Annotations

See the related annotations listed in § 1[b].

Encyclopedias and Texts

6 Am Jur 2d, Assault and Battery §§ 53, 55; 40 Am Jur 2d, Homicide § 5; 54 Am Jur 2d, Military and Civil Defense §§ 229, 246, 252, 254; 79 Am Jur 2d Weapons and Firearms §§ 1, 2

Practice Aids

2 Am Jur Pl & Pr Forms (Rev), Assault and Battery, Forms 1 et seq.
24 Am Jur Proof of Facts 2d 515, Defense to Charges of Sex Offense
36 Am Jur Trials 241, Defending Assault and Battery Cases

Digests and Indexes

L Ed Digest, Armed Forces §§ 38, 40; Assault and Battery § 1; Homicide § 1
ALR Digest, Armed Forces § 5; Assault and Battery §§ 28, 34; Courts-Martial § 2; Homicide § 1; Weapons and Firearms § 1
ALR Index, Acquired Immunodeficiency Syndrome; Armed Forces; Assault and Battery; Blood Tests; Court-Martial; Deadly Weapon; Homicide; Sentence and Punishment; Sex and Sexual Matters; Sodomy

Auto-Cite®

Cases and annotations referred to herein can be further researched through the Auto Cite® computer-assisted research service. Use Auto-Cite to check citations for form, parallel references, prior and later history, and annotation references.

RESEARCH SOURCES

The following are the research sources that were found to be helpful in compiling this annotation:

Encyclopedias

6 Am Jur 2d, Assault and Battery §§ 53, 55
40 Am Jur 2d, Homicide § 5
54 Am Jur 2d, Military and Civil Defense §§ 229, 246, 252, 254
79 Am Jur 2d, Weapons and Firearms §§ 1, 2
6 CJS, Armed Services §§ 157 (n 56.5), 164, 179, 181
6A CJS, Assault and Battery §§ 77, 79 (fn 58.5), 80
40 CJS, Homicide § 37
94 CJS, Weapons § 1

continues

Exhibit 2–31   continued

Law Reviews

Blumberg, Transmission of the AIDS Virus Through Criminal Activity, 25 Crim L Bull 454 (1989)

Criminalizing the Sexual Transmission of HIV: An International Analysis, 15 Hastings Intl & Comp L Rev (1992)

Deadly and Dangerous Weapons and AIDS: The Moore Analysis is Likely to be Dangerous, 74 Iowa L Rev 951 (1989)

Brock v State: The AIDS Virus as a Deadly Weapon, 24 John Marshall L Rev 677 (1991)

Gostin, The Politics of AIDS: Compulsory State Powers, Public Health, and Civil Liberties, 49 Ohio St L J 1017 (1989)

Herrmann, Criminalizing Conduct Related to HIV Transmission, 9 St. Louis U Pub L Rev 351 (1990)

Electronic Search Query

HIV or (AIDS w/5 virus) w/10 crim! or sentence or assault or battery or homocide or murder or (dangerous pre/1 weapon or instrument!) or rape

West Digest Key Numbers

Assault and Battery 48, 49, 52–56, 58–60, 77, 78, 85, 89, 92(1), 92(2), 92(3), 92(4), 92(5), 96(5), 96(6), 96(7), 96(8), 96(9), 100

Criminal Law 304(1), 304(2), 304(3), 627.5(3), 627.5(6), 982.5(1), 982.5(2), 986.2(1), 986.2(4), 986.6(3), 1208.1(5), 1208.6(1), 1208.6(2), 1208.6(3)

Health and Environment 23

Homicide 2–5, 15, 16, 18(4), 58, 59, 63, 87, 89(1), 89(3), 90, 134, 135(1), 141(1), 141(7), 141(9), 142(7), 142(8), 156(1), 161, 164, 171(1), 171(2), 173,

Source: A.L.R.5th, vol. 13, pp. 630–631 (Total Client-Service Library References). Permission has been granted by the current copyright holder, West Group. Further reproduction of any kind is strictly prohibited. For additional information, please contact West Group Customer Services representative at 1-800-328-4880.

When we turn to section 9 in the annotation to learn more about attempted murder in this context, we find an extensive examination of cases relevant to this issue. (See Exhibit 2–32.) Principal discussion is devoted to a case decided by the Indiana Court of Appeals, *State v. Haines*, 545 N.E.2d 834 (Ind. App. 1989), but other cases are discussed and cited that also promise to be of interest. When we check the pocket supplement, we find additional cases on point and learn that the Texas case that was cited in the main volume, *Weeks v. State*, 834 S.W.2d 559 (Tex. App. 1992), appears to have begun its passage through the federal court system as *Weeks v. Collins*, 867 F. Supp. 544 (S.D. Tex. 1994). (See Exhibit 2–33.) We also review section 5 of the annotation to learn whether the act of biting and spitting by offenders who are HIV-positive has been recognized as assault with a deadly weapon.

The issues presented in our other cases are researched similarly. For Case 1, the words we consult in the A.L.R. Index might include **"Confrontation," "Televi-**

Exhibit 2–32

---

§ 9. Attempted murder

In the following cases involving attempts by HIV-positive defendants to expose others, the courts held supportable convictions for attempted murder.

The court reversed the trial judge's grant of judgment on the evidence in favor of the defendant, after the jury had convicted him of three counts of attempted murder based on his efforts to transmit the AIDS virus to those trying to help him after a suicide attempt, in State v Haines (1989, **Ind** App) 545 NE2d 834. When police officers arrived at the defendant's apartment in response to a radio call, they found him unconscious, lying face down in a pool of blood. The officer who tried to revive the defendant saw that his wrists were slashed and bleeding. When he heard paramedics arriving, the defendant revived and screamed that he should be left to die because he had AIDS. He told the officer that he wanted to f____ him and "give it to him" and that he would "use his wounds," after which he began jerking his arms, causing blood to spray into the officer's mouth and eyes. He repeatedly yelled that he had AIDS, that he could not deal with it, and that he was going to make the officer deal with it. He struggled with the paramedics, threatening to infect them with AIDS and spitting at them. When the police officer grabbed him, the defendant scratched, bit, and spit at him, at one point grabbing a blood-soaked...

...virus as a result of the defendant's conduct. The court distinguished United States v Moore (1988,

CA8 Minn) 846 F2d 1163 (§ 5[b]), on the ground that in that case there had been insufficient evidence that the defendant's conduct could transmit the AIDS virus. Here, stated the court, it could only be concluded that the defendant had knowledge of his disease and unrelentingly, unequivocally sought to kill those helping him by infecting them, and took a substantial step toward killing them by his conduct, believing that he could do so, all of which amounted to more than a theoretical or speculative chance of transmitting the disease. The court instructed the trial judge to reinstate the jury's verdict and to sentence the defendant accordingly.

See Re Anonymous (1989, 4th Dept) 156 App Div 2d 1028, 549 NYS2d 308 (§ 3[a]), affd 76 NY2d 766, 559 NYS2d 976, 559 NE2d 670, in which the court approved an order requiring a blood test for use in an attempted murder prosecution arising from the biting of three police officers by an AIDS carrier.

In Weeks v State (1992, **Tex** App Eastland) 834 SW2d 559, petition for discretionary review ref (Oct 14, 1992), the court, affirming an attempted murder conviction, rejected the defendant's contention that his conduct was not shown to be reasonably capable of harming the victim. The defendant inmate had spit twice in the face of a prison guard. The court noted the evidence that the defendant previously had told the guard that the defendant knew he was HIV-positive and that the defendant was going to "take somebody with him when he went."

Source: A.L.R.5th, vol. 13, pp. 654–656, sec. 9. Permission has been granted by the current copyright holder, West Group. Further reproduction of any kind is strictly prohibited. For additional information, please contact West Group Customer Services representative at 1-800-328-4880.

---

sion," "**Child**," "**Witness**," "**Testimony**," and others pertaining to the issue. We discover at least two annotations that may be of interest: "Confrontation of witnesses, closed circuit television, 80 ALR3d 1212; 61 ALR4th 1155." (See Exhibit 2–34.) The annotation, "**Closed-Circuit Television Witness Examination**," 61 A.L.R. 4th 1155, promises to provide a good overview of the law, including an extensive discussion of the admissibility of testimony over closed-circuit televi-

Exhibit 2–33

13 ALR5th 628–683

§ 5. Aggravated assault—with deadly weapon or dangerous instrument
[a] Offense established

See People v Dempsey (1993, 5th Dist) 242 Ill App 3d 568, 182 Ill Dec 784, 610 NE2d 208, § 10.

Defendant, who was HIV positive and bit correctional officer, was properly convicted of aggravated assault and attempted murder where, although there was some evidence that defendant knew that HIV could not be transmitted through bite, defendant had frequently threatened to spit at and/or bite officers to give them "what I have" and, in one instance, to kill officer. State v Smith (1993) 262 NJ Super 487, 621 A2d 493, certif den (NJ) 634 A2d 523.

§ 9. Attempted murder

See State v Smith (1993) 262 NJ Super 487, 621 A2d 493, certif den (NJ) 634 A2d 523, § 5[a].

Attempted murder conviction of HIV positive defendant who spat twice in face of prison guard was supported by evidence where record showed that (1) defendant knew he was HIV positive, (2) defendant had stated that he was going to take as many with him as he could, (3) defendant believed that he could kill victim by spitting on him, and (4) experts had not entirely ruled out possibility of transmitting HIV through saliva. Weeks v State (1992, Tex App Eastland) 834 SW2d 559, petition for discretionary review ref (Oct 14, 1992).

Conviction of defendant for attempted murder in connection with his spitting on prison guard was supported by evidence that (1) on date of spitting, defendant was in advanced stage of infection, HIV-4, (2) some HIV-positive patients have virus growing in their saliva, (3) there is greater possibility of virus being present in saliva if blood were in saliva, (4) blood would more likely be in saliva if defendant needed dental work or had just eaten, (5) shortly before spitting incident, defendant had eaten lunch, (6) defendant had been to dentist 2 months prior to incident and needed additional dental work, (7) virus could be transmitted through saliva, especially where saliva comes in contact with mucous membrane, (8) spit hit guard in face and got up inside his nose, and (9) nose is lined by mucous membrane. Weeks v Collins (1994, SD Tex) 867 F Supp 544 (applying Tex law).

Source: A.L.R.5th, vol. 13 (Supp. 1995), pp. 5–6, sec. 9 of 13 ALR 5th 628–683. Permission has been granted by the current copyright holder, West Group. Further reproduction of any kind is strictly prohibited. For additional information, please contact West Group Customer Services representative at 1-800-328-4880.

sion by "**Children or minors**." In Exhibit 2–35, we display a portion of this annotation's pocket supplement, in which this same issue is addressed.

You should also have a hand at constructing a search strategy for A.L.R. annotations relating to Case 3, which, you recall, presents the issue of whether a prisoner's involuntary exposure to second-hand tobacco smoke can violate the right to be free from cruel and unusual punishment. If you were to look in the A.L.R. Index, including its pocket part, under topics such as "**Prison and prisoners**" and "**Cruel and unusual punishment**," you would find scores of annotations, two of which seem to be especially relevant to our issue. The index describes one as covering "conditions as amounting to cruel and unusual punishment, 51 ALR3d 111," and the other as addressing "Conditions of confinement as constituting cruel and unusual punishment in violation of the Federal Constitution's Eighth Amendment, 115 L.Ed.2d 1151." Investigating the latter

Exhibit 2–34

ALR INDEX

TELEVISION
For digest treatment, see Title **Radio and Television in ALR Digest**
Generally, as to telecommunications, see topic **Telecommunications** in this index
Advertising and advertisements, measure of damages, to advertiser, for radio or television station's breach of wrongful termination of contract, 90 ALR2d 1199
Antennas, standing of owner of property adjacent to zoned property, but not within territory of zoning authority, to attack zoning, 69 ALR3d 805, §§ 4, 7[b]
Attorneys, advertising as ground for disciplinary action, 30 ALR4th 742, § 10
Broadcast, liability for personal injury or death allegedly resulting from television or radio broadcast, 20 ALR4th 327

Burglary charge, maintainability where entry into television shop is made with consent, 58 ALR4th 335, § 8
Cable Television (this index)
Closed Circuit Television (this index)
Communications Act, construction and application of Communications Act statute of limitations (47 USCS § 415(b)) relating to recovery from carrier of damages not based on overcharges, 81 ALR Fed 700
Competition, enforceability of covenant not to compete involving radio or television personality, 36 ALR4th 1139
Confrontation of witnesses, closed circuit television, 80 ALR3d 1212; 61 ALR4th 1155
Copyrights
— co-owners of copyright, rights and remedies of, 3 ALR3d 1301

Source: A.L.R. Index, ALR 2d-3d-4th-5th, Federal, LEd2d (T-Z), p. 67, "Television: Confrontation of witnesses, closed circuit television." Permission has been granted by the current copyright holder, West Group. Further reproduction of any kind is strictly prohibited. For additional information, please contact West Group Customer Services representative at 1-800-328-4880.

reference permits us to make use of an annotation in the Lawyers' Edition 2d case reports.

The annotation, which begins at page 1151 in volume 115 of Lawyers' Edition 2d, refers to a Supreme Court decision, printed in the same volume, that involves a prisoner's federal lawsuit based on a claim of cruel and unusual punishment, *Wilson v. Seiter*, 501 U.S. 294, 111 S. Ct. 2321, 115 L. Ed. 2d 271 (1991). The annotation focuses on U.S. Supreme Court decisions that have addressed whether different conditions of confinement amount to cruel and unusual punishment. When we locate this annotation, we note that its section 10 discusses the very issue in which we are interested, "Exposure to environmental tobacco smoke." From our previous research, we know we should examine the numerous tips regarding related references at the start of the annotation, read the other sections of the annotation for general principles and background, and then focus on section 10. Section 10 summarizes the holding in *Helling v. McKinney*, 509 U.S. 25, 113 S. Ct. 2475, 125 L. Ed. 2d 22 (1993), a Supreme Court decision that apparently relates directly to our topic. The annotation reports that in *Helling v. McKinney* the Supreme Court ruled that a lower court correctly refused to dismiss a state prisoner's lawsuit alleging a violation of his right to be free from cruel and un-

Exhibit 2–35

---

61 ALR4th 1155–1171

§ 12. Children or minors
[a] Held to be constitutional

In determining whether a state has made an adequate showing of the necessity of using a special procedure which permits a child witness in a child abuse case to testify at trial against a defendant in the absence of face-to-face confrontation with the defendant, without a violation of the confrontation clause of the Federal Constitution's Sixth Amendment—which clause provides that, in all criminal prosecutions, the accused has the right to be confronted with the witnesses against the accused—a trial court's requisite finding of necessity must be a case-specific one; where the special procedure involves testimony by one-way closed-circuit television, the trial court must (1) hear evidence and determine whether use of the closed-circuit television procedure is necessary to protect the welfare of the particular child witness who seeks to testify, and (2) if the court is to find the procedure necessary, find that (a) the child witness would be traumatized not by the courtroom generally, but by the presence of the defendant, and (b) the emotional distress suffered by the child witness in the presence of the defendant would be more than mere nervousness or excitement or some reluctance to testify. Maryland v Craig (1990, US) 111 L Ed 2d 666, 110 S Ct 3157, 30 Fed Rules Evid Serv 1, on remand 322 Md 418, 588 A2d 328.

In prosecution for murder of daughter, son was properly allowed to testify by closed circuit television where court considered expert testimony as to likely effect of testifying and concluded child would be unable to testify in open court due to presence of defendant and would suffer emotional trauma from testifying. United States v Quintero (1994, CA9 Ariz) 21 F3d 885, 94 CDOS 2436, 94 Daily Journal DAR 4651.

In prosecution for sexual abuse involving two young girls aged 8 and 7, District Court's order allowing child witnesses to testify by closed-circuit television was supported by its . . .

---

[b] Held to be unconstitutional

In state court prosecution for capital sexual battery in which five-year-old victim was permitted to testify by closed circuit television, state trial court failed to make sufficient individualized findings about possibility of harm to victim to reliably conclude that it was necessary for her to testify outside of defendant's presence in violation of his rights under confrontation clause to meet face to face all those who appear and give evidence at trial. No one at trial appeared to have considered defendant's confrontation clause rights. State's motion for use of closed circuit television did not mention necessity, nor even request that, victim testify outside defendant's presence, and there was no discussion during hearing on state's motion about whether there would be danger of significant traumatization to victim if defendant stayed in courtroom during her testimony. There was nothing to indicate that she was afraid of defendant or that testifying by closed circuit television would enhance protection she needed. Thus, there was no support in record for proposition that state demonstrated, or trial court articulated, any reason to justify victim's separation from defendant during her testimony. Cumbie v Singletary (1993, CA11 Fla) 991 F2d 715, 7 FLW Fed C 341.

In murder prosecution, prosecutor would not be permitted to have four-year-old son of victim testify outside courtroom by use of closed-circuit television procedure in which prosecutor and defense counsel would question child outside courtroom and process would be viewed . . .

Source: A.L.R.4th, vol. 61 (1995 Supp.) pp. 60–64, 61 ALR 4th 1151–1171, sec. 12a, 12b (excerpts). Permission has been granted by the current copyright holder, West Group. Further reproduction of any kind is strictly prohibited. For additional information, please contact West Group Customer Services representative at 1-800-328-4880.

usual punishment based on the prison administrators' deliberate indifference to the health risks associated with his exposure to environmental tobacco smoke. (See Exhibit 2–36.) Since the annotation in Lawyers' Edition 2d confines itself to an examination of U.S. Supreme Court decisions, we want to be sure to follow up by checking other references, including the related annotation to which we were referred by the A.L.R. Index. This annotation, "Prison conditions as amounting to cruel and unusual punishment," 51 A.L.R.3d 111, considers a wide range of lower court decisions that may relate to our issue.

Computer-Assisted Legal Research and A.L.R.

All of the Lawyers Cooperative annotation series except Lawyers' Edition 1st and A.L.R.1st are accessible through the LEXIS® computer database system. In addition, Lawyers Cooperative has made available CD-ROM versions of its an-

---

Exhibit 2–36

WILSON v SEITER
Reported p 271, supra

§ 10. Exposure to environmental tobacco smoke

The United States Supreme Court has held that a state prison inmate stated a cause of action under the Federal Constitution's Eighth Amendment by alleging that prison personnel have, with deliberate indifference, exposed the inmate to levels of environmental tobacco smoke that posed an unreasonable risk of serious damage to the inmate's future health.

Thus, a state prison inmate was held in Helling v McKinney (1993, US) 125 L Ed 2d 22, 113 S Ct 2475, to have stated a cause of action under the Federal Constitution's Eighth Amendment by alleging that prison personnel had, with deliberate indifference, exposed the inmate to levels of environmental tobacco smoke (ETS) that posed an unreasonable risk of serious harm to the inmate's future health. In affirming a Federal Court of Appeals' judgment which reversed a directed verdict in favor of the defendant prison personnel, the Supreme Court rejected the view that only deliber-

ate indifference to current serious health problems is actionable under the Eighth Amendment. The Supreme Court noted, however, that the inmate, in order to prevail on remand, would have to show (1) that he himself was being exposed to unreasonably high levels of ETS, as to which determination the fact that the inmate had been moved to another prison and was no longer the cellmate of a five-pack-a-day smoker would be relevant; (2) that the risk he complained of was not one that contemporary society chose to tolerate, as the determination whether the inmate's conditions of confinement violated the Eighth Amendment required, in addition to a scientific and statistical inquiry into the seriousness of the potential harm and the likelihood that such an injury to health would actually be caused by exposure to ETS, an assessment whether society considered the risk complained of to be so grave that it violated contemporary standards of decency to expose anyone unwillingly to such a risk; and . . .

Source: United States Supreme Court Reports, Lawyers Edition 2d, vol. 115, pp. 1166–1167, sec. 10. Reprinted with the permission of LEXIS-NEXIS, a division of Reed Elsevier Inc. LEXIS and NEXIS are registered trademarks of Reed Elsevier Properties Inc. FREESTYLE, KWIC, SuperKWIC and MEGA are trademarks of Reed Elsevier Properties Inc. SHEPARD'S and SHEPARDIZE are registered trademarks of Shepard's Company, a Partnership.

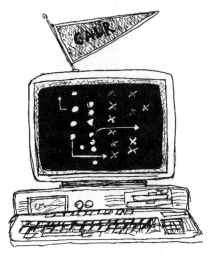

notation series. However, we will limit our discussion to the online versions available through the LEXIS system.

General Computer-Assisted Legal Research Strategies. In Chapter 1 we touched briefly on some of the advantages and disadvantages of using computer databases for legal research. With the assistance of our hypothetical cases, we can provide some concrete examples of what we meant. Much of law, especially case law, involves the application of established legal principles to new factual circumstances or to different social phenomena. In determining whether **computer-assisted legal research (CALR)** has advantages over research using printed sources, it can be helpful to separate the legal principles from the unique factual circumstances of a case. For example, in Case 1 the general legal principle involves a defendant's right to confront his accusers. The unique factual circumstance related to this principle is the closed-circuit televising of an alleged child-victim's testimony. In Case 2, the legal principles involve attempted murder and assault with a deadly weapon, and the unique factual circumstances are the use of HIV or AIDS as the means of inflicting the harm, as well as the transmission method—biting or spitting. In Case 3, one of the legal principles involves cruel and unusual punishment, and the unique facts are double celling and second-hand smoke.

CALR is likely to prove advantageous in situations where the factual circumstances are unique or so new that they are not well covered by the printed references. Fifteen years ago it would have been difficult to find any reference to HIV or AIDS in the sources mentioned above. Even today, it may prove easier to find relevant documents by using key word searching in the computerized databases rather than using the subject approach in the printed indexes. We will use our hypothetical cases to illustrate strategies for using the computerized databases to get information similar to what we retrieved through the printed sources.

Before going through these examples, we shall discuss different standards for gauging the quality of results obtained through computerized searching. These standards are **recall** and **precision**. "Recall" is measured by comparing the number of relevant documents retrieved from a database to the total number of relevant documents in the database. Thus, if a search query retrieves 90 relevant documents from a database containing 100 relevant documents, the recall rate is 90 percent. "**Precision**" refers to the ratio of relevant documents retrieved as measured against the total number of documents retrieved. Thus, if 50 documents are retrieved but only 20 are relevant to the search topic, the precision rate is 40 percent. Ideally, a search strategy will produce all the relevant documents in a database, and will retrieve only the relevant documents, resulting in a score of 100 percent on both recall and precision. As illustrated below, designing a strategy requires a balancing of recall and precision. Unfortunately, computer research is a bit like baseball: There are no perfect 1.000 hitters in either enterprise. However, with a little thought and some practice, your average in CALR can be considerably higher than the .300 batting average benchmark.

Here are some tips for improving your CALR average. First, remember you are doing key word or key phrase searching. Unlike the subject searching demonstrated with the printed sources, key word or phrase searching does not restrict the context of the word to a particular issue. Accordingly, you must be careful how you use words that have different meanings in different contexts. For example, in the second hypothetical, you must be careful how you enter the term "AIDS" as an acronym for acquired immune deficiency syndrome. The computer will retrieve documents not only where "AIDS" is used as an acronym, but also where the word "aids" appears in other contexts, such as "Computer *aids* researcher in search for relevant documents." The way to get around this problem in LEXIS is to put the word in capital letters—that is, "**AIDS**." LEXIS reads terms in all capital letters as acronyms or initialisms. You should also include synonyms or closely related terms in your search query. For example, in Case 2 you should use both "human immunodeficiency virus" and "acquired immune deficiency syndrome," as well as HIV and AIDS. In WESTLAW, acronyms and initialisms are distinguished by placing periods between the letters: **A.I.D.S.** or **H.I.V.**

Fortunately, both WESTLAW and LEXIS include features in their respective software that enable the researcher to improve his or her average when doing keyword searching. Documents in each system are broken down into parts called **fields** in WESTLAW and **segments** on LEXIS. Search queries can be restricted to designated fields or segments. This feature usually does not come into play when searching the annotations so we will delay a detailed discussion.

Another feature is the use of **proximity requirements** or **Boolean operators** to provide more precision for search results. Proximity requirements or Boolean operators link two or more search terms or phrases. The commonly used Boolean operators are **and**, **or**, and **not**. The **and** operator requires that the terms on either side of the **and** appear in the same document, or the same field if you are using field restrictions. Thus, the query "**rifle and shotgun**" would retrieve documents in which both "**rifle**" and "**shotgun**" appear. The **or** operator requires that one of the terms on either side of the "**or**" appears in the retrieved documents, and thus expands the number of documents retrieved. Thus, the query "**rifle or shotgun**" would retrieve documents containing either the word "**rifle**" or the word "**shotgun**," as well as documents containing both terms. The **not** operator retrieves documents containing one term but not another. Thus, the query "**rifle not shotgun**" would retrieve all documents containing the term "**rifle**" except for those that also contain the term "**shotgun**." Figure 2–1 provides a graphic illustration of how these three basic operators affect search results. The shaded area in each **Venn** diagram indicates the portion of documents retrieved.

Both WESTLAW and LEXIS have added to these basic operators another set of proximity operators that greatly expand the flexibility of searching the respective database systems. The additional operators used in LEXIS are illustrated in our examples below. Those used on WESTLAW will be discussed later.

Three comparable search methods exist for using both WESTLAW and LEXIS systems. The most sophisticated form of searching is called **terms and connectors** searching on WESTLAW, and **Boolean searching** on LEXIS. This search method requires knowledge of the proximity requirements and the fields or segments used in the systems. The least sophisticated form of searching is called

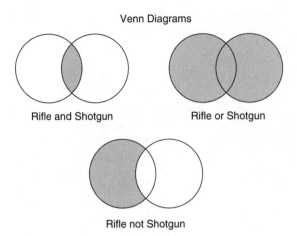

Figure 2–1 Venn Diagrams

natural language searching on WESTLAW and **FREESTYLE™ searching** on LEXIS, and does not require knowledge of fields or proximity requirements. We will be discussing and providing examples of this latter method of searching later in this chapter. The third method relies on menus to guide the user through the system. The menu system is called **EZ Access** on WESTLAW and **Easy Search** on LEXIS.

Both WESTLAW and LEXIS are large and complex database systems, and each has well over 1,000 databases accessible through it. For this reason, we recommend training in the use of the systems before you attempt to search them. Both systems have tutorial packages as well as online help information, so basic techniques can be acquired if formal training is not possible. Both systems have Windows and DOS versions of their programs, but the basic search commands and capabilities are the same in both versions.

    Searching A.L.R. Electronically: Designing the Search Query. With this much by way of background, we are ready to search for A.L.R. annotations by using LEXIS. We already have noted that the more recent printed versions of annotations provide within their Total Client-Service Library References a description of the **Electronic Search Query** that can be used to retrieve the annotation. Exhibit 2–31 depicts the electronic search query that could be used to retrieve the annotation related to our first hypothetical case: **"HIV or (AIDS w/5 virus) w/ 10 crim! or sentence or assault or battery or homicide or murder or (dangerous pre/1 weapon or instrument!) or rape."** (Please note that we have changed "hom*o*cide" to "hom*i*cide.") To those unfamiliar with the Boolean operators and other features of the LEXIS search system, this strategy may look like so much gibberish. Novice searchers who are unable to translate the strategy into English should be wary of using these suggested queries. We will explain this search strategy, but first we want to construct our own LEXIS search strategy for our initial hypothetical case. We will be using the Boolean searching method.

Searching the electronic version of the annotations involves a process different from searching the print version. In the print version we select a term or phrase describing a key element or concept in our topic and then look up that key word or phrase in the A.L.R. Index. In the electronic version, we take all of the terms or phrases describing all our concepts and include them in our initial search query. An advantage of using the electronic version is that if we have properly identified our key terms or phrases and also have properly designed our query, then our results should focus narrowly on the issue we are researching. One disadvantage of using the electronic version is that by focusing on retrieving only annotations dealing with all the concepts in our topic we may lose some of the serendipitous references that ultimately might prove helpful even if they are not precisely on point.

We will use the hypothetical case involving our HIV-positive defendant, Andrew Adams, who has been charged with attempted murder or assault with a deadly weapon for biting and spitting on Officer Fiegel. We previously have identified at least some of the key terms describing the elements of the research issue: **"attempt," "murder," "assault," "deadly weapon," "AIDS," "HIV."** We will add to this list **"acquired immunodeficiency syndrome"** and **"human immunodeficiency virus."** As a general rule, it is a good idea to search not only initialisms or acronyms referring to a concept, but also the fully spelled out words or phrase to which the acronym or initialism refers. This is particularly appropriate in electronic searching, where you are usually searching by key words appearing in the text of the documents rather than by subject descriptors assigned to the documents.

Our next step is to group together the related terms. We join **"assault"** and **"murder"** in one group and put in a second group the related terms **"AIDS," "HIV," "acquired immunodeficiency syndrome,"** and **"human immunodeficiency virus." "Attempt"** would stand by itself, as would **"deadly weapon,"** since we have not identified related terms for either of them. We next have to use Boolean operators or proximity connectors to link the terms together. The basic Boolean and proximity operators used in the LEXIS system are consistent with the general principles we described above. They are:

> **and** retrieves documents containing at least one of the terms appearing on both sides of the connector.

> **or** retrieves documents that include one of the terms appearing on either side of the operator.

> **and not** Documents containing the term(s) following this operator will be excluded.

> **w/p** retrieves documents where the term(s) on both sides of the operator appear in the same paragraph.

> **w/s** retrieves documents where the term(s) on both sides of the operator appear in the same sentence.

> **w/n** retrieves documents where the term(s) appearing on both sides of the operator must appear within a designated number of words of each

other. For example, "**cruel w/5 punishment**" would retrieve documents where the word "**cruel**" appears within five words of the word "**punishment**."

pre/n retrieves documents where the term(s) to the left of the operator must precede the term(s) on the right side of the operator and be within *n* words of each other. For example, "**cruel pre/5 punishment**" would retrieve documents where "**cruel**" precedes and is within five words of "**punishment**."

If no operator is inserted between words, then the grouped words will be interpreted as a phrase. For example, "**death penalty**" will retrieve documents in which the phrase "**death penalty**" appears, but will not retrieve a document with the following wording appearing in the text: "a sentence of death is an appropriate penalty."

There are other proximity operators used in LEXIS®, but those listed above should be sufficient for most research topics in the annotation databases.

We will want to link all related concepts with the **or** operator:

murder or assault

HIV or AIDS or acquired immunodeficiency syndrome or human immunodeficiency virus

Our next step is to designate the proximity operator(s) we want to use to link each of the four separate elements of our search query:

Attempt

deadly weapon

murder or **assault**

HIV or AIDS or acquired immunodeficiency syndrome or human immunodeficiency virus

We do not have to use the same operator between all the elements, but some caution is needed in selecting the operators because they affect not only the results retrieved but also the order in which the computer searches the various elements of our query. In LEXIS, the computer first searches for terms connected by the **or** operator, followed by the **w/n** operator, and finally the **and** operator. For our example the order of searching is not as important as the proximity requirements. The **and** operator is probably too broad a proximity requirement for this search as it would require only that the various elements all appear in the same document. A more precise operator might be **w/p**, which would require the elements to appear in the same paragraph. If we use the **w/p** operator, our search query will look like this:

HIV or AIDS or acquired immunodeficiency syndrome or human immunodeficiency virus w/p deadly weapon w/p murder or assault w/p attempt!

The "!" is the truncation symbol in LEXIS, which means that using this symbol allows us to retrieve all forms of a word with a common root but with potentially different endings. Thus, "**attempt!**" will retrieve not only "**attempt**," but "**attempted**," "**attempting**," and "**attempts**." This same query will retrieve annotations containing (1) either "**HIV**," "**AIDS**," "**acquired immunodeficiency syndrome**," or "**human immunodeficiency virus**," and (2) in the same paragraph, the phrase "**deadly weapon**," and (3) in the same paragraph either "**murder**" or "**assault**," and (4) in the same paragraph any word beginning with the root form "**attempt**." We can compare the search query we have constructed with the Electronic Search Query suggested in Exhibit 2–31:

> **HIV or (AIDS w/5 virus) w/10 crim! or sentence or assault or battery or rape or homicide or murder or (dangerous pre/1 weapon or instrument!) or rape**

Note that the use of parentheses rearranges the order in which LEXIS searches for terms. The part of the query within the parentheses will always be searched first. Therefore, in this query the computer will look for documents containing the word "**AIDS**" within five words of "**virus**," and "**dangerous**" immediately preceding and within one word of either "**weapon**" or a word beginning with the root form "**instrument**." The computer then will look for HIV as an optional term for AIDS within five words of "**virus**," and for any of the words "**sentence**," "**assault**," "**battery**," "**homicide**," "**murder**," "**rape**," or a word beginning with the root "**crim**" as alternatives to "**dangerous weapon**" or "**dangerous instrument(s)**." Essentially, there are two groupings of words: **HIV** or **AIDS** within five words of "**virus**" constitutes the first group, and all of the other terms or phrases are in the second group. The **w/10** connector requires that one of the terms or phrases to the left of the connector appears within 10 words of one of the terms to the right of the connector in any document retrieved. Whether this search strategy is more effective than the one we designed is problematical. The differences are that we required "**attempt**" to be in the documents we retrieved, and we treated "**deadly weapon**" as a separate concept, whereas the suggested query includes "**dangerous weapon**" or "**dangerous instrument**" as part of a string of related concepts. There is no perfect query for most research problems, and there usually are several ways of designing a query that will retrieve similar, if not identical, results.

Running the Search. The databases in LEXIS are first arranged by **libraries**, which are broad groupings of materials by jurisdiction, subject matter, or types of materials. Within each library there are individual **files**. Each library and file has its own alphabetical code. An online directory can be used to find the proper file to be searched. However, novice users may want to determine the identifier for their library and file prior to signing on by consulting the printed **LEXIS-NEXIS® Directory of Online Services**. Both the library and the file containing the A.L.R. annotations are named "ALR." As you sign on to LEXIS, **enter** "**ALR**" at the Libraries screen, as depicted in Exhibit 2–37. You will come to a screen asking you to enter the name of the file you want. **Enter "ALR,"** as depicted in Exhibit 2–38.

Exhibit 2–37

ALR

LIBRARIES — Page 1 of 3

Please ENTER the NAME <only one> of the library you want to search.
—For more information about a library, ENTER its page <PG> number.
—To see a list of additional libraries, press the NEXT PAGE key.

```
- - - - - - - - -  Types  - - - - - - - -   - - - - - - - -  Topics  - - - - - - - -
    General                      Public   BUSFIN    2      Intellect         Medical
    News                Legal    Records  CMPCOM    2      Property     GENMED   14
  NEWS      1   CODES     6   DOCKET  5   ENERGY    9      COPYRT   8   MEDLNE   14
  REGNWS    1   LAWREV   11   INCORP  5   ENTERT    2      PATENT  12
  TOPNWS    1   LEGNEW    1   LEXPAT  5   ENVIRN    9      TRDMRK  13       Political
                MEGA      6   LIENS   5   INSURE   10                   CMPGN     3
                MODEL    11   VERDCT  5   MARKET    1          Legal    EXEC      3
                                         PEOPLE    2      BANKNG   7   LEGIS     3
                                         SPORTS    2      FEDSEC  10
                                         TRANS    13      GENFED   6        Tax
    Financial                    Reference               LABOR   11   FEDTAX   10
  ACCTG     2                  BUSREF    2               PUBCON  12   STTAX    12
  COMPNY    2                  LEXREF   11               STATES   6
  NAARS     2


  - - - - - -  Int'l  - - - - - -      Assists
  ASIAPC    4   MDEAFR    4   EASY     14
  CANADA   19   NSAMER    4   GUIDE    14
  DUTCH     4   WORLD     4   PRACT    14
  EUROPE    4   TXTLNE   18   TERMS    14
  GERMAN    4                 CATALOG  14
```

Source: LEXIS Libraries screen (ALR). Reprinted with the permission of LEXIS-NEXIS, a division of Reed Elsevier Inc. LEXIS and NEXIS are registered trademarks of Reed Elsevier Properties Inc. FREESTYLE, KWIC, SuperKWIC and MEGA are trademarks of Reed Elsevier Properties Inc. SHEPARD'S and SHEPARDIZE are registered trademarks of Shepard's Company, a Partnership.

Exhibit 2–39 depicts the query entry screen where we already have typed in our search strategy. The summary results of the strategy are depicted in Exhibit 2–40. Note that nine A.L.R. annotations meet the requirements of our search request. We have several options for browsing through our results. One option is to look at the bibliographic citations for the nine documents. To get a list of citations we **enter ".ci,"** as shown in the upper left corner of Exhibit 2–40.

Exhibit 2–41 displays the citations to the first five documents retrieved. Notice that the third citation refers to the same annotation we found using the print version of A.L.R. We can go to the full text of that annotation by entering its list number, in this case "3," and we will come to the first screen of the annotation.

Exhibit 2–38

ALR_

Please ENTER the abbreviated NAME of the file you want to search. To see a description of a file, type its page number and press the ENTER key.

FILES—PAGE 1 of 1

NAME	PG	DESCRIP	NAME	PG	DESCRIP
ALR	1	American Law Reports Federal; Lawyer's Edition Second; American Law Reports Fifth; American Law Reports Fourth; American Law Reports Third; American Law Reports Second	GUIDE	1	LEXIS<R> Product Guide

Enter file name; .gu for file content & coverage. Example: ALR;.gu

Source: LEXIS Files screen (ALR). Reprinted with the permission of LEXIS-NEXIS, a division of Reed Elsevier Inc. LEXIS and NEXIS are registered trademarks of Reed Elsevier Properties Inc. FREESTYLE, KWIC, SuperKWIC and MEGA are trademarks of Reed Elsevier Properties Inc. SHEPARD'S and SHEPARDIZE are registered trademarks of Shepard's Company, a Partnership.

Exhibit 2–39

AIDS or HIV or acquired immunodeficiency syndrome or human immunodeficiency virus w/p assault or murder w/p deadly weapon w/p attempt!

Please type your search request then press the ENTER key.
What you enter will be Search Level 1.

Type .fr to enter a FREESTYLE<TM> search.

For further explanation, press the H key <for help> and then the ENTER key.

Source: LEXIS Query entry screen (AIDS or HIV . . .). Reprinted with the permission of LEXIS-NEXIS, a division of Reed Elsevier Inc. LEXIS and NEXIS are registered trademarks of Reed Elsevier Properties Inc. FREESTYLE, KWIC, SuperKWIC and MEGA are trademarks of Reed Elsevier Properties Inc. SHEPARD'S and SHEPARDIZE are registered trademarks of Shepard's Company, a Partnership.

(See Exhibit 2–42.) The information in the LEXIS version is exactly the same as in the print version, with the exception that the online version may include recent updates to the annotation that are not yet available in printed form. For this reason, it is a good idea to update your search in LEXIS even if you started your

Exhibit 2–40

.ci
AIDS OR HIV OR ACQUIRED IMMUNODEFICIENCY SYNDROME OR HUMAN IMMUNODEFI-
CIENCY VIRUS W/P ASSAULT OR MURDER W/P DEADLY WEAPON W/P ATTEMPT!

Your search request has found 9 ALR ANNOTATIONS through Level 1.
To DISPLAY these ALR ANNOTATIONS press either the KWIC, FULL, CITE or SEGMTS key.
To MODIFY your search request, press the M key <for MODFY> and then the ENTER key.

For further explanation, press the H key <for HELP> and then the ENTER key.

Source: LEXIS Search request results screen (9 ALR ANNOTATIONS. . .). Reprinted with the permission of
LEXIS-NEXIS, a division of Reed Elsevier Inc. LEXIS and NEXIS are registered trademarks of Reed Elsevier Properties Inc. FREESTYLE, KWIC, SuperKWIC and MEGA are trademarks of Reed Elsevier Properties Inc. SHEPARD'S
and SHEPARDIZE are registered trademarks of Shepard's Company, a Partnership.

Exhibit 2–41

LEVEL 1–9 ALR ANNOTATIONS

1. ANNOTATION: ADMISSIBILITY OF STATEMENTS MADE FOR PURPOSES OF MEDICAL
DIAGNOSIS OR TREATMENT AS HEARSAY EXCEPTION UNDER RULE 803(4) OF THE UNI-
FORM RULES OF EVIDENCE, 38 A.L.R.5th 433, ALR 5th; Copyright (c) 1996 Lawyers Cooperative
Publishing Co.

2. ANNOTATION: VALIDITY, CONSTRUCTION, AND EFFECT OF "HATE CRIMES" STAT-
UTES, "ETHNIC INTIMIDATION" STATUTES, OR THE LIKE, 22 A.L.R.5th 261, ALR 5th; Copyright (c) 1995 Lawyers Cooperative Publishing Co.

3. ANNOTATION: TRANSMISSION OR RISK OF TRANSMISSION OF HUMAN IMMUNODEFI-
CIENCY VIRUS (HIV) OR ACQUIRED IMMUNODEFICIENCY SYNDROME (AIDS) AS BASIS
FOR PROSECUTION OR SENTENCING IN CRIMINAL OR MILITARY DISCIPLINE CASE, 13
A.L.R.5th 628, ALR 5th; Copyright (c) 1995 Lawyers Cooperative Publishing Co.

4. ANNOTATION: EXCESSIVENESS OR INADEQUACY OF PUNITIVE DAMAGES AWARDED
IN PERSONAL INJURY OR DEATH CASES, 12 A.L.R.5th 195, ALR 5th; Copyright (c) 1995 Lawyers
Cooperative Publishing Co.

5. ANNOTATION: CONSTRUCTION AND APPLICATION OF STATUTES JUSTIFYING THE
USE OF FORCE TO PREVENT THE USE OF FORCE AGAINST ANOTHER, 71 A.L.R.4th 940, ALR
4th; Copyright (c) 1995 Lawyers Cooperative Publishing Co.

Source: LEXIS Search results display screen (LEVEL 1–9 ALR ANNOTATIONS—ANNOTATIONS 1–5). Reprinted with the permission of LEXIS-NEXIS, a division of Reed Elsevier Inc. LEXIS and NEXIS are registered trademarks of Reed Elsevier Properties Inc. FREESTYLE, KWIC, SuperKWIC and MEGA are trademarks of Reed Elsevier
Properties Inc. SHEPARD'S and SHEPARDIZE are registered trademarks of Shepard's Company, a Partnership.

Exhibit 2–42

LEVEL 1–3 OF 9 ALR ANNOTATIONS

ALR 5th; Copyright (c) 1995 Lawyers Cooperative Publishing Co.

ANNOTATION

TRANSMISSION OR RISK OF TRANSMISSION OF HUMAN IMMUNODEFICIENCY VIRUS (HIV) OR ACQUIRED IMMUNODEFICIENCY SYNDROME (AIDS) AS BASIS FOR PROSECU-TION OR SENTENCING IN CRIMINAL OR MILITARY DISCIPLINE CASE

Alan Stephens, J.D.

13 A.L.R.5th 628

I. Preliminary Matters

1. Introduction
   [a] Scope
   [b] Related annotations

2. Summary

Source: LEXIS Annot., 13 A.L.R.5th 628 screen. Reprinted with the permission of LEXIS-NEXIS, a division of Reed Elsevier Inc. LEXIS and NEXIS are registered trademarks of Reed Elsevier Properties Inc. FREESTYLE, KWIC, SuperKWIC and MEGA are trademarks of Reed Elsevier Properties Inc. SHEPARD'S and SHEPARDIZE are registered trademarks of Shepard's Company, a Partnership.

research with the printed A.L.R. series. There are several ways of browsing through your search results in LEXIS. Exhibit 2–43 depicts the "Help" function available on LEXIS that lists all the basic commands. You can use these commands to work your way through your search results.

To look at the full text of the annotation, **enter .fu** at the first screen. To go to a particular section listed in the contents of the annotation, **enter p*** followed by the number of the section. For example, to go to section 9, as we did with the printed version, enter **p*9**, and the screen as depicted in Exhibit 2–44 will appear. Notice the reference to *State v. Haines*, and the symbol in front of the cite, **<=59>**. This symbol indicates that there is a hypertext connection to the document following the cite. To go to the text of *State v. Haines*, **enter =59**. To return from the text of *State v. Haines* to the annotation, **enter .es**.

At this point you may want to try your hand at constructing LEXIS queries for Case 1 and Case 3. Try your strategies to see if you retrieve references to the same annotations that we located using the printed sources. Above all, remember that LEXIS, like WESTLAW, is a sophisticated online database system and that it takes time to learn to use these systems effectively. It is hoped that you will have both the printed and online versions of the annotations available so that you can decide which one is more to your liking. With practice and patience, you should be able to use either the printed or electronic versions of A.L.R. and other legal authorities.

Exhibit 2–43

	HELP				
Browse	b	Help	h	Print Doc	.pr
Change Lib	.cl	KWIC (TM)	.kw	Request	r
Change File	.cf	Mail-It	.mi	Screen Print	.sp
Cite	.ci	Modify	m	Segments	.se
Disp Dif Level	.dl	New Client	c	Select Serv	.ss
Exit FOCUS (TM)	.ef	New Search	.ns	Sign Off	.so
Exit Serv	.es	Next Doc	.nd	Sort	s
First Doc	.fd	Next Page	.np	Star Paging	p*
First Page	.fp	Pages	p	Time	t
FOCUS	.fo	Prev Doc	.pd	Var KWIC	.vk
Full	.fu	Prev Page	.pp		

Source: LEXIS HELP screen. Reprinted with the permission of LEXIS-NEXIS, a division of Reed Elsevier Inc. LEXIS and NEXIS are registered trademarks of Reed Elsevier Properties Inc. FREESTYLE, KWIC, SuperKWIC and MEGA are trademarks of Reed Elsevier Properties Inc. SHEPARD'S and SHEPARDIZE are registered trademarks of Shepard's Company, a Partnership.

Exhibit 2–44

13 A.L.R.5th 628, *9

[*9] Attempted murder

In the following cases involving attempts by HIV-positive defendants to expose others, the courts held supportable convictions for attempted murder.

The court reversed the trial judge's grant of judgment on the evidence in favor of the defendant, after the jury had convicted him of three counts of attempted murder based on his efforts to transmit the AIDS virus to those trying to help him after a suicide attempt, in <=59> State v Haines (1989, Ind App) 545 NE2d 834. When police officers arrived at the defendant's apartment in response to a radio call, they found him unconscious, lying face down in a pool of blood. The officer who tried to revive the defendant saw that his wrists were slashed and bleeding. When he heard paramedics arriving, the defendant revived and screamed that he should be left to die because he had AIDS. He told the officer that he wanted to f____ him and "give it to him" and that he would "use his wounds," after which he began jerking his arms, causing blood to spray into the officer's mouth and eyes. He repeatedly yelled that he had AIDS, that he could not deal with it, and that he was going to make the officer deal with it. He struggled with the paramedics, threatening to infect them with AIDS and spitting at them. When the police officer grabbed him, the defendant scratched, bit, and spit at him, at one point grabbing a blood-soaked wig and hitting the officer in the face with it, causing blood to splatter onto his eyes, mouth, . . .

Source: LEXIS Annot., 13 A.L.R.5th 628, sec. 9 screen. Reprinted with the permission of LEXIS-NEXIS, a division of Reed Elsevier Inc. LEXIS and NEXIS are registered trademarks of Reed Elsevier Properties Inc. FREESTYLE, KWIC, SuperKWIC and MEGA are trademarks of Reed Elsevier Properties Inc. SHEPARD'S and SHEPARDIZE are registered trademarks of Shepard's Company, a Partnership.

Legal Periodicals

One of the most rewarding experiences in legal research is finding a law review article on the subject of your search. A good law review article can cut to the quick of the issues you are examining, alert you to nuances and facets that had not previously occurred to you, pull together literally hundreds of other secondary and primary authorities related to the topic, and enhance your understanding of an issue as no other single reference can. Unlike most other secondary legal authorities, the articles you will find in legal periodicals usually do not stop with a description of the law. Rather, they push beyond basic description and critically analyze legal issues. Thus, they examine the implications of court decisions and legislative enactments; question legal policies; and frequently criticize judicial opinions, legislation, and other sources of the law. A law review article's detailed examination of laws, legal theory, and administration can provide a comprehensive and eye-opening analysis of the issues you are researching.

Notice that we have reserved these accolades for *good* law review articles. Unfortunately, not all law review articles merit this classification; and more than a few of those published will fall short of enlightening you. Each of the hundreds of law schools in the country publishes a law review, and many publish two or more. Private companies and other organizations also publish legal periodicals. The articles published may be authored by anyone ranging from Supreme Court Justices and law professors to practicing attorneys and second- and third-year law students. Some authors strive to be objective, but others have an axe to grind. Although law review articles are rigorously scrutinized by those students who staff their editorial boards, these writings are unlike articles published in many other scholarly periodicals, which first must undergo anonymous peer review by experts in the field. In short, you should be grateful for law review articles that are good, but ever vigilant lest what you read and rely on does not measure up to those standards.

Printed Sources

You can gain entry into the rich world of law review literature in different ways. The **Index to Legal Periodicals (ILP)**, which recently was renamed the **Index to Legal Periodicals and Books**, is the standard printed indexing source; the **Current Law Index** is another. The **Current Index to Legal Periodicals**, which is printed weekly, provides information about very recently published articles that have not yet been entered in either the Index to Legal Periodicals or the Current Law Index. The Index to Legal Periodicals is available on CD-ROM and through various online database systems, as well as in print. The Current Index to Legal Periodicals can be searched electronically on CD-ROM or as an online database. The Current Law Index has a CD-ROM version called **LegalTrac** and an online version called **Legal Resource Index**. WESTLAW and LEXIS are among the database vendors that provide access to these indexes through their online systems. In addition, WESTLAW and LEXIS have databases that contain the full text of law review and other legal periodical articles. The full-text databases can be searched using WESTLAW's and LEXIS' own searching techniques. We begin by exploring the ILP.

The ILP is divided into sections. Most of the ILP is comprised of a **Subject and Author Index**, which groups law review articles by subject matter; provides their title, author, and citation; and lists the authors of published articles and what they have written. Other helpful indexes appear toward the back of the ILP. The **Table of Cases** lists court decisions by name and identifies law review articles that give the cited case prominent consideration. This table is especially handy when you know the name of a leading case and want commentary about that case or the involved issues. The **Table of Statutes** provides a similar service by listing legislative acts (both federal and state) by name and identifying law review articles that have focused on them. The **Book Review Index** lists book reviews that are published in legal periodicals. At the very front of the ILP you will find the names of the periodicals indexed and the abbreviations used for them, as well as a list of the hundreds of subject headings (from "Abandonment of family" and "Abatement and removal" to "Yugoslavia/Civil War, 1991–" and "Zoning"), which are used to classify articles in the subject and author index.

Since approximately 1980 a new hardbound volume of the ILP has been issued annually, collecting articles published from September through August of each succeeding year. Before 1980, individual ILPs generally covered three-year periods. Paperback issues of the ILP are published during the course of the year to keep up with recently published articles. The multivolume nature of the ILP, which is a necessary feature, also is its biggest inconvenience. To make sure that you have not overlooked potentially useful articles, you have to search each and every volume of the ILP spanning a reasonable time period. For example, you may have to examine the subject headings in both the paper issues of the ILP, and in several hardbound volumes, such as the September 1995–August 1996 ILP, the September 1994–August 1995 ILP, and on and on for several years. You doubtless have done more difficult things in your life than perusing a half dozen or more ILPs, and your diligence is likely to be rewarded through being referred to some very interesting law review articles.

We begin by searching for law review articles that will help us analyze the confrontation issue presented in Case 1. We will be making use of volume 34 of the ILP, which covers the period September 1994 through August 1995. We can quickly examine the list of subject headings in the front of the ILP to confirm that we will not be wasting our time by looking under "**Confrontation**" in the subject and author index. We find that "**Confrontation clause**" is used as a subject heading for articles and we also are referred to the heading "**Witnesses**." (See Exhibit 2–45.)

In the subject-author index section under "Confrontation clause," we find citations to more than 20 law review articles, which deal with a variety of issues. (See Exhibit 2–46.) Note that the only information listed here is each article's title and author; the citation to the periodical, including its page length; and the publication date. There are no abstracts or other clues provided about the contents of the articles, which puts a premium on having the titles of law review articles be as descriptive as possible. Titles that are cryptic, obscure, or too cute to describe what is covered in an article may result in the article being overlooked by a potential reading audience. Student authors are identified as such, to distinguish their submis-

Exhibit 2–45

LIST OF SUBJECT HEADINGS

Computer law *See* Computers

Computer programs *See* Computer Software

Computer simulations as evidence

Computer software

    See also Artificial intelligence; Computer viruses; Computers; Copyright/Computer software; Information systems; Patents/Computer software

Computer viruses

Computerized information *See* Information systems

Computers

    See also Automation; Computer crimes; Computer data as evidence; Computer simulations as evidence; Computer software; Digital sampling; Information systems; Virtual reality

Computers and crime *See* Computer crimes

Computers and privacy

Concessions (International law)

Conciliation

Concurrent jurisdiction

Condemnation of land *See* Eminent domain

Conditional sales

    See also Liens

Condominium and cooperative buildings

    See also Cooperatives

Conferences, International *See* International conferences

Confessions

    See also Admissions; Custodial interrogation; Self-incrimination

Confidential informants *See* Informers

Confidentiality

    See also Informers

Confiscation

Conflict of laws

    See also Comity; Domicile and residence; Enforcement of judgments abroad; Full faith and credit; International law; Pleading and proof of foreign law; Recognition of foreign judgments

Conflict of laws/Contracts

Conflict of laws/Corporations

Conflict of laws/Domestic relations

Conflict of laws/Inheritance and succession

Conflict of laws/Torts

Confrontation clause ◀

    See also Witnesses

Congestion in courts *See* Court congestion

Conjugal violence *See* Domestic violence

Conscientious objectors

    See also Civil disobedience

Source: Index to Legal Periodicals, vol. 34, p. xxxvii, "Confrontation clause" (Sept. 1994–Aug. 1995), © H.W. Wilson Company, used with permission.

sions from articles written by professors, judges, lawyers, and others who no longer occupy student status.

We can eliminate some of the articles listed under "Confrontation clause" in the ILP as not being relevant to our issue, but there is no magic in choosing which of the potentially relevant articles should be consulted, or in what order. There is no harm in jotting down the cites to all articles that might prove to be helpful, and it will be less burdensome to prune marginally relevant articles later than if we have to retrace our steps because we chose only a few articles and those are not

Exhibit 2–46

INDEX TO LEGAL PERIODICALS & BOOKS

Conflict of laws—Torts—*cont.*
. . . interest analysis in Cooney v. Osgood Machinery, Inc. A. D. Twerski; Cooney v. Osgood Machinery, Inc.: a less than complete "contribution". L. J. Silberman. 59 *Brook. L. Rev.* 1323–83 Wint '94

Conflict resolution in the workplace enters a new era; ALI-ABA video law review study materials: January 27, 1994, live via satellite to 70+ cities. American Law Institute-American Bar Assn. Com. on Continuing Professional Educ. 1994 390p
LC 94–163788

Confrontation clause
See also
Witnesses
The admission of hearsay evidence where defendant misconduct causes the unavailability of a prosecution witness. P. T. Markland, student author. 43 *Am. U. L. Rev.* 995–1021 Spr '94

Admitting confessions of codefendants: has Lee v. Illinois [106 S. Ct. 2056 (1986)] created an additional hearsay exception? C. Noworyta, student author. 48 *U. Miami L. Rev.* 435–50 N '93

Broadening the scope of counselor-patient privilege to protect the privacy of the sexual assault survivor. A. Y. Joo. 32 *Harv. J. on Legis.* 255–99 Wint '95

Constitutional hearsay: requiring foundational testing and corroboration under the confrontation clause. C. R. Nesson, Y. Benkler. 81 *Va. L. Rev.* 149–74 F '95

Constitutional law: everything you say can and will be used against you in a court of law . . . or will it? [State v. Rakestraw, 871 P.2d 1274 (Kan. 1994)] T. J. Parascandola, student author. 34 *Washburn L.J.* 174–92 Fall '94

Department of Social Services v. Brock [499 N.W.2d 752 (Mich. 1993)]: videotaped testimony in lieu of live testimony, H. Jefferson, student author. 1994 *Det. C.L. Rev.* 897–926 Summ '94

Determining reliability factors in child hearsay statements: Wright [Idaho v. Wright, 110 S. Ct. 3139 (1990)] and its progeny confront the psychological research. D. B. Lord, student author. 79 *Iowa L. Rev.* 1149–79 Jl '94

Due process concerns in video production of defendants. P. Raburn-Remfry. 23 *Stetson L. Rev.* 805–41 Summ '94

Evidence—the prosecution is not required to produce the four year old victim of a sexual assault at trial or to have the trial court find the victim was unavailable for testimony before the out-of-court statements of the child are admitted under the spontaneous declaration and medical examination exceptions to the hearsay rule—White v. Illinois, 112 S. Ct. 736 (1992). E. A. Delagardelle, student author. 43 *Drake L. Rev.* 209–18 '94

Facing the accuser: ancient and medieval precursors of the confrontation clause. F. R. Hermann, B. M. Speer. 34 *Va. J. Int'l L.* 481–552 Spr '94

Justice for our children: New Jersey addresses evidentiary problems inherent in child sexual abuse cases. D. M. Enea, student author. 24 *Seton Hall L. Rev.* 2030–56 '94

A law and economics analysis of the right to face-to-face confrontation post-Maryland v. Craig [110 S. Ct. 3157 (1990)]: distinguishing the forest from the trees. P. T. Wendel. 22 *Hofstra L. Rev.* 405–94 Wint '93

Losing the right to confront: defining waiver to better address a defendant's actions and their effects on a witness. D. J. Tess, student author. 27 *U. Mich. J.L. Ref.* 877–918 Spr/Summ '94

Maryland v. Craig [110 S. Ct. 3157 (1990)]: the Supreme Court clarifies when a child protective statute which allows a child witness to testify outside the presence of the accused will violate the confrontation clause. T. A. Cotton, student author. 19 *T. Marshall L. Rev.* 309–32 Spr '94

continues

Exhibit 2–46   continued

Ohio rule of Evidence 807: does it go far enough in protecting the confrontation clause rights of the accused? T. A. Ballato, student author. 20 *Ohio N.U. L. Rev.* 981–1006 '94

Remaking confrontation clause and hearsay doctrine under the challenge of child sexual abuse prosecutions. R. P. Mosteller. 1993 *U. Ill. L. Rev.* 69–807 '93

Stephens v. Miller [13 F.3d 998 (1994)]: restoration of the rape defendant's sixth amendment rights. L. M. Dillman, student author. 28 *Ind. L. Rev.* 97–114 '94

The true value of the confrontation clause: a study of child sex abuse trials. J. M. Beckett, student author. 82 *Geo. L.J.* 1605–42 Ap '94

Two critical evidentiary issues in child sexual abuse cases: closed-circuit testimony by child victims and exceptions to the hearsay rule. A. C. Goodman, student author. 32 *Am. Crim. L. Rev.* 855–82 Spr '95

Illinois

People v. Fitzpatrick [633 N.E.2d 685 (Ill. 1994)]: the path to amending the Illinois Constitution to protect child witnesses in criminal sexual abuse cases. T. Conklin, student author. 26 *Loy. U. Chi. L.J.* 321–50 Wint '95

Pennsylvania

Constitutional law—criminal procedure—child testimony via videotape or closed circuit television—defendant's right to confront witnesses—the Supreme Court of Pennsylvania held that a statute allowing children to testify outside the physical presence of a defendant by means of video-tape or closed circuit television violates the defendant's constitutional right to confront witnesses face-to-face. The court further held that Article I, section 9 of the Pennsylvania Constitution requires a face-to-face confrontation between a defendant and a witness, and allows exceptions only when the defendant has previously had the opportunity to physically confront and cross-examine the witness. M. L. Bell, student author. 33 *Duq. L. Rev.* 361–76 Wint '95

Confronting sexual assault; a decade of legal and social change; edited by Julian V. Roberts and Renate M. Mohr. University of Toronto Press 1994 355p il pa ISBN 0-8020-5928-7; 0-8020-6868-5 LC 94-188417

Congestion in courts *See* Court congestion

Source: Index to Legal Periodicals, vol. 34, p. 156, "Confrontation clause" (Sept. 1994–Aug. 1995), © H.W. Wilson Company, used with permission.

useful. For illustrative purposes, let us begin by examining the **Hofstra Law Review** article written by P.T. Wendel. Incidentally, the correct form for citing this article is Peter T. Wendel, *A Law and Economics Analysis of the Right to Face-to-Face Confrontation post*-Maryland v. Craig: *Distinguishing the Forest from the Trees,* 22 Hofstra L. Rev. 405 (1993). (See **The Bluebook: A Uniform System of Citation** 113 (16th ed. 1996).) We find this article in volume 22 of the Hofstra Law Review beginning at page 405. We reproduce excerpts from its initial pages to demonstrate the hefty, reference-laden footnotes that commonly are found in law review articles. The footnotes complement the narrative in the article's text and can be an excellent source for finding other authorities related to your issue. (See Exhibit 2–47.)

Since we have run across the Supreme Court's decision in *Maryland v. Craig* so frequently while conducting research for Case 1, we might want to find law

Exhibit 2–47

A LAW AND ECONOMICS ANALYSIS OF THE RIGHT TO FACE-TO-FACE
CONFRONTATION POST-*MARYLAND V. CRAIG*: DISTINGUISHING THE
FOREST FROM THE TREES

Peter T. Wendel*

I. INTRODUCTION

In a criminal case, should the defendant have the right to confront, face-to-face, the witnesses who appear at trial and testify against the defendant?[1] In analyzing this issue, the courts have repeatedly recognized that there are costs and benefits associated with a right to face-to-face confrontation.[2] The principal cost is the potential anxiety, in some cases even trauma, which the witness may experience from having to confront the defendant face-to-face.[3] The principal benefit is the reduced risk of an erroneous conviction.[4] In comparing the costs and benefits, the courts traditionally and generally have held that the benefits of face-to-face confrontation (the savings from the reduced risk of an erroneous judgment) outweigh the costs (the potential anxiety, and even trauma, to the witness)[5]—until *Mary-* . . .

1. Although the right to face-to-face confrontation and the right to cross-examination will be one and the same for most purposes, *see* Ohio v. Roberts, 448 U.S. 56, 63–64 (1980), the right to face-to-face confrontation as an independent right is the focus of this Article.

2. *See* Coy v. Iowa, 487 U.S. 1012, 1014, 1020-21 (1988) (reversing the Iowa Supreme Court's decision, which allowed the use of a screen between the defendant and a child witness, and thereby disallowing the State's assertion that the appellant's right to confrontation was outweighed by the State's interest in protecting the sexual abuse victim); *Roberts*, 448 U.S. at 64 (recognizing that "*competing interests* if 'closely examined' . . . may warrant dispensing with confrontation at trial") (emphasis added) (quoting Chambers v. Mississippi, 410 U.S. 284, 295 (1973)); State v. Dolen, 390 So. 2d 407, 409-10 (Fla. Dist. Ct. App. 1980) (remanding a case so that its outcome may properly be based upon "whether the potential detrimental effect upon the witness outweighs the interest or benefit to the defendant"); *see also The Supreme Court, 1987 Term—Leading Cases,* 102 Harv. L. Rev. 143, 157 n.46 (1988) ("The [Supreme] Court has balanced costs and benefits in several confrontation clause cases.").

3. *See Coy*, 487 U.S. at 1020 ("[F]ace-to-face [confrontation] may, unfortunately, upset the truthful rape victim."); State v. Vincent, 768 P.2d 150, 162 (Ariz. 1989) ("To whatever extent the law insists on face-to-face confrontation, it heightens the anxiety, and perhaps the trauma, of those who are willing to bear witness against crime."); *see also* Michael H. Graham, *Indicia of Reliability and Face to Face Confrontation: Emerging Issues in Child Sexual Abuse Prosecutions,* 40 U. Miami L. Rev. 19, 83 (1985) ("Witnesses who testify in open court often suffer some emotional distress. Many, if not most, rape victims suffer severe emotional distress or trauma while testifying, especially when face to face with the accused. Presumably, so do many other groups of victims.").

continues

Exhibit 2–47   continued

4. *See Coy,* 487 U.S. at 1019-20 (comparing the right to face-to-face confrontation with the right to cross-examine the witness and finding that both serve much the same purpose in that both ensure the integrity of the factfinding process. "[F]ace-to-face presence . . . may confound and undo the false accuser, or reveal the child coached by a malevolent adult."); United States v. Leonard, 494 F.2d 955, 987 (D.C. Cir. 1974) ("Elaboration and application of the rules of evidence and the Confrontation guarantee are themselves directed to a fully pragmatic concern; they are designed to provide some reasonable assurance that defendants found guilty are guilty."); *Vincent,* 768 P.2d at 162 ("To whatever extent the law cushions a witness against the crucible of confrontation, it diminishes a fundamental courtroom test of truth."); Herbert v. Superior Court, 172 Cal. Rptr. 850, 855 (Ct. App. 1981) ("A witness's reluctance to face the accused may be the product of fabrication rather than fear or embarrassment."); *see also Coy,* 487 U.S. at 1019 ("It is always more difficult to tell a lie about a person 'to his face' than 'behind his back.'"); *Roberts,* 448 U.S. at 63 n.6 (reiterating the difficulty of lying when in the presence of an innocent defendant).

5. *See Coy,* 487 U.S. at 1020-22 (holding that the defendant's confrontation right was violated when the lower court allowed a child witness to testify from behind a screen); Barber v. Page, 390 U.S. 719, 724-26 (1968) (refusing to affirm the lower court's decision due to the prosecution's failure to produce a witness residing in a federal prison); United States v. Benfield, 593 F.2d 815, 817, 821 (8th Cir. 1979) (ruling that a procedure which allowed a kidnapping victim to testify via pretrial deposition under circumstances that allowed the defendant to be present at the deposition, but outside of the victim's view, was violative of the defendant's confrontation right); Britton v. Maryland, 298 F. Supp. 641, 647 (D. Md. 1969) (holding that the State did not make a good faith effort to produce the witness, who was in the armed services, and therefore use of the deposition was not allowed); Hochheiser v. Superior Court, 208 Cal. Rptr. 273, 278 & n.2 (Ct. App. 1984) (overruling the trial court's decision to allow children to testify via video, and explaining that physical confrontation falls within the scope of the Sixth Amendment and "[t]he closed-circuit television order . . . raise[d] significant and complex federal and state constitutional issues, potentially affecting petitioner's fundamental rights to a public trial, confrontation of witnesses against him and due process") (footnote omitted); *Herbert,* 172 Cal. Rptr. at 853 (agreeing with the defendant that a seating arrangement violated his right to confront accusatory witnesses and stating that hearsay rules indicate that "a personal view of the witness by the defendant at some point is part of the right of confrontation"); Keshishian v. State, 386 A.2d 666, 667 (Del. 1978) (de- . . .

Source: Peter T. Wendel, *A Law and Economics Analysis of the Right to Face-to-Face Confrontation Post-Maryland v. Craig: Distinguishing the Forest from the Trees,* 22 HOFSTRA L. REV. 405 (1993). Reprinted with permission from the Hofstra Law Review.

review articles that focus discussion on the *Craig* case. We do so by turning to the table of cases in the rear of the ILP, where case names are arranged alphabetically. Three articles are identified as addressing *Maryland v. Craig,* including the Hofstra Law Review article we have just examined and two others that also appear in the subject-author index that we just consulted. (See Exhibit 2–48.) Since we have only checked the September 1994 through August 1995 volume of the ILP, we must replicate our subject-author index and table of cases search in other recent volumes to avoid overlooking additional law review articles that might be informative.

Similar techniques allow us to find law review articles addressing the issues raised in our other hypothetical cases. Several subject headings in the ILP may reference articles that will help us with Case 2, including "**AIDS**," "**Assault and battery**," "**Attempt, criminal**" (which directs us to "**Criminal responsibility**"), "**HIV infection**" (which directs us to "**AIDS**"), "**Homicide**," and "**Weapons**." For instance, if we examine recent ILPs under those topics, volume 32 (September 1992–August 1993) will direct us to an article that looks interesting: K.A. Harris (who is identified as a student author), *Death at First Bite*: *A Mens Rea Approach in Determining Criminal Liability for Intentional HIV Transmission*, 35 Arizona

---

Exhibit 2–48

TABLE OF CASES   977

Maritime Delimitation in the Area between Greenland and Jan Mayen (Denmark v. Norway) I.C.J. Rep 1993, 38
   43 *Int'l & Comp. L.Q.* 678-96 Jl '94
Markiewicz; United States v., 978 F.2d 786 (1992)
   17 *Suffolk Transnat'l Rev.* 539-50 Spr '94
Marks v. Stinson, No. CIV. A. 93-6157, 1994 WL 146113 (1994)
   99 *Dick. L. Rev.* 501-20 Wint '95
Mars; Bryant v., 830 S.W.2d 869 (Ark. 1992)
   16 *U. Ark. Little Rock L.J.* 313-25 '94
Marsh; Ben-Shalom v., 881 F.2d 454 (1980)
   2 *Law & Sex.* 209-36 '92
Marshall v. Southampton & S.W. Hampshire Area Health Auth., [1986] 2 W.L.R. 780
   57 *Mod. L. Rev.* 859-79 N '94
Marshall v. Southampton & S.W. Hampshire Area Health Auth. (No. 2), [1988] I.R.L.R. 325
   15 *D.U.L.J. (n.s.)* 173-88 '93
   18 *Fordham Int'l L.J.* 641-83 D '94
Martin; Commonwealth v., 626 A.2d 556 (Pa. 1993)
   67 *Temp. L. Rev.* 861-82 Summ '94

Martin v. Richards, Docket 91-0016 Wis. S. Ct. May 4, 1995
   68 *Wis. Law* 12-14+ Je '95
Martin v. Warden, Atlanta Penitentiary, 993 F.2d 824 (1993)
   18 *Suffolk Transnat'l Rev.* 347-58 Wint '95
Martin, In re 923 F.2d 504 (1991)
   26 *Tax Adviser* 228-35 Ap '95
Martineau; R. v., [1990] 2 S.C.R. 633
   52 *U. Toronto Fac. L. Rev.* 379-404 Spr '94
Martinez-Hidalgo; United States v., 993 F.2d 1052 (1993)
   18 *Tul. Mar. L.J.* 401-14 Summ '94
Maryland v. Craig, 110 S. Ct. 3157 (1990)
   32 *Am. Crim. L. Rev.* 855-82 Spr '95
   22 *Hofstra L. Rev.* 405-94 Wint '93
   19 *T. Marshall L. Rev.* 309-32 Spr '94
Maryland Casualty Co.; Lloyd E. Mitchell, Inc. v., 595 A.2d 469 (Md. 1991)
   22 *U. Balt. L. Rev.* 167-82 Fall '92

Source: Index to Legal Periodicals, vol. 34, p. 977, "Maryland v. Craig" (Sept. 1994–Aug. 1995), © H.W. Wilson Company, used with permission.

Law Review 237–264 (Spring 1993). (See Exhibit 2–49.) This article begins with a discussion of a case from Indiana, *State v. Haines*, 545 N.E.2d 834 (Ind. Ct. App. 1989), which resembles the hypothetical fact situation we have presented in Case 2. (This is the same case we encountered in our previous discussion of A.L.R. annotations.) The article promises to cite numerous related sources as it analyzes the involved issues of *Haines*. (See Exhibit 2–50.)

Imagine taking the steps—or, better yet, go to the library and actually take the steps—that lead you to law review articles pertinent to our hypothetical Case 3. When you check the subject headings "**Cruel and unusual punishment**" and "**Prisons and prisoners**" in recent ILPs, you will find several articles relating to exposure to environmental tobacco smoke in prisons as a form of cruel and un-

---

Exhibit 2–49

SUBJECT AND AUTHOR INDEX

AIDS (Disease)—*cont.*

A case study on interest group behaviour in a federal system: the AIDS lobby groups in Canada. E. Cipparone, student author. 50 *U. Toronto Fac. L. Rev.* 131-60 Spr '92

Causation in transfusion-associated AIDS cases. R. K. Jenner, S. Schupak. 29 *Trial* 60-5 My '93

Commentary: AIDS testing of health care workers. L. J. Frankel. 16 *Nova L. Rev.* 1161-9 Spr '92

The conflict between Illinois Rule 1.6(b) and the AIDS Confidentiality Act. S. H. Isaacman. 25 *J. Marshall L. Rev.* 727-36 Summ '92

Constitutional challenges to the criminalization of same-sex sexual activities: state interest in HIV-AIDS issues. J. K. Strader. 70 *Denv. U. L. Rev.* 337-57 '93

Constitutional issues surrounding the mass testing and segregation of HIV-infected inmates. J. F. Horner, Jr., student author. 23 *Mem. St. U. L. Rev.* 369-98 Wint '93

Constitutional law/search and seizure/equal protection. The portion of a corrections statute requiring HIV testing of convicted prostitutes is constitutional under the U.S. and Illinois Constitutions. People v. Adams, 597 N.E.2d 574 (Ill. 1992). D. A. Iannicola, Jr. 81 *Ill. B.J.* 159-62 Mr '93

The constitutionality of parole departments disclosing the HIV status of parolees. D.E. Post, student author. 1992 *Wis. L. Rev.* 1993-2030 '92

The constitutionality of police protection statutes. F. J. Schlosser, student author. 16 *S. Ill. U. L.J.* 707-25 Spr '92

Control of childbearing by HIV-positive women: some responses to emerging legal policies. S. Sangree. 41 *Buff. L. Rev.* 309-449 Spr '93

Controlling HIV-positive women's procreative destiny: a critical equal protection analysis. J. S. Weiss, student author. 2 *Seton Hall Const. L.J.* 643-718 Spr '92

Death at first bite: a mens rea approach in determining criminal liability for intentional HIV transmission. K. A. Harris, student author. 35 *Ariz. L. Rev.* 237-64 Spr '93

Do the orthodox rules of lawyering permit the public interest advocate to "do the right thing?": A case study of HIV-infected prisoners. J. Mosoff. 30 *Alta. L. Rev.* 1258-75 '92

Source: Index to Legal Periodicals, vol. 32, p. 21, AIDS (Disease) (excerpts, including "Death at first bite . . . 35 Ariz. L. Rev. 237-64 Spr '93"), © H.W. Wilson Company, used with permission.

Exhibit 2–50

DEATH AT FIRST BITE: A *MENS REA* APPROACH IN DETERMINING CRIMINAL LIABILITY FOR INTENTIONAL HIV TRANSMISSION

Kimberly A. Harris

Joseph Haines[1] had an advanced case of Acquired Immune Deficiency Syndrome (AIDS) related complex.[2] On August 6, 1987, he slashed his wrists in his apartment.[3] Police officers Dennis and Hayworth, responding to the radio call of a possible suicide, arrived at the scene to find Haines face down in a pool of blood.[4] Dennis attempted to revive Haines and told Haines that they were there to help him.[5]

When Haines heard the paramedics approaching he stood up, ran at Dennis, and screamed that he should be left to die because he had AIDS.[6] As the officers attempted to subdue him, Haines repeatedly shouted that he had AIDS, that he could not deal with it, and that he was going to make Officer Dennis know what it meant to suffer from AIDS.[7] He told Dennis he would "use his wounds" and jerked his arms at Dennis, spraying blood into the officer's eyes and mouth.[8]

Haines then struggled with the paramedics, threatening to infect them with AIDS and spitting at them.[9] Haines bit a paramedic on the upper arm, breaking the skin, and said he was going to show everyone what it was like to . . .

---

1. State v. Haines, 545 N.E.2d 834 (Ind. Ct. App. 1989.)

2. AIDS is the name given to a complex of opportunistic infections that develop when a person's immune system has broken down. Kenneth Vogel, *Discrimination on the Basis of HIV Infection: An Economic Analysis*, 49 OHIO ST. L.J. 965, 967-68 (1988-89). The virus that causes AIDS has various names in the scientific community: Human T-Lymphotropic Virus Type III (HTLV-III), lymphadenopathy-associated virus (LAV), or AIDS-related complex (ARC). *Id.* at 967 n.18. For simplicity, this Note collectively refers to these strains as AIDS. AIDS occurs in people who have been infected by the Human Immunodeficiency Virus (HIV).

A positive HIV antibody test indicates the individual has produced antibodies in reaction to exposure to HIV. Antibodies usually develop within 6–12 weeks following exposure to the virus. Exchanging of body fluids, sharing of contaminated needles, and transfusion of infected blood are the primary means of transmission. The average life expectancy of a patient who has contracted one of the opportunistic diseases associated with AIDS is approximately fifteen months after diagnosis. Rhonda R. Rivera, *Lawyers, Clients, and AIDS: Some Notes from the Trenches,* 49 OHIO ST. L.J. 883, 884–85 n.14 (1988–89).

3. *Haines*, 545 N.E.2d at 835.

4. *Id.*

5. *Id.*

6. *Id.*

7. *Id.*

8. At one point, Haines struck Dennis in the face with a blood-soaked wig splattering blood onto Dennis' eyes, mouth, and skin. *Haines,* 545 N.E.2d at 835.

9. *Id.*

Source: Kimberly A. Harris, "Death at First Bite: A Mens Rea Approach in Determining Criminal Liability for Intentional HIV Transmission," *Arizona Law Review,* Vol. 35, p. 237. Copyright © 1993 by the Arizona Board of Regents. Reprinted by permission.

usual punishment. Many of the titles of these articles name the U.S. Supreme Court case we uncovered in A.L.R., *Helling v. McKinney*, 509 U.S. 25, 113 S. Ct. 2475, 125 L. Ed. 2d 22 (1993). (Remember that you also can expect to find law review articles discussing *Helling v. McKinney* by using the Table of Cases index included in the rear of ILPs.)

Searching for Law Review Articles on CD-ROM

Print references provide the immense reassurance and comfort of allowing the researcher to manipulate the pages of a book and to browse whenever and wherever a whim strikes, even if it is distinctly illogical. These attributes of control can be compromised when mechanical classification and identification systems replace human touch and vision. **LegalTrac**, WESTLAW, and LEXIS have many convenient features that in some ways make them more efficient guides to legal periodicals than printed versions of sources such as the ILP. Still, their advances come at some cost to those who like the feel of hand on paper and who like to see the research paths not chosen as well as the ones that are pursued.

LegalTrac is an index to law review articles, bar journals, and legal newspapers that is stored on a CD-ROM. The index begins with materials published in 1980 and is cumulative thereafter. One important feature of most digital bibliographic databases, whether they are online databases or CD-ROMs such as LegalTrac, is the ability to complete a single search that covers a span of years. As we have discussed, a researcher using the printed ILP would have to check a series of individual volumes to search for law reviews dating from 1980 to the present. (Of course, the ILP's CD-ROM and online versions also permit searches over several years.)

There are differences between LegalTrac and the ILP. Like the ILP, the references collected in LegalTrac are indexed by subject, author, and the names of cases and statutes. However, there are two significant differences between the subject indexing system used in the ILP and that used in LegalTrac. First, the ILP makes use of broad subject headings but very few subheadings. Thus, you may have to scan through scores or even hundreds of articles under a broad heading in order to find the few that deal with the particular issue in which you are interested. This process becomes particularly demanding when you are searching the electronic version and considering a broad span of years. This disadvantage of the ILP is partially ameliorated in the electronic versions, where it is possible to combine a subject search with a key word in the title or to look for documents that have two or more specific subject headings assigned to them. LegalTrac makes extensive use of subheadings and therefore facilitates the search for documents dealing with a particular aspect of a topic.

The second major difference is that the subject headings used in the ILP are chosen specifically for this index and rely heavily on legal terminology. These headings are not problematic, and, as long as the researcher is comfortable with the legal lexicon, may be advantageous. LegalTrac's subject indexing relies heavily on Library of Congress subject headings, which is the same system used by most academic libraries to provide access to their books. The Library of Congress subject headings include not only unique legal terminology but also terms that are more likely to be used by the layperson.

LegalTrac covers many more legal periodicals than the ILP. However, for serious legal researchers, this more extensive coverage is a mixed blessing at best. The ILP restricts its coverage to scholarly articles, which are usually lengthy and documented by reference to numerous sources of authority. LegalTrac covers legal newspapers such as the **National Law Journal** and the **Los Angeles Daily Journal** as well as law reviews. Search results may include plentiful references to these news articles, typically brief descriptive accounts of legal news items, which may be interesting but usually are not very helpful as a form of legal authority. In light of the proliferation of legal newspaper articles, it can be a tedious process separating the wheat from the chaff in the citations provided by LegalTrac.

The search entry screen for LegalTrac appears as shown in Exhibit 2–51. Note that there are three basic search methods: **Subject Guide**, **Key Word Search**, and **PowerTrac**. The **Subject Guide** search limits your search results to documents that have particular subject headings. The **Key Word Search** searches for words or phrases from the titles, authors, names, or subject headings assigned to the documents. **PowerTrac** is a more powerful search option that allows you to search for documents containing several concepts. The **Subject Guide** search works best if you already know the Library of Congress subject heading(s) for your topic. These subject headings can be found in a multivolume set entitled **Library of Congress (LC) Subject Headings**, which is arranged alphabetically. We refer to our first hypothetical to illustrate how to use this search technique. In

---

Exhibit 2–51

InfoTrac EF	LegalTrac	Start a Search

EasyTrac provides two simple ways to search: Subject Guide and Key Word. Try the Subject Guide first.

Enter word(s) for SUBJECT GUIDE and press Enter

[   ]

Subject Guide   Browse listings of subjects, personal names, or companies that include the word(s) you type, e.g., rap music, censorship or Boris Yeltsin.

Key Word Search   Search for article references by combining words or phrases from titles, authors or subjects, e.g., family values AND Dan Quayle.

↓ Search using key words   Esc   Exit

                                   F1   Help   F10   PowerTrac (advanced search)

Source: © 1997 Information Access Company, LegalTrac™ Search Entry screen.

Case 1 John Winston has been accused of sexually abusing a child, and the question is whether allowing the child to testify via closed circuit television violates Winston's constitutional right to confront his accuser.

Subject Guide Searches. Some digging into the LC Subject Headings will produce those headings that appear relevant to our case: **"children as witnesses,"** **"child sexual abuse,"** and **"closed-circuit television."** To perform a subject search, highlight **Subject Guide** and then type in your subject terms. For example, in Exhibit 2–52 we have typed in the subject, **"children as witnesses,"** and Exhibit 2–53 displays the results of our search. In this case we have retrieved 669 documents that have the subject heading **"children as witnesses"** assigned to them. Please note that if you try to replicate this search, you are certain to retrieve more than 669 documents, as more documents are being added to the database on a regular basis. (At the time we completed our search the database contained citations to 661,147 documents from the time period 1980 through August 1995). We could start browsing through the 669 documents by pressing **enter**, but our search results appear to indicate that we have a "precision" problem—that is, we have probably retrieved a high percentage of documents that do not focus on the specific elements of our topic.

We try to narrow our search results either by looking at the subheading under **"children as witnesses"** or by adding another subject phrase to the search query.

Exhibit 2–52

InfoTrac EF   LegalTrac   Start a Search

EasyTrac provides two simple ways to search: Subject Guide and Key Word.
Try the Subject Guide first.

Enter word(s) for SUBJECT GUIDE and press Enter

children as witnesses

Subject Guide   Browse listings of subjects, personal names, or companies that include the word(s)
                     you type, e.g., rap music, censorship or Boris Yeltsin.
Key Word Search   Search for article references by combining words or phrases from titles, authors or
                     subjects, e.g., family values AND Dan Quayle.

↓ Search using key words   Esc   Erase entry line

                                F1   Help   F10   PowerTrac (advanced search)

Source: © 1997 Information Access Company, LegalTrac™ Search Entry screen ("children as witnesses").

Exhibit 2–53

InfoTrac EF	LegalTrac	Subject Guide

Subjects containing the words: children as witnesses

	Rec.'s
Children as Witnesses	669

▶ (32) subdivisions

Press **Enter** to view the citation(s) for the highlighted subject

Esc	Return to start
F1	Help
F2	Start over
F3	Print
F4	Mark

Source: © 1997 Information Access Company, LegalTrac™ Search Results screen ("children as witnesses").

We decide on the latter approach. Press **Esc** to return to the query entry screen. We attempt to add the subject phrase "**closed-circuit television**" to our query by typing in the operator **and**, followed by "**closed-circuit television**," a query that would retrieve a set of documents that include both headings as subject descriptors. Unfortunately, the query entry screen is not large enough to allow us to enter all of the characters of both of these lengthy subject descriptors, so we shorten our second descriptor to "**television**" and add this word to our query, as illustrated in Exhibit 2–54. Subject Guide searching is designed so that if a word or phrase typed in the entry box does not appear as a subject descriptor, LegalTrac will automatically look for the word in the title of citations or in the abstract that accompanies some citations. Our search strategy is now designed to retrieve documents including the subject descriptor "children as witnesses" and the key word "television." Exhibit 2–55 illustrates the second page of the results. By combining two concepts in our search as subject headings, we have skipped over the summary results screen (see Exhibit 2–53) and have gone directly to the citations to the retrieved documents. At the top center of the screen in Exhibit 2–55, the phrase "4 of 27" tells us that there were 27 documents retrieved, and the citation currently highlighted is document number 4. The documents are listed in reverse chronological order, beginning with the most current and working back through older articles.

Exhibit 2–54

InfoTrac EF   LegalTrac   Start a Search

EasyTrac provides two simple ways to search: Subject Guide and Key Word.
Try the Subject Guide first.

Enter word(s) for SUBJECT GUIDE and press Enter

children as witnesses and television

Subject Guide   Browse listings of subjects, personal names, or companies that include the word(s)
                       you type, e.g., rap music, censorship or Boris Yeltsin.
Key Word Search   Search for article references by combining words or phrases from titles, authors or
                       subjects, e.g., family values AND Dan Quayle.

↓ Search using key words   Esc   Erase entry line

                                  F1   Help   F10   PowerTrac (advanced search)

Source: © 1997 Information Access Company, LegalTrac™ Search Entry screen ("children as witnesses and television").

Exhibit 2–55

InfoTrac EF   LegalTrac   Brief Citations

Key Words: children as witnesses and television
——————————————————— 4 of 27 ———————————————————

2   Constitutional law—criminal procedure—child testimony via videotape or closed circuit televi-
      sion—defendant's right to confront witnesses. (Case Note) Michael L. Bell. Duquesne Law Review,
      Wntr 1995 33 n2 p361–376.
3   Testimony via closed circuit television after Gonzales v. State: is the Sixth Amendment right to
      confront adverse witnesses at stake in Texas criminal trials? (Case Note) Stacey L. Jones. Baylor Law
      Review, Fall 1992 44 n4 p957–971.
4   A conflict of interests: the constitutionality of closed-circuit television in child sexual abuse cases.
      (Indiana) John Paul Serketich. Valparaiso University Law Review Fall 1992 27 n1 p217–255.
      Enter

Display   Narrow   Explore   Esc   Return to start

Display extended citation   F1   Help   F2   Start over   F3   Print   F4   Mark

Source: © 1997 Information Access Company, LegalTrac™ Search Results screen ("children as witnesses and television—p. 2").

Exhibit 2–56 illustrates the full citation to document 4. The "**Subjects**" portion of the screen can be particularly useful for determining related subject headings to incorporate in your search strategy. Some of our documents, such as the one we have highlighted, are clearly on point, but we have to be concerned with our "re-call" rate; that is, did we retrieve all or nearly all of the relevant documents in the database? One way of determining the quality of a search strategy is to check the retrieved documents to see if they include references to known articles on the topic. In our case we know that P.T. Wendel wrote a relevant article that appeared in the Hofstra Law Review. If we scan our 27 documents, we will find that the Wendel article is not listed. This omission indicates either that the citation to the article is not in the database or that our search query was not structured to pick it up. At this point we try alternative search strategies, such as searching the subject phrase "**Child sexual abuse and confrontation**." Indeed, if you try this strategy as a "Subject Guide" search, you will find a cite to the Wendel article. (This revision process should convince you that thorough database searching requires you to be flexible and willing to try various approaches.)

An alternative approach to using the Subject Guide mode is to look for references to a particular court decision or the popular name of a statute. This strategy parallels the use of the Table of Cases and Table of Statutes indexes in the ILP. In our example, we have previously identified *Maryland v. Craig* as an important

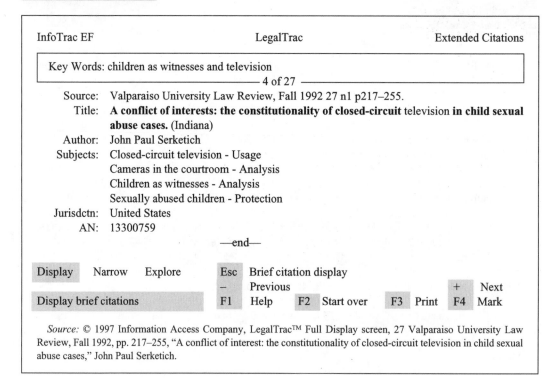

Exhibit 2–56

InfoTrac EF	LegalTrac	Extended Citations

Key Words: children as witnesses and television
———— 4 of 27 ————

Source:	Valparaiso University Law Review, Fall 1992 27 n1 p217–255.
Title:	**A conflict of interests: the constitutionality of closed-circuit** television **in child sexual abuse cases.** (Indiana)
Author:	John Paul Serketich
Subjects:	Closed-circuit television - Usage
	Cameras in the courtroom - Analysis
	Children as witnesses - Analysis
	Sexually abused children - Protection
Jurisdctn:	United States
AN:	13300759

—end—

Display   Narrow   Explore   Esc Brief citation display
                              –   Previous   + Next
Display brief citations   F1 Help   F2 Start over   F3 Print   F4 Mark

Source: © 1997 Information Access Company, LegalTrac™ Full Display screen, 27 Valparaiso University Law Review, Fall 1992, pp. 217–255, "A conflict of interest: the constitutionality of closed-circuit television in child sexual abuse cases," John Paul Serketich.

case. We would type in the name of the case, as illustrated in Exhibit 2–57. Exhibit 2–58 indicates that 41 documents were retrieved using this strategy. Exhibit 2–59 shows a portion of the list of the articles retrieved, with the article authored by P.T. Wendel highlighted. The full citation to the document appears in Exhibit 2–60. Note that the citation includes a field **Cases** that makes reference to the name of our case.

Key Word Searches. The Key Word Searching option is somewhat more flexible than the Subject Guide option in that it looks for the search words or phrases appearing in titles' or authors' names as well as in the subject headings assigned to each document. If you already have looked up the Library of Congress subject headings, you simply can enter the same search strategy that we used in the Subject Guide. Alternatively, you can just enter key words or phrases that you think describe the concepts related to your topic. For example, in Exhibit 2–61 we have entered the following search query: **"children and witnesses and television."** The query will retrieve only those citations and documents that include all three of the specific words. One thing to be wary of in key-word searching is that the computer will retrieve words only in the precise form they are entered. Therefore, in our example it will retrieve documents containing the word **"children,"** but not those in which the word "child" appears. To retrieve a word in its various forms, you must place an asterisk at the end of the root form of the word. For example, if we had entered **"child*"** instead of **"children,"** we would have found references

---

Exhibit 2–57

| InfoTrac EF | LegalTrac | Start a Search |

EasyTrac provides two simple ways to search: Subject Guide and Key Word. Try the Subject Guide first.

Enter word(s) for SUBJECT GUIDE and press Enter

    Maryland v. Craig

Subject Guide   Browse listings of subjects, personal names, or companies that include the word(s) you type, e.g., rap music, censorship or Boris Yeltsin.

Key Word Search   Search for article references by combining words or phrases from titles, authors or subjects, e.g., family values AND Dan Quayle.

↓ Search using key words   Esc   Erase entry line

    F1   Help   F10   PowerTrac (advanced search)

Source: © 1997 Information Access Company, LegalTrac™ Search Entry screen ("Maryland v. Craig").

Exhibit 2–58

InfoTrac EF	LegalTrac	Subject Guide

Subjects containing the words: maryland v. Craig

Rec.'s

Maryland v. Craig   1
Maryland v. Craig   41
▸ (3) subdivisions

Press Enter to view the   Esc   Return to start
highlighted subdivisions.
                              F1   Help   F2 Start over   F3   Print   F4   Mark

Source: © 1997 Information Access Company, LegalTrac™ Search Results screen ("Maryland v. Craig").

Exhibit 2–59

InfoTrac EF	LegalTrac	Brief Citations

Subject: maryland v. craig
—— 4 of 41 ——

4   A law and economics analysis of the right to face-to-face confrontation post-Maryland v. Craig: distinguishing the forest from the trees. Peter T. Wendel. Hofstra Law Review, Wntr 1993 22 n2 p405–494.
        Enter
5   The collision of policy and history. (Case Note) Timothy F. Cullen. New England Journal on Criminal & Civil Confinement. Wntr 1993 19 n1 p141–173.
6   Beyond Maryland v. Craig: can and should adult rape victims be permitted to testify by closed-circuit television? (Case Note) Lisa Hamilton Thielmeyer. Indiana Law Journal, Summer 1992 67 n3 p797–816.
7   Children as witnesses. (Case Note) Gail D. Cecchettini-Whaley. Southern California Law Review, May 1992 65

Display   Narrow   Explore   Esc   Return to subject list

Display extended citation   F1   Help   F2 Start over   F3   Print   F4   Mark

Source: © 1997 Information Access Company, LegalTrac™ Search Results screen ("Maryland v. Craig" search results 4–7).

Exhibit 2–60

InfoTrac EF   LegalTrac   Extended Citations

Subject: maryland v. craig

———————————— 4 of 41 ————————————

Source: Hofstra Law Review, Wntr 1993 22 n2 p405–494.
Title: **A law and economics analysis of the right to face-to-face confrontation post-Maryland v. Craig: distinguishing the forest from the trees.**
Author: Peter T. Wendel
Subjects: Law and economics—Analysis
Confrontation (Criminal law)—Litigation
Child sexual abuse—Litigation
Cases: Maryland v. Craig—497 U.S. 836 (1990)
Jurisdctn: United States
AN: 16384619

—end—

Display   Narrow   Explore   Esc   Brief citation display
   –   Previous   +   Next
Display brief citations   F1   Help   F2   Start over   F3   Print   F4   Mark

Source: © 1997 Information Access Company, LegalTrac™ Full Display screen, 22 Hofstra Law Review, Winter 1993, pp. 405–494, "A law and economics analysis of the right to face-to-face confrontation post-Maryland v. Craig: distinguishing the forest from the trees," Peter T. Wendel.

Exhibit 2–61

InfoTrac EF   LegalTrac   Start a Search

EasyTrac provides two simple ways to search: Subject Guide and Key Word.
Try the Subject Guide first.

Enter word(s) for KEY WORD SEARCH and press Enter

children and witnesses and television

Subject Guide   Browse listings of subjects, personal names, or companies that include the word(s) you type, e.g., rap music, censorship or Boris Yeltsin.
Key Word Search   Search for article references by combining words or phrases from titles, authors or subjects, e.g., family values AND Dan Quayle.

↑ Browse the Subject Guide   Esc   Erase entry line
    F1   Help   F10   PowerTrac (advanced search)

Source: © 1997 Information Access Company, LegalTrac™ Search Entry screen ("children and witnesses and television").

to "**child**," "**children**," "**childhood**" and other citations that contain the various endings of "child."

The use of truncation symbols requires caution. If a very common root is used, it will retrieve many irrelevant citations. For example, "**tele***" would produce not only "**television**" but "**telecommunication**," "**telephone**," and other words beginning with the root form "**tele**." In Exhibit 2–62 we see the first page of citations retrieved with the query used for Case 1. Note that we have retrieved 28 documents, and that we have been provided with the options **Narrow** and **Explore** at the bottom of the screen. These options, which we will not further illustrate here, allow us to make additional refinements in our search strategy.

In key word searching, it is extremely important to attempt different combinations of relevant terms or phrases. Remember that this is not subject searching and that some effort must be spent to identify the various words or phrases that could be used in the titles of relevant documents. Furthermore, as in the **Subject Guide** mode, only a limited number of characters or words can be typed into the entry box. Exhibit 2–63 presents an alternative search strategy, "**child sexual abuse and confrontation**," which will retrieve citations containing both the phrase "**child sexual abuse**," and the word "**confrontation**." Exhibit 2–64 presents a screen retrieved using this strategy. Note that many more documents (106) are retrieved, including the highlighted Wendel article.

Exhibit 2–62

InfoTrac EF	LegalTrac	Brief Citations

Key Words: children and witnesses and television
—————————————— 1 of 28 ——————————————

1   CCTV for child defence witnesses. (closed circuit television) (Scotland) Robert Shiels. SCOLAG, April 30, 1995 n222 p51(1).
        Enter

2   Constitutional law - criminal procedure - child testimony via videotape or closed circuit television - defendant's right to confront witnesses. (Case Note) Michael L. Bell. Duquesne Law Review, Wntr 1995 33 n2 p361–376.

3   Testimony via closed circuit television after Gonzales v. State: is the Sixth Amendment right to confront adverse witnesses at stake in Texas criminal trials? (Case Note) Stacey L. Jones. Baylor Law Review, Fall 1992 44 n4 p957–971.

4   A conflict of interests: the constitutionality of closed-circuit television in child sexual abuse cases.

Display	Narrow	Explore		Esc	Return to start			

Display extended citation   F1   Help   F2   Start over   F3   Print   F4   Mark

Source: © 1997 Information Access Company, LegalTrac™ Search Results screen ("children and witnesses and television" search results 1–4).

Exhibit 2–63

InfoTrac EF	LegalTrac	Start a Search

EasyTrac provides two simple ways to search: Subject Guide and Key Word.
Try the Subject Guide first.

Enter word(s) for KEY WORD SEARCH and press Enter

child sexual abuse and confrontation

Subject Guide   Browse listings of subjects, personal names, or companies that include the word(s) you type, e.g., rap music, censorship or Boris Yeltsin.

Key Word Search   Search for article references by combining words or phrases from titles, authors or subjects, e.g., family values AND Dan Quayle.

↑ Browse the Subject Guide   Esc   Erase entry line
                                  F1   Help   F10   PowerTrac (advanced search)

Source: © 1997 Information Access Company, LegalTrac™ Search Entry screen ("child sexual abuse and confrontation").

Exhibit 2–64

InfoTrac EF	LegalTrac	Brief Citations

Key Words: child sexual abuse and confrontation
————————————— 11 of 106 —————————————

Law Review, Wntr 1993 22 n1 p189–206.

11   A law and economics analysis of the right to face-to-face confrontation post-Maryland v. Craig: distinguishing the forest from the trees. Peter T. Wendel. Hofstra Law Review, Wntr 1993 22 n2 p405–494.

        Enter

12   The child victim/witness: balancing of defendant/victim rights in the emotional caldron of a criminal trial. (Kansas) Thomas D. Haney. The Journal of the Kansas Bar Association, Jan 1993 62 n1 p38(6).

13   The collision of policy and history. (Case Note) Timothy F. Cullen. New England Journal on Criminal & Civil Confinement, Wntr 1993 19 n1 p141–173.

Display   Narrow   Explore   Esc   Return to start

Display extended citation   F1   Help   F2   Start over   F3   Print   F4   Mark

Source: © 1997 Information Access Company, LegalTrac™ Search Results screen ("child sexual abuse and confrontation" search results 11–13).

PowerTrac. **PowerTrac** should be used in situations where you are combining several legal concepts and the **Subject Guide** and **Key Word Search** options do not provide sufficient space to type in all of your search terms. Our present example is a good one for demonstrating PowerTrac because it involves a combination of several concepts. To enter the PowerTrac search mode, press the **F10** function key from the main search query screen. The next screen will have a search entry box at the top. Enter the word or phrase that describes your first concept, such as "**child or children**," as represented in Exhibit 2–65. Alternatively, we could have used the truncation symbol (*) and typed in "**child***." Exhibit 2–66 indicates that the search strategy retrieves 16,792 documents ("**Rec.'s**" on the righthand side of the query line).

Note the **R1** that appears to the left of our search terms. This reference will come in handy as we proceed with our search strategy. We obviously must add additional terms to reduce the number of documents retrieved. Follow the direction at the top of the screen, **Press S to Search**, and you will be allowed to type in a new set of terms in the query entry box.

Exhibit 2–67 displays the second step of our process, in which we are looking for citations containing any of the three words "**witness**," "**witnesses**," or "**testimony**." Exhibit 2–68 indicates that the second step retrieves 5,722 documents.

Exhibit 2–65

InfoTrac EF	LegalTrac	Search

Enter search expression and press Enter

child or children

Display   Search   Choose   Esc   Erase entry line   F6   Fields/Indexes

F1   Help   F2   Start over

Source: © 1997 Information Access Company, LegalTrac™ PowerTrac Search Entry screen ("child or children").

Exhibit 2–66

InfoTrac EF	LegalTrac	Review

Press S to Search

		Rec.'s
R1 child or children		16792

Display   Search   Choose   Esc   Exit PowerTrac   F6   Fields/Indexes

Display highlighted result   F1   Help   F2   Start over   F3   Print

Source: © 1997 Information Access Company, LegalTrac™ PowerTrac Search Results screen ("child or children").

Exhibit 2–67

InfoTrac EF	LegalTrac	Search

Enter search expression and press Enter

witness or witnesses or testimony

	Rec.'s
R1 child or children	16792

Display   Search   Choose   Esc   Erase entry line   F6   Fields/Indexes

F1   Help   F2   Start over

Source: © 1997 Information Access Company, LegalTrac™ PowerTrac Search Results screen ("witness or witnesses or testimony R1 child or children").

Exhibit 2–68

InfoTrac EF	LegalTrac	Review

Press S to Search

		Rec.'s
R1	child or children	16792
R2	witness or witnesses or testimony	5722

Display	Search	Choose	Esc	Exit PowerTrac	F6	Fields/Indexes

| Display highlighted result | F1 | Help | F2 | Start over | F3 | Print |

Source: © 1997 Information Access Company, LegalTrac™ PowerTrac Search Results screen ("R1 child or children R2 witness or witnesses or testimony").

We could continue with this process by typing terms or phrases at each step that describe a different aspect of our topic. Thus, in Exhibit 2–69 we see that in step three **(R3)** of the process we retrieve 3,592 documents that make reference either to television or videotape. Likewise, in step 4 **(R4)** we retrieve 1,381 documents where the citation includes the phrase "**sexual abuse**," and in step 5 **(R5)**, a total of 498 document citations contain the term "**confrontation**." Once we have completed these separate searches we can start combining the various search results. In Exhibit 2–69 we have typed in "**r1 and r2**," which will retrieve documents containing at least one of the terms in **R1** and at least one of the terms in **R2**. Exhibit 2–70 displays the results of combining the different steps in our search process. We have highlighted **R8**, which combines steps 1, 4, and 5. Step 8 contains a set of 107 citations to documents that include either "**child**" or "**children**," the phrase "**sexual abuse**," and the word "**confrontation**."

Exhibit 2–71 displays a screen from the list of the 107 citations. We have highlighted document 11, which is the reference to the Wendel article.

Conclusion. We have provided an abbreviated introduction to LegalTrac. We do not claim that the search strategy we have used is necessarily the best one for the topic we are investigating, but we hope that it provides some indication of the ways in which LegalTrac can be useful. As always, we invite you to replicate the search we have just completed and to explore more efficient and effective search strategies for this topic and others of your choosing.

Exhibit 2–69

InfoTrac EF   LegalTrac   Search

```
                    Enter search expression and press Enter
  rl and r2
```

	Rec.'s
R1 child or children	16792
R2 witness or witnesses or testimony	5722
R3 television or videotape	3592
R4 sexual abuse	1381
R5 confrontation	498

Display   Search   Choose   Esc Erase entry line   F6 Fields/Indexes

                              F1   Help   F2 Start over

Source: © 1997 Information Access Company, LegalTrac™ PowerTrac Search Results screen ("rl and r2 R1—child or children R2—witness or witnesses or testimony R3—television or videotape R4—sexual abuse R5—confrontation").

Exhibit 2–70

InfoTrac EF   LegalTrac   Review

```
                         Press S to Search
```

	Rec.'s
R1 child or children	16792
R2 witness or witnesses or testimony	5722
R3 television or videotape	3592
R4 sexual abuse	1381
R5 confrontation	498
R6 rl and r2	857
R7 rl and r3 and r5	24
R8 rl and r4 and r5	107

Display   Search   Choose   Esc Exit PowerTrac   F6 Fields/Indexes

Display highlighted result   F1   Help   F2 Start over   F3 Print

Source: © 1997 Information Access Company, LegalTrac™ PowerTrac Search Results screen ("R1 child or children . . . R8 rl and r4 and r5").

Exhibit 2–71

InfoTrac EF	LegalTrac	Brief Citations

R8   r1 and r4 and r5

———————————— 11 of 107 ————————————

10   Defining the contours of unavailability and reliability for the confrontation clause. (response to article by Eleanor Swift in this issue, p. 145) (A Symposium on Current Trends of the Confrontation Clause) Barbara Rook Snyder. Capital University Law Review, Wntr 1993 22 n1 p189–206.

11   A law and economics analysis of the right to face-to-face confrontation post-Maryland v. Craig: distinguishing the forest from the trees. Peter T. Wendel. Hofstra Law Review, Wntr 1993 22 n2 p405–494.
     Enter

12   The child victim/witness: balancing of defendant/victim rights in the emotional caldron of a criminal trial. (Kansas) Thomas D. Haney. The Journal of the Kansas Bar Association, Jan 1993 62 n1 p38(6).

Display   Search   Review   Esc   Return to Review mode

Display extended citation   F1   Help   F2   Start over   F3   Print   F4   Mark

Source: © 1997 Information Access Company, LegalTrac™ Results screen ("R8 r1 and r4 and r5 search results 10–12").

Using WESTLAW and LEXIS to Find Law Review Articles

Both WESTLAW and LEXIS have files that contain the full text of law review articles. The date at which the coverage of individual law reviews begins varies on the two systems, as does whether all or just selected articles from each law review are included. This information is provided in the printed guides that accompany each system, and also is available online within the systems. The law review databases can be searched by using any of the systems' different search modes. We will illustrate how to use the **natural language** (Nat Lang) search mode on WESTLAW, which is similar to the **FREESTYLE** mode of LEXIS. Neither of these search modes requires the use of Boolean operators or field restrictions.

The first step is to locate the appropriate database to search, along with its **identifier**. The **identifier** is a letter code representing the database. WESTLAW has a printed list of the available databases and corresponding identifiers, by which we determine that WESTLAW has a database that contains the full text of law review and other legal periodical articles, and that its identifier is **JLR**, for "Journal & Law Reviews." (On LEXIS the library is **LAWREV** and the file is **ALLREV**.) If the printed list is not available, once you enter the WESTLAW system a series of menu screens appears that quickly allows you to find the law review database and its **identifier**. You can simply follow the menu screens to reach the law review database or shortcut the process by **entering JLR** from any of the menu screens.

As we have previously discussed, you can search WESTLAW by using either DOS or Windows versions. We will demonstrate using the Windows version. When we **enter** the **JLR** database, a search query screen appears. (See Exhibit 2–72.) The top line and the third line on the screen indicate that we are in the **Terms and Connectors** search mode, which is the method that uses Boolean operators and field restrictions. To switch to "natural language" simply click the **Nat Lang** button toward the bottom right side of the screen. The search query screen reappears, indicating that the search method is now **Nat Lang**.

Exhibit 2–72

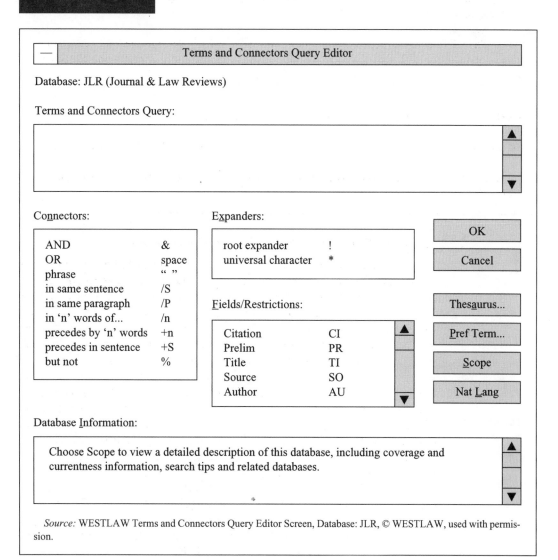

Terms and Connectors Query Editor		

Database: JLR (Journal & Law Reviews)

Terms and Connectors Query:

Connectors:

AND	&
OR	space
phrase	" "
in same sentence	/S
in same paragraph	/P
in 'n' words of...	/n
precedes by 'n' words	+n
precedes in sentence	+S
but not	%

Expanders:

root expander	!
universal character	*

Fields/Restrictions:

Citation	CI
Prelim	PR
Title	TI
Source	SO
Author	AU

OK

Cancel

Thesaurus...

Pref Term...

Scope

Nat Lang

Database Information:

Choose Scope to view a detailed description of this database, including coverage and currentness information, search tips and related databases.

Source: WESTLAW Terms and Connectors Query Editor Screen, Database: JLR, © WESTLAW, used with permission.

Entering the query. You can enter your query either by typing a sentence that describes the search topic, or by selecting and entering the key terms related to your issue. We will again use Case 1 as our example. Since a limited number of lines can be entered in the query box, it is best to use key words or phrases instead of full sentences when you have a topic involving several concepts. There are four concepts involved in our issue, expressed as the phrases "**right to confront witnesses**," "**child sexual abuse**," "**closed-circuit television**," and "**emotional trauma**." We type those phrases in the query box, as illustrated in Exhibit 2–73. Note that **Nat Lang** automatically truncates the end form of words. For example,

---

Exhibit 2–73

—	Natural Language Description Editor

Database: JLR (Journal & Law Reviews)
Maximum Result: 20
Natural Language Description:

right to confront witnesses child sexual abuse closed circuit television emotional trauma   ▲
▼

Restrictions Specified: Restrictions:

| | Date... | DA | | OK |
| | Added Date... | AD | | Cancel |

Thesaurus...

Scope

Remove   Control Concepts   Term Srch

Database Information:

Choose Scope to view a detailed description of this database, including coverage and currentness information, search tips and related databases. WIN natural language is protected by U.S. Patent No. 5,265,065 and 5,418,948.   ▲
▼

Source: WESTLAW Natural Language Description Editor Screen, Database: JLR, "right to confront witnesses child sexual abuse closed circuit television emotional trauma," © WESTLAW, used with permission.

by typing the word "**child**" we also get references to words such as "**children**," so we do not have to be concerned about using plural or other end forms of words.

It is important to remember that we are doing key word searching, not subject searching. In key word searching it is essential to include synonymous words or phrases for the terms entered. WESTLAW facilitates this process by having a built-in thesaurus, as shown by the button on the righthand side of the screen. After you have typed in your initial query but before you actually send it, click the thesaurus button to look for synonymous terms. A screen appears that lists your key terms in a box. If you click on one of your terms, the box to the right will display synonyms for that term, or what WESTLAW calls "**Related Terms**." In Exhibit 2–74 we show the related concepts that appear when we click the word

Exhibit 2–74

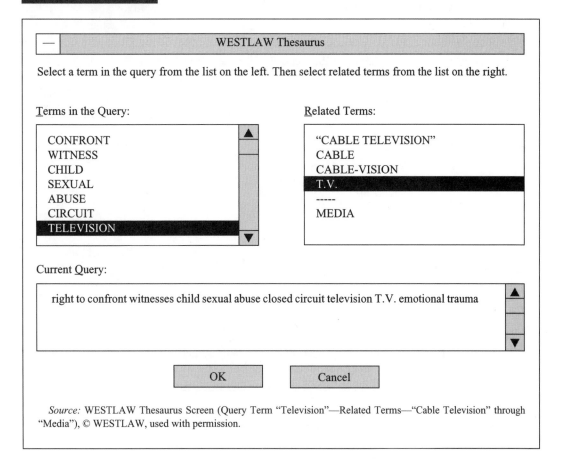

Source: WESTLAW Thesaurus Screen (Query Term "Television"—Related Terms—"Cable Television" through "Media"), © WESTLAW, used with permission.

"**television**." We add the abbreviation "**T.V.**" as a synonym for "television," click "**T.V.**," and press **enter**. When we have finished selecting **Related Terms** we return to our search query screen (see Exhibit 2–75) and see that "**T.V.**" has been added to our search query and is placed in parentheses after the word "**television**." The parentheses in **natural language** are read as an **or**, so our revised search strategy will now retrieve documents containing either "**television**" or "**T.V.**" We can add our own "**Related Terms**" even if they do not appear in the thesaurus by inserting them within parentheses after the appropriate word. For example, if we

---

Exhibit 2–75

—	Natural Language Description Editor

Database: JLR (Journals & Law Reviews)
Maximum Result: | 20 |
Natural Language Description:

right to confront witnesses child sexual abuse closed circuit television (T.V.) emotional trauma ▲ ▼

Restrictions Specified:

Restrictions:

Date...   DA
Added Date...   AD

OK

Cancel

Thesaurus...

Remove   Control Concepts   Scope

Term Srch

Database Information:

Choose Scope to view a detailed description of this database, including coverage and currentness information, search tips and related databases. WIN natural language is protected by U.S. Patent No. 5,265,065 and 5,418,948. ▲ ▼

Source: WESTLAW Natural Language Description Editor Screen, Database: JLR, "right to confront witnesses child sexual abuse closed circuit television (T.V.) emotional trauma," © WESTLAW, used with permission.

wanted to search for the word "**videotape**" as an alternative to "**television**" or "**T.V.**" we could insert it within parentheses immediately after "**T.V.**" (Other options appear on the search query screen that can be used to refine your search even more, but we will not discuss these other features now since you have more than enough to continue your search.

Interpreting and browsing your search results. WESTLAW's **natural language** search mode includes a statistical program which uses two criteria to retrieve relevant documents. The statistical program first looks for documents in the designated database that contain the most occurrences of the terms or phrases included in the search query. It also assigns a weighted value to each term or phrase in the query. A term or phrase that appears in relatively few documents receives a higher value than a term or phrase that appears in more documents. Documents that include the most occurrences of the terms or phrases with higher weighted values and that have the most occurrences of all the search terms are considered the most relevant. The retrieved documents are supposedly ranked in order of relevancy, with the most relevant presented first. You can set the number of documents to be retrieved from 1 to 100. (The default option retrieves 20 documents, and we will use that option in our example.) *Warning:* Whatever number of documents you set for retrieval, the system will retrieve that number regardless of whether all your search terms appear in those documents. If there are not 20 documents that contain the terms in your search query, the system will start dropping terms from the query in order to retrieve the designated number. Thus, when you use **Nat Lang**, it is essential that you actually browse the text of the retrieved documents in order to determine their relevancy. We discuss how to browse documents below.

As your search is running through the system, you will see the screen illustrated in Exhibit 2–76. Under the line "**Your description is:**" you will see your search query as you actually typed it. Below that you will see another line, "**Concepts in your description:**". Below this line you see the actual words and phrases the system has taken from your search strategy and is using to retrieve documents. There is not much difference between the two sets of words and phrases in our example, except that where quotation marks surround a phrase that entire phrase is being searched, as well as the individual words within it. Greater differences would have been apparent if we had entered the query in sentence form, in that commonly used words such as pronouns would have been omitted.

Exhibit 2–77 shows the first page of the first document retrieved, which is the one in the database that best meets the specifications of our search strategy. As with the printed indexes, there is no attempt to assess the quality of the articles, which are selected exclusively according to the criteria mentioned above. Note the five pieces of information appearing at the top of the screen: the WESTLAW cite for the document (15 NILULR 719), the rank of the document (in this case, the first of 20 documents retrieved), the page (or, more appropriately, the screen number) of the document on view, the identifier for the database (**JLR**), and the search mode (**Term**). The search mode tells you how you will be browsing the text of the document. The **Term** mode indicates that each time you hit the **enter** key you will go to the next screen in the document where one of your search terms

Exhibit 2–76

Your description is:
RIGHT TO CONFRONT WITNESSES CHILD SEXUAL ABUSE CLOSED
CIRCUIT TELEVISION (T.V.) EMOTIONAL TRAUMA

Concepts in your description:
RIGHT CONFRONT WITNESSES CHILD "SEXUAL ABUSE" "CLOSED
CIRCUIT" TELEVISION (T.V.) EMOTIONAL TRAUMA

Your database is JLR
Your search is proceeding.
To cancel your search, type X and press ENTER

Source: WESTLAW Search Screen ("RIGHT TO CONFRONT WITNESSES CHILD
SEXUAL ABUSE CLOSED CIRCUIT TELEVISION (T.V.) EMOTIONAL TRAUMA,"
Database: JLR), © WESTLAW, used with permission.

Exhibit 2–77

Copr. © West 1996 No claim to orig. U.S. govt. works
AUTHORIZED FOR EDUCATIONAL USE ONLY

Citation	Rank (R)	Page (P)	Database	Mode
15 NILULR 719	R1 OF 20	P 1 OF 69	JLR	Term

(Cite as: 15 N. Ill. U. L. Rev. 719)

Northern Illinois University Law Review
Summer, 1995

Comment

*719 ILLINOIS' CONFRONTATION WITH THE USE OF CLOSED CIRCUIT
TESTIMONY IN CHILD SEXUAL ABUSE CASES: A LEGISLATIVE
APPROACH TO THE SUPREME COURT DECISION OF PEOPLE V.
FITZPATRICK

Michael G. Clarke

Copyright (c) 1995 by the Board of Regents for Northern Illinois University;
Michael G. Clarke

WESTLAW LAWPRAC INDEX
COA — Court Automation: Computer Hardware & Software

Source: WESTLAW Search Results Screen (Comment, "Illinois' Confrontation with the
Use of Closed Circuit Testimony in Child Sexual Abuse Cases: A Legislative Approach to the
Supreme Court Decision of People v. Fitzpatrick," 15 N. Ill. U.L. Rev. 719 (1995)), ©
WESTLAW, used with permission.

appears. When all the search terms in the first document are identified, you auto-matically are referred to the first page of the next document, or R2 (second-ranked document). You can cancel the **Term** mode and browse page by page by entering "**p**" (for "page"), which takes you through the document page by page, irrespec-tive of whether your search terms appear. (The mode at the top of the screen now indicates "**Page**". Switch back to the **Term** mode by entering "**t**" for term.)

You will be provided with a listing of all 20 retrieved documents by entering "**l**" for "**list**." This command produces bibliographic information on the first 20 docu-ments, which takes up several screens. You can go to the full text of any of the documents by entering the number to the left of the citation (see Exhibit 2–78). To return to your original document **enter** either "**gb**" (for go back) or "**map**," which will provide you with the history of your search session and allow you to return to whatever step you choose.

Remember that **natural language** will retrieve 20 documents even if there are not that many documents in the database that fully meet the requirements of the query and that the documents are ranked in order of relevancy. The easiest way to determine if all 20 documents are relevant is to go to the last document retrieved by entering *r20* (**r** for "rank"). If that document is not relevant, you may want to

---

Exhibit 2–78

Copr. © West 1996 No claim to orig. U.S. govt. works
AUTHORIZED FOR EDUCATIONAL USE ONLY
CITATIONS LIST (Page 1)   Search Result Documents: 20
Database: JLR

1. 15 N. Ill. U. L. Rev. 719 Summer, 1995 Comment ILLINOIS' CONFRON-TATION WITH THE USE OF CLOSED CIRCUIT TESTIMONY IN CHILD SEXUAL ABUSE CASES: A LEGISLATIVE APPROACH TO THE SUPREME COURT DECISION OF PEOPLE V. FITZPATRICK Michael G. Clarke

2. 27 Val. U. L. Rev. 217 Fall, 1992 Note A CONFLICT OF INTERESTS: THE CONSTITUTIONALITY OF CLOSED-CIRCUIT TELEVISION IN CHILD SEXUAL ABUSE CASES   John Paul Serketich

3. 15 U. Puget Sound L. Rev. 913 Spring, 1992 Comment WASHINGTON'S CLOSED-CIRCUIT TESTIMONY STATUTE: AN EXCEPTION TO THE CONFRONTATION CLAUSE TO PROTECT VICTIMS IN CHILD ABUSE PROSECUTIONS   Karen R. Hornbeck [FNa]

4. 24 Seton Hall L. Rev. 2030 1994 Comment JUSTICE FOR OUR CHIL-DREN: NEW JERSEY ADDRESSES EVIDENTIARY PROBLEMS IN-HERENT IN CHILD SEXUAL ABUSE CASES   Dione Marie Enea

Source: WESTLAW WESTMATE [Session] Screen, Citations List (Page 1) of Search Re-sults, Database: JLR, 1 through 4, © WESTLAW, used with permission.

work your way back by entering **r10**, and so on, until you reach a relevant document. If the twentieth-ranked document is relevant, you may decide to search for additional documents in the database. You can do this by entering the command **next.** The **next** command will retrieve the next 10 most relevant documents in the database. In our example it will retrieve documents ranked 21 through 30, which we can browse in the same manner as the original 20 retrieved documents. We can continue to use the **next** command in the **Nat Lang** mode until we reach a limit of 100 documents retrieved.

One way to evaluate our strategy is to check our search results to see if they contain references to documents that we previously have determined are relevant. For example, we previously searched the ILP and discovered the article by Peter Wendel that was published in the Hofstra Law Review. That article appears as the nineteenth-ranked document in our result list. (See Exhibit 2–79.)

Exhibit 2–79

Copr. © West 1996 No claim to orig. U.S. govt. works
AUTHORIZED FOR EDUCATIONAL USE ONLY
CITATIONS LIST (Page 5)   Search Result Documents: 20
Database: JLR

17. 50 U. Pitt. L. Rev. 1187 Summer, 1989 Comment COY v. IOWA: SHOULD CHILDREN BE HEARD AND NOT SEEN? Sharon Parker Brustein

18. 8 T.M. Cooley L. Rev. 389 Trinity Term, 1991 Casenote SOMETHING MORE THAN A GENERALIZED FINDING: [FN1] THE STATE'S INTEREST IN PROTECTING CHILD SEXUAL ABUSE VICTIMS IN MARYLAND v. CRAIG [FN2] OUTMUSCLES THE CONFRONTATION CLAUSE Eldonna M. Ruddock

19. 22 Hofstra L. Rev. 405 Winter 1993 A LAW AND ECONOMICS ANALYSIS OF THE RIGHT TO FACE-TO-FACE CONFRONTATION POST-MARYLAND V. CRAIG: DISTINGUISHING THE FOREST FROM THE TREES Peter T. Wendel [FNa]

20. 26 Loy, U. Chi. L.J. 321 Winter, 1995 Note PEOPLE v. FITZPATRICK: THE PATH TO AMENDING THE ILLINOIS CONSTITUTION TO PROTECT CHILD WITNESSES IN CRIMINAL SEXUAL ABUSE CASES Thomas Conklin

YOU ARE AT THE END OF THE CITATION LIST. PLEASE ENTER YOUR NEXT COMMAND.

Source: WESTLAW WESTMATE [Session] Screen, Citations List (Page 5) of Search Results, Database: JLR, 17 through 20, © WESTLAW, used with permission.

As we have pointed out previously, you will get different results if you later try to replicate our current search strategy, because new documents are constantly being added to the database. You can get a more accurate replication by including a date restriction, which we will explain in Chapter 3. Remember that "relevancy ranking" in WESTLAW relies on a statistical package that measures the frequency of the occurrence of certain words and/or phrases, not on any attempt to assess the *quality* or *authority* of the article. For this reason, it is extremely important not to settle for reviewing the first few documents retrieved via WESTLAW, just as you would not settle for consulting the first articles listed in a printed index. The better practice is to obtain as comprehensive a list of relevant articles as possible, and then assess which are likely to be the most *authoritative* and *helpful* ones to read.

We still have only scratched the surface of some of the capabilities of CALR. In Chapters 3 and 4 we discuss additional features of CALR, such as the "hypertext" functions and how to run the same search strategy in a different database. Nevertheless, you now have enough basic information and commands at your disposal to get started. Below is a summary list of the commands discussed:

t	Term mode; displays the next page containing a search term.
p	Page mode; displays the next page of a document or a specified page if followed by a number; e.g., p4.
l	Displays a list of citations to retrieved documents.
r	Displays the next ranked document, or a specified document when followed by the document rank number; e.g., **r3**.
gb	Means go back; WESTLAW command to return to previous location.
map	Displays search steps.
next	Retrieves the next 10 documents using the same search strategy in **Nat Lang** searching.

Social Science References

Although this book focuses on legal authorities and related research techniques, we must keep in mind that the law does not exist in a vacuum. In Chapter 1 we briefly discussed how social science research contributes to the assessment and analysis of legal issues related to criminal justice. We also mentioned some of the major reference sources to social science literature. You are likely to be familiar with many of these references, and an exhaustive account of them would cause us to stray too far from our path. We will, however, take the opportunity to use our hypothetical cases to illustrate a few ways that social science references can complement your legal research objectives.

Let us start with Case 1, which involves John Winston's right to confront a child witness he has been accused of sexually abusing. The prosecution has argued that the child should be allowed to testify via closed-circuit television in order to avoid being intimidated and psychologically traumatized by Mr. Winston's presence. Underlying the legal issues is a key assumption that may require supporting evidence. The assumption is that child victims of sexual abuse will be psychologically traumatized if confronted by the alleged abuser in the courtroom. A related assumption is that the resulting psychological trauma will render the victim unable to provide credible testimony regarding the alleged abuse. To help determine the validity of these assumptions, the court may rely on the oral or written testimony of experts who have studied these issues. Indeed, the leading Supreme Court case that we have located on this topic, *Maryland v. Craig*, 497 U.S. 836, 110 S. Ct. 3157, 111 L. Ed. 2d 666 (1990), makes several references to social science research. For example, the following reference is cited in support of the proposition that child victims who testify in court suffer psychological trauma: G. Goodman et al., *Emotional Effects of Criminal Court Testimony on Child Sexual Assault Victims, Final Report to the National Institute of Justice* (presented as conference paper at annual convention of American Psychological Assn., August 1989). Maryland v. Craig, 497 U.S. at 855. The majority opinion also makes reference to an article presenting research in support of the contention that where there is face-to-face confrontation with the accused, the emotional trauma suffered by a child witness may prevent the child from effectively testifying. Id., 497 U.S. at 857, citing Goodman & Helgeson, *Child Sexual Assault: Children's Memory and the Law*, 40 U. Miami L. Rev. 181, 203–204 (1985).

We might want to examine whether more recent research on this topic has been reported. Our first step is to determine what social science discipline or field of study is likely to have conducted the relevant research. In this instance this is not a difficult task. The primary discipline dealing with psychological trauma obviously is psychology. If you are unsure about what reference source to consult for citations to literature on your topic, ask the reference librarian at your institution's library and try to be as specific as possible regarding your information needs.

Psychological Abstracts is the major reference source providing citations to books and journal articles in the discipline of psychology. There is a CD-ROM version of Psychological Abstracts, called **PsycLIT**, which most academic libraries are likely to carry. We recommend searching the CD-ROM version for information for several reasons. First, PsycLIT allows you to search several years at the same time, rather than plowing through annual volumes of the printed version. Second, the CD-ROM version quickly permits retrieval of sets of documents containing more than one concept. In the printed volumes, you are limited to searching one concept or subject at a time. Third, the CD-ROM version allows you to search for key words or phrases appearing either in the title of a document or in an abstract summarizing the contents of a document. It also allows you to search for documents containing specific subject terms, which in PsycLIT is somewhat less important because this reference source has excellent subject indexing. Before searching PsycLIT you should consult the **Thesaurus of Psychological Index Terms**, which lists, describes, and cross-references the subject descriptors assigned to documents referenced in PsycLIT.

The thesaurus produces the following subject descriptors: "**sexual abuse**," "**child abuse**," "**emotional trauma**," and "**crime victims**." Several different vendors produce different versions of PsycLIT on CD-ROM. We will be using the

Silverplatter version. The other versions differ somewhat in their search techniques, but the basic search strategy should be similar to our Silverplatter example. The particular section of PsycLIT we will be searching contains only journal articles indexed from January 1990 through June 1996. Other sections of PsycLIT contain references to older journal articles and references to books and chapters in books. We enter our query phrases one at a time and retrieve the results depicted in Exhibit 2–80.

When two or more terms are entered without connectors between them, the Silverplatter software searches the words as a phrase. The phrase could appear in the title of the document, the abstract accompanying the document, or as a subject descriptor assigned to the document. Search result #3 indicates that 2,587 documents were in the database where the term "**sexual abuse**" appears in one of these fields. When terms are entered without a connector between them, Silverplatter also retrieves all documents containing the individual words that appear in any of the fields mentioned above. Thus, in Exhibit 2–80, result #1 indicates that there are 10,696 documents in the database containing the word "**sexual**" in one of the fields, and result #2 indicates there are 12,678 documents where the word "**abuse**" appears. We next repeat the process and search for the term "**child abuse**." Result #6 indicates that there are 2,803 documents in the database where "**child abuse**" appears in one of the three fields. The search also yields the number of documents in which each individual word appears.

We enter our third search term, "**emotional-trauma**," with a hyphen between the words. In Silverplatter, placing the hyphen between the words in a query or after a single word in a query will limit the search result to documents in which

---

Exhibit 2–80

PsycLIT Journal Articles 1991–6/97

No.	Records	Request
#1:	10696	SEXUAL
#2:	12678	ABUSE
#3:	2587	SEXUAL ABUSE
#4:	19113	CHILD
#5:	12678	ABUSE
#6:	2803	CHILD ABUSE
#7:	795	EMOTIONAL-TRAUMA
#8:	275	CRIME-VICTIMS
#9:	1	#3 and #6 and #7 and #8

FIND:
To Show records, press F4.

Source: Psyclit (cd-rom version of Psychological Abstracts, SilverPlatter) Query Entry Screen "Sexual," "Abuse," . . . "Crime-Victims," © 1986–1997 American Psychological Association, © 1991–1997 SilverPlatter International, N.V.

the phrase is used as a subject descriptor for the document. Thus, result #7 indicates that 795 documents in the database have the subject descriptor "**emotional-trauma**" attached to them. We also search our final term "**crime-victims**," as a subject descriptor. Result #8 indicates that 275 documents included "**crime-victims**" as a subject descriptor.

Here we begin to see the advantage of searching electronically rather than using the printed indexes. If we had begun our search with the subject indexes to Psychological Abstracts for the years 1990 through June 1996 by looking for references to "**emotional trauma**" as a subject heading, we would have located at least 795 documents (and in the printed indexes we also would have found references to books and book chapters, which we have excluded in our example). We then would have to go through each of these references individually to determine which ones were also likely to relate to the other aspects of our topic. The electronic version allows us to combine all aspects of our search topic, as we will demonstrate below.

By placing the connector **and** between the numbers of the search results we want to link, we combine our results to get a set of documents that includes all four search elements. The combined query "**#3 and #6 and #7 and #8**" retrieves a set of documents where "**sexual abuse**" and "**child abuse**" and the subject descriptors "**emotional-trauma**" and "**crime-victim**" all appear in the reference to the document. Result #9 indicates that only one document in the database meets all four elements of our search query. To view the document we press **F4** and retrieve the reference illustrated in Exhibit 2–81.

Each two-letter code to the left of the screen indicates a separate field of the document. Thus, "**TI**" indicates the title of the article being referenced; "**JN**" is followed by the name of the journal in which the article appears; "**AB**" indicates the abstract for the article; and "**DE**" lists each of the subject descriptors assigned to the document. Note that the subject descriptors assigned to a document produced by this initial search result can help you revise your query. In this instance the Brannon article produces an additional subject descriptor, "**legal-testimony**," that appears to be particularly relevant, and a second descriptor, "**children-**," may help us as well. Once we have chosen additional terms to search, we can return to our search query screen either by entering **f** (for "Find") or highlighting the FIND: key at the bottom of the screen and pressing **enter**. Exhibit 2–82 depicts our revised search result.

We first search for the subject descriptor "**legal-testimony**" and retrieve 162 documents. Then we search for the subject descriptor "**children-**" and retrieve 4,314 documents, as indicated in result #11. We next look for documents that contain three subject descriptors, "**emotional-trauma**," "**legal-testimony**," and "**children-**," by entering the query "**#7 and #10 and #11**." Result #12 indicates that five documents in the database include all three of these subject descriptors. Exhibit 2–83 depicts the reference to one of the documents found in result #12. We could continue with this process by trying different combinations of search terms and revising our search strategy until we are confident that we have retrieved all of the relevant documents.

Our second hypothetical raises the factual issue of whether the human immunodeficiency virus can be transmitted by biting or spitting. Research dealing with the transmission of HIV is more likely to be found in the medical or public health

Exhibit 2–81

PsycLIT Journal Articles 1991–6/97

TI DOCUMENT TITLE: The trauma of testifying in court for child victims of sexual assault v. the accused's right to confrontation.
AU AUTHOR(S): Brannon, -L. -Christine
JN JOURNAL NAME: Law-and-Psychology-Review; 1994 Spr Vol 18 439–460
IS   ISSN: 00985961
LA   LANGUAGE: English
PY   PUBLICATION YEAR: 1994

AB   ABSTRACT: Reviews emotional effects on child victim witnesses of testifying against the accused, including the beneficial effects of the legal process, fear and trauma stemming from confronting the accused, and the effect of fear and trauma on the child's willingness and ability to testify. The accused's right to confrontation, and balancing the competing interests of the accused and the child witness are discussed. Developments in the law after the Maryland v. Craig case are addressed, including the enactment of the Child Victims' and Child Witnesses' Rights statute. (PsycLIT Database Copyright 1995 American Psychological Assn, all rights reserved)
KP KEY PHRASE: trauma of testifying in court vs accused's right to confrontation & development in law; child victims of <u>sexual abuse</u>
DE DESCRIPTORS: LEGAL-TESTIMONY; <u>EMOTIONAL-TRAUMA</u>; <u>CHILD-ABUSE</u>; CIVIL-RIGHTS; LAWS-; <u>SEXUAL ABUSE</u>; LEGAL-PROCESSES; FEAR-; DEFENDANTS-; CHILDREN-; <u>CRIME-VICTIMS</u>; WITNESSES-
CC CLASSIFICATION CODE(S): 4230; 3230; 42; 32
PO POPULATION: Human
AG COMPOSITE AGE GROUP: Child
UD UPDATE CODE: 9503
AN PSYC ABS. VOL. AND ABS. NO.: 82-11788
JC JOURNAL CODE: 2002

Source: Psyclit (cd-rom version of Psychological Abstracts, SilverPlatter) Search Result Screen, TI: The trauma of testifying in court for child victims of sexual assault v. the accused's right to confrontation. AU: Brannon,-L.-Christine. JN: Law-and Psychology-Review; 1994 Spr Vol 18 439–460 . . . Reprinted with permission of The American Psychological Association, publisher of The Psyc INFO and Psyc LIT database, all rights reserved. © 1986–1997 American Psychological Association, © 1991–1997 SilverPlatter International, N.V.

---

literature than in social science journals. Similarly, in the third hypothetical there is an assertion that exposure to second-hand tobacco smoke can lead to serious health problems. Whether this assertion is valid is best evaluated through research that is most likely to be reported in the medical or public health literature. However, another dimension of these questions is likely to be reported in the social science literature. We should investigate references addressing the public policy and regulatory actions of government bodies relating to these factual issues. Specifically, in Case 2 we want to know whether government bodies have adopted policy positions regarding the methods by which the HIV can be transmitted. If they have, the policies could be used to help evaluate the underlying assumption that HIV can be transmitted by biting or spitting. In Case 3 we would want evi-

Exhibit 2–82

PsycLIT Journal Articles 1991–6/97

No.	Records	Request
#1:	10696	SEXUAL
#2:	12678	ABUSE
#3:	2587	SEXUAL ABUSE
#4:	19113	CHILD
#5:	12678	ABUSE
#6:	2803	CHILD ABUSE
#7:	795	EMOTIONAL-TRAUMA
#8:	275	CRIME-VICTIMS
#9:	1	#3 and #6 and #7 and #8
#10:	162	LEGAL-TESTIMONY
#11:	4314	CHILDREN-
#12:	5	#7 and #10 and #11

FIND:
To Show records, press F4.

Source: Psyclit (cd-rom version of Psychological Abstracts, SilverPlatter) Search Result Screen, "#1 10512 SEXUAL" through "#12 5 #7 and #10 and #11, © 1986–1987 American Psychological Association, © 1991–1997 SilverPlatter International, N.V.

---

dence of government action regulating second-hand tobacco smoke based on the health threat imposed on nonsmokers who share the same physical environment.

Public policy issues are addressed to some extent in the law review literature we have previously discussed. An additional source for references to public policy literature is **Public Affairs Information Service Bulletin International (PAIS)**. **PAIS** contains references to journal articles, books, government documents, research reports, and other materials relevant to public and social policy. It is available in print, on CD-ROM, and through online database vendors.

In Exhibit 2–84 we display a search strategy for our third hypothetical entered into the Silverplatter CD-ROM version of PAIS. We enter three search phrases: **"second-hand smoke," "environmental tobacco smoke or ets,"** and **"passive smok*."** When you are using the Silverplatter software, recall that words entered without connectors between them will generate search results for each individual word as well as for the words as a phrase. Thus, result #4 indicates that three references in the database contain the phrase **"second hand smoke."** Result #9 indicates that 20 references contain either the phrase **"environmental tobacco smoke"** or its initialism **"ets."** An asterisk is used as the truncation symbol in the Silverplatter system that retrieves the different end forms of a word root. Thus, we use **"smok*"** to retrieve **"smoke"** or **"smoker"** or **"smoking."** Result #12 indicates that 14 documents in the database include a reference to **"passive smoke,"**

Exhibit 2–83

TI   DOCUMENT TITLE: Evaluation of child witnesses for confrontation by criminal defendants.
AU   AUTHOR(S): Small, -Mark-A.; Melton, -Gary-B.
IN   INSTITUTIONAL AFFILIATION OF FIRST AUTHOR: Southern Illinois U, Ctr for Study of Crime, Delinquency & Corrections, Carbondale, US
JN   JOURNAL NAME: Professional-Psychology-Research-and-Practice; 1994 Aug Vol 25 (3) 228-233
IS   ISSN: 07357028
LA   LANGUAGE: English
PY   PUBLICATION YEAR: 1994
AB   ABSTRACT: States have passed legislation governing the procedures by which children may testify in cases of child abuse. In Maryland v. Craig (1990), the US Supreme Court reviewed the constitutionality of these procedures. As a result, psychologists may be asked to perform evaluations regarding the potential trauma a child faces in confronting a defendant in a criminal case. Specifically, opinions of psychologists may be sought as to the potential trauma a child may endure as a result of a face-to-face confrontation with a defendant. Research relevant to this assessment is reviewed, and it is suggested that psychologists may be of most help in preparing children to testify. Careful attention should be given to ethical and legal issues, and psychologists should be careful not to overstep the limits of their expertise. (PsycLIT Database Copyright 1994 American Psychological Assn, all rights reserved)
KP   KEY PHRASE: issues in evaluation of child's potential trauma resulting from testimony confrontation with defendants in child abuse cases; psychologists
DE   DESCRIPTORS: CLINICAL-JUDGMENT-NOT-DIAGNOSIS; EMOTIONAL-TRAUMA; LE-GAL-TESTIMONY; CHILD-ABUSE; PSYCHOLOGISTS-; DEFENDANTS-; CHILDREN-
CC   CLASSIFICATION CODE(S): 4230; 3230; 42; 32
PO   POPULATION: Human
AG   COMPOSITE AGE GROUP: Child
UD   UPDATE CODE: 9411
AN   PSYC ABS. VOL. AND ABS. NO.: 81-43476
JC   JOURNAL CODE: 1604

Source: Psyclit (cd-rom version of Psychological Abstracts, SilverPlatter) Search Result Screen, TI: Evaluation of child witnesses for confrontation by criminal defendants. AU: Small,-Mark-A.; Melton,-Gary-B. IN: Southern Illinois U, Ctr for Study of Crime, Delinquency & Corrections, Carbondale, US JN: Professional-Psychology-Research-and-Practice; 1994 Aug Vol 25(3) 228–233. Reprinted with permission of The American Psychological Association, publisher of The Psyc INFO and Psyc LIT database, all rights reserved. © 1986–1997 American Psychological Association, © 1991–1997 SilverPlatter International, N.V.

"**passive smoker**," or "**passive smoking**." Result #13 shows that we have retrieved 34 documents that include at least one of our search phrases.

The reason for the discrepancy between the 34 documents indicated in result #13 and the 37 documents obtained by adding the individual search results displayed in lines 4, 9, and 12 is that some documents contain more than one of our search phrases and thus are counted in more than one set. However, the documents containing terms from more than one search query will only be counted once when the **or** command is applied to the different search results. For example, if a reference contains both the phrase "**second hand smoke**" and "**environmental tobacco smoke**," it would be counted in both search result #4 and search result

Exhibit 2–84

PAIS International 1972–4/97

No.	Records	Request
#1:	2082	SECOND
#2:	303	HAND
#3:	116	SMOKE
#4:	3	SECOND HAND SMOKE
#5:	13689	ENVIRONMENTAL
#6:	532	TOBACCO
#7:	116	SMOKE
#8:	10	ETS
#9:	20	ENVIRONMENTAL TOBACCO SMOKE OR ETS
#10:	137	PASSIVE
#11:	562	SMOK*
#12:	14	PASSIVE SMOK*
#13:	34	#4 or #9 or #12

FIND:
To Show records, press F4.

Source: P.A.I.S. (Public Affairs Information Service Bulletin International, Silverplatter cd-rom version) Search Entry Screen, "#1: 1977 SECOND" through "#13: 30 #4 or #9 or #12," © 1991–1996 Public Affairs Information Services, Inc., © 1991–1996 SilverPlatter International, N.V.

#9. Duplicates are discarded when the results are combined with the **or** command in search result #l3, and the document will be counted only once.

We can view the documents retrieved by pressing **F4**. Exhibits 2–85 and 2–86 display two of the 34 references from our search result. As in PsycLIT, the two-letter codes in PAIS at the left of the screens are labels for the various fields or pieces of information comprising the references. It appears that the article cited in Exhibit 2–85 will be of interest to Warden Miller, the defendant in hypothetical Case 3. This article apparently criticizes a study linking environmental tobacco smoke to health problems, and it might be used to suggest that any policy decisions based on the study also are flawed. The article cited in Exhibit 2–86 indicates that the U.S. Supreme Court has issued a decision on this issue, which would be especially important information if we had started our research with PAIS. The article apparently also discusses the legal and policy issues stemming from the decision. Correctional administrators and/or policymakers might find the discussion of the legal and policy issues helpful in designing regulations consistent with the decision of the Supreme Court.

Many other reference sources to the social science literature may be useful to your research. Unfortunately, a detailed discussion of these resources is beyond the scope of this book. The above illustrations of how social science materials can be relevant to issues of law remind you not to exclude other reference sources when you are confronted with a legal research problem. The social science and

Exhibit 2–85

PAIS International 1972–4/97

12 of 34

AN ACCESSION NUMBER: 94-0205734
TI TITLE: Smoke and mirrors: the EPA's flawed study of <u>environmental tobacco smoke</u> and lung cancer.
AU AUTHOR: Huber, -Gary-L.; and-others
SO SOURCE: Regulation-(Cato-Inst); 16:44-54 no 3 1993
PY PUBLICATION YEAR: 1993
NT NOTE: Critical of the 1992 US study on <u>passive smoking</u>.
DE DESCRIPTORS: United-States-Environmental-policy; Smoking-Research; Cancer-Research; Indoor-air-pollution-United-States; Public-health-United-States
LA LANGUAGE: E; English
IS INTERNATIONAL STANDARD SERIAL NUMBER: 0147-0590
SP SPECIAL FEATURES: bibl(s) il(s) table(s) chart(s)
PT PUBLICATION TYPE: P; Periodical

Source: P.A.I.S. (Public Affairs Information Service Bulletin International, SilverPlatter cd-rom version) Search Result Screen, "AN: 94-0205734 TI: Smoke and mirrors: the EPA's flawed study of environmental tobacco smoke and lung cancer." AU: Huber,-Gary-L. and others SO: Regulation-(Cato-Inst); 16:44–45 no 3 1993 . . . , © 1991–1996 Public Affairs Information Services, Inc. © 1991–1996 SilverPlatter International, N.V.

policy references you produce can greatly enhance your appreciation of law-related issues, and you should not hesitate to consult them when appropriate.

Statistical Sources

In the nineteenth century, British Prime Minister Benjamin Disraeli reportedly observed: "There are three kinds of lies: lies, damned lies, and statistics." The Prime Minister clearly did not put much faith in statistics. However, statistics that are methodically collected and properly presented can shed considerable light on social and legal issues. Courts have not been reluctant to rely on statistical evidence to help support their decisions. See, for example, *Michigan Department of State Police v. Sitz,* 496 U.S. 444, 110 S. Ct. 2481, 110 L. Ed. 2d 412 (1990) (upholding police sobriety checkpoints against Fourth Amendment challenge, relying in part on statistics demonstrating magnitude of drunk-driving problem and efficacy of roadblocks in combatting this problem); *Thompson v. Oklahoma,* 487 U.S. 815, 108 S. Ct. 2687, 101 L. Ed. 2d 702 (1988) (ruling that execution of offender who was 15 years old at the time of his crime would be cruel and unusual punishment and citing statistics describing the incidence of homicides and death sentences for adult and juvenile offenders); *Tennessee v. Garner,* 471 U.S. 1, 105 S. Ct. 1694, 85 L. Ed. 2d 1 (1985) (holding that police may not use deadly force to apprehend a nondangerous fleeing felon and citing arrest statistics); *Olim v. Wakinekona,* 461 U.S. 238, 103 S. Ct. 1741, 75 L. Ed. 2d 813 (1983) (finding that

Exhibit 2–86

PAIS International 1972–4/97

AN ACCESSION NUMBER: 94-0108170
TI   TITLE: Legal and policy issues from the Supreme Court's decision on smoking in prisons.
AU AUTHOR: Vaughn,-Michael-S.; del-Carmen,-Rolando-V.
SO SOURCE: Federal-Probation; 57:34-9 S 1993
PY PUBLICATION YEAR: 1993
NT NOTE: Considers the 1993 decision which held that exposure to <u>environmental tobacco smoke</u> may sometimes be cruel and unusual punishment; US.
DE DESCRIPTORS: United-States-Supreme-court-Decisions; Prisoners-Legal-status,-laws,-etc.; Smoking-Legal-aspects; Environmental-health-United-States; Correction-penology-United-States; Prisons-Legal-aspects
LA LANGUAGE: E; English
IS   INTERNATIONAL STANDARD SERIAL NUMBER: 0014-9128
PT PUBLICATION TYPE: P; Periodical

Source: P.A.I.S. (Public Affairs Information Service Bulletin International, SilverPlatter cd-rom version) Search Result Screen, "AN: 94-0108170 TI: Legal and policy issues from the Supreme Court's decision on smoking in prisons. AU: Vaughn,-Michael-S.; del-Carmen,-Rolando-V. SO: Federal-Probation; 57:34 S 1993 . . . , © 1991–1996 Public Affairs Information Services, Inc. © 1991–1996 SilverPlatter International, N.V.

interstate prison transfer did not violate prisoner's due process rights, citing statistics describing frequency of such transfers and percentage of prisoners incarcerated in the proximity of their homes). Fortunately, the U.S. government, through the U.S. Department of Justice, maintains excellent data collection and statistical reporting services that are of interest to criminal justice researchers.

The Federal Bureau of Investigation (FBI) conducts the **Uniform Crime Reports (UCR)** program. Under the UCR program, the FBI collects information on eight "index crimes" reported to law enforcement authorities throughout the United States: homicide, forcible rape, robbery, aggravated assault, burglary, larceny-theft, motor-vehicle theft, and arson. Arrest information is reported for an additional 21 crime categories. The UCR summarizes crime information for the whole country, and also provides information for particular regions, states, counties, cities, and towns. The UCR findings are published in an annual report, **Crime in the United States**. The annual report also includes information on crimes cleared, the age, sex, and race of persons arrested, law enforcement personnel either murdered or assaulted while on duty, and the characteristics of homicides (including age, sex, and race of offenders and victims, the relationship between the offender and victim, and types of weapons used). The UCR in addition publishes special reports based on the data collected through the program. Almost all college and university libraries contain at least the most recent volumes of **Crime in the United States**. Research libraries that have been designated as depository libraries for federal government publications should have not only **Crime in the United States** but also the special reports generated through the UCR program.

The federal government is increasingly relying on electronic means of distributing its information. Some of the UCR program information is accessible on a CD-ROM disk entitled **NESE**, for the National Economic, Social & Environmental data bank. In addition, the FBI has its own web site on the Internet, **http://www.fbi.gov.** Each year, prior to issuing the more complete annual report, the FBI publishes a preliminary report, which is available through the Internet site. The full annual reports, beginning with the 1995 issue, are also accessible through the Internet site.

The Bureau of Justice Statistics, another agency within the U.S. Department of Justice, conducts the **National Crime Victims Survey (NCVS)**. The NCVS collects information on crimes suffered by individuals and households, whether or not those crimes have been reported to law enforcement. The information is based on interviews conducted with all members (12 years of age or older) of a representative sample of U.S. households. Information is collected about the number and nature of several crimes—including rape, sexual assault, personal robbery, aggravated and simple assault, household burglary, theft, and motor vehicle theft. The survey includes information about the victims of crime (age, sex, race, ethnicity, marital status, income, and education level), known characteristics of offenders (sex, race, approximate age, and victim-offender relationship), and the crimes (time and place of crime, weapons used, nature of injury, and economic consequences). The Bureau of Justice Statistics publishes an annual report, *Criminal Victimization in the United States*, which is available in most college and university libraries. It also publishes special reports on topics such as crimes against women, the use of guns in the commission of crimes, and crime in urban areas. These specialized reports should be available at research libraries that serve as depositories for federal government publications. Many of them also will be available on the **NESE** CD-ROM and on the Bureau of Justice Statistics web site, **http://www.ojp.usdoj.gov/bjs/**, which provides access to Internet users.

The U.S. Department of Justice Bureau of Justice Statistics also publishes an annual compendium of crime-related statistics, *The Sourcebook of Criminal Justice Statistics (Sourcebook)*. The **Sourcebook** contains not only statistical information taken from the UCR and NCVS, but also many additional crime-related statistics adopted from both government and nongovernment sources. The **Sourcebook** has a subject index and also includes a section that provides bibliographic information and descriptions of the sources of the statistical information.

Some research issues require statistical information that is not strictly crime-related. For example, in our third hypothetical, concerning the prisoner who is seeking relief from a smoking cellmate, we might want to look for statistical evidence supporting the contention that second-hand smoke can lead to physical ailments. The federal government collects and reports statistical information on a wide variety of topics in an annual compendium, the *Statistical Abstract of the United States*. Most libraries have a copy of this book in their reference collection. Most of the information in the **Statistical Abstract** is presented in tables. References accompanying each table direct the user to the source of the information. The information is arranged by broad subject area, with a brief narrative introducing each section. A subject index also is provided in the back of each volume.

If you do not find helpful information in the ***Statistical Abstract of the United States***, you may want to consult the ***American Statistics Index (ASI)***. **ASI** is a commercial publication of the Congressional Information Service (CIS). Ten issues are published each year, with annual and five-year cumulations. There are two parts to **ASI**, the Index volume and the Abstract volume. **ASI** claims to provide references to all U.S. government publications that contain statistical information. Users should start with the Index volume, which includes a subject index as well as several others. Once a relevant entry is found in the Index volume, the researcher will be cross-referenced to an entry in the Abstract volume. The entry in the Abstract volume contains bibliographic information and a narrative description of the contents of the source document. Many of the documents referenced in **ASI**, but not all, will be contained in depository collections of federal documents. CIS also sells collections of the documents referenced in **ASI** on microfiche.

CIS publishes two other indexes to statistical information, **Statistical Reference Index (SRI),** and **International Statistics Index (ISI)**. **SRI** provides references to state government and nongovernment publications that contain statistical information. **ISI** primarily provides references to foreign publications with statistical information. Both are structured and accessed similarly to **ASI**. CIS also sells a CD-ROM version of these indexes, **Statistical Masterfile**.

Experienced researchers may want to create their own statistical studies based on the NCVS and UCR data files. Those data files are available through the **National Archive of Criminal Justice Data (NACJD)** at the University of Michigan. The NACJD was established in 1978 by the Bureau of Justice Statistics and the Inter-University Consortium for Political and Social Research (ICPSR). It includes many crime-related data files in addition to the NCVS and UCR files. A printed guide to this collection, Guide to Resources and Services, should be included in most research libraries' reference collections. ICPSR also has a website, **http://www.icpsr.umich.edu/**. Researchers who are affiliated with an institution belonging to ICPSR can obtain data files and code books from ICPSR at no charge. Other researchers must pay to obtain the data files and accompanying information. Data files and code books can be downloaded from the ICPSR ftp (file transfer protocol) site via the Internet.

CONCLUSION

Secondary legal authorities and the related social science and statistical materials we have reviewed in this chapter serve a number of different functions. The most limited secondary legal authorities are law dictionaries and the multivolume series, Words and Phrases. They primarily are useful to define legal terms. Words and Phrases extracts definitions of terms from case law. By citing the decisions on which these definitions are based, this reference work also helps direct the researcher to an important form of primary legal authority. Legal encyclopedias, treatises, casebooks, and A.L.R. annotations give helpful descriptions of legal issues and are useful as finding aids because of the numerous secondary and primary authorities they rely upon and cite. Court briefs filed by parties and *amici curiae* often are overlooked as a legal research tool. Because briefs typically are prepared by persons, organizations, or government bodies interested in the outcome of a case, they must be used with appropriate caution. Still, they can help illuminate the strengths and weaknesses of different positions taken on legal issues, and they often pull together scores of important and relevant references. A few good law review articles can be a tremendous boon to legal research. These writings, like no other secondary authorities, combine the descriptive, analytical, and finding aid functions that generally make it advisable to begin a research project by using secondary legal authorities.

The law does not exist in isolation from the criminal justice institutions that administer it, or from the crime-related problems to which lawmakers and law enforcement officials must respond. Thus, as you conduct legal research on criminal justice issues, you frequently will find useful social scientific and other empirical information related to crime, police, courts, corrections, and other aspects of the criminal justice system. For these reasons, we have provided brief descriptions of social science references and indexes, and statistical sources that may contribute to your research endeavors.

We also have commented in this chapter on matters that generally are important to the process of legal research—matters such as defining the issue you are investigating with appropriate breadth and specificity, being willing to practice and experiment with the described references by going to a library and actually using the books and computer databases, and having the grace to accept successes in your research tasks with equanimity and the doggedness not to be discouraged if your efforts do not meet with success as quickly as you would have liked. In Chapter 3 we explore the process of locating and using primary legal authorities. Secondary legal authorities cannot be cited and considered as binding sources of the law, but they can be indispensable for promoting your understanding of legal issues, by describing and analyzing them, and directing you to primary sources of law.

3 Primary Legal Authorities and Related Research Tools

INTRODUCTION

In this chapter we consider primary legal authorities and the tools and methods that will enable you to find these sources of law and confirm that they remain reliable and up to date. As we discussed in Chapter 1, primary legal authorities *are* "the law." They include administrative regulations, statutes, constitutions, and judicial decisions. Each of these sources of law is important in its own right. However, legal research generally cannot be brought to a conclusion until you have uncovered relevant judicial decisions, or case law. This attribute of the legal research process relates to the very nature of laws and the legal system in the United States.

Disputes about the proper application and interpretation of administrative regulations, statutes, and constitutional provisions are settled by the courts. Constitutions are written in purposefully broad and sometimes majestic language, as exemplified by the rights to due process and equal protection of the law, and freedom from cruel and unusual punishments. Administrative rules and statutes often are written imprecisely or have uncertain application to novel circumstances. Their exact contours typically take shape only after the courts have spoken. Legal research normally does not end until judicial decisions have been produced because these decisions tend to be the clearest and most definitive statement about what other sources of the law mean. Judges also "make law" independent of regulations, statutes, and constitutions by applying common law principles and building on established judicial precedent. Case law obviously must be researched when common law adjudication defines legal doctrine.

A legal research project can be started by directly consulting cases and other forms of primary legal authority. However, we hasten to remind you that beginning your research with primary authorities can be ill-advised, especially if you know relatively little about a topic. Such a strategy also can be very inefficient. As you contemplate finding out about the administrative regulations, statutes, constitutional provisions, and case law applicable to the issues you are investigating, do not ignore what you have just learned about secondary legal authorities. Law dictionaries, the Words and Phrases series, legal encyclopedias, treatises, case books, briefs, American Law Reports, and law review articles not only can en-

141

hance your understanding of primary authorities but also can help you locate them. Keep in mind that it often makes sense to begin your research by consulting secondary legal authorities.

We also consider important legal research tools in this chapter, including case digests and Shepard's citators. These materials are not legal authorities and should never be cited as such. Case digests organize and report brief excerpts or summaries of the principles of law established in judicial decisions. You use them to find leads to cases that address a particular legal issue. Then, you must look up the cases you find promising and read the decisions in full. Shepard's case citators allow you to learn about the judicial history of cases in which you are interested, to confirm that case decisions have not been overturned or eroded to the point that they no longer are good law, and to identify all subsequent decisions that have cited the cases you are Shepardizing. The latter service is extremely useful to help locate the most recent decisions discussing the issues in which you are interested. Other kinds of Shepard's citators serve analogous functions for different kinds of legal authorities, including administrative regulations, statutes, constitutions, and law review articles, among others.

We describe case digests and Shepard's in much greater detail later in the chapter. For now, it will suffice that you are aware that these references exist and that although they are indispensable for conducting legal research, neither is considered a form of secondary or primary legal authority. We begin by describing the different sources of law, or primary legal authorities, and then explain how to go about finding those authorities that are relevant to particular research problems. In the process, we continue to make use of the three case scenarios that we introduced in Chapter 2.

PRIMARY LEGAL AUTHORITIES AND RELATED FINDING TOOLS

Administrative Regulations

Federal, state, and local governments create administrative agencies to perform specific functions, such as regulating industries (e.g., the Food and Drug Administration), monitoring compliance with governmental policies (e.g., the Equal Employment Opportunities Commission), and resolving disputes and adjudicating special types of claims (e.g., the Veterans Administration). You probably recognize most federal administrative agencies by the initials by which they are known, such as the EEOC, EPA, FCC, FDA, FTC, HUD, ICC, and many, many more. State agencies also abound, including athletic commissions, health departments, motor vehicle bureaus, and gambling and lottery commissions. Locally created agencies and commissions include school boards, housing and zoning authorities, and water boards, among others.

The agencies in charge of these public services customarily develop rules to govern the performance of their duties. The rules of federal agencies are accessible in the references we discuss. Tremendous variations exist among the states and among different state agencies concerning the publication of operating rules and regulations, and the same is true at a local level. Our discussion will focus on federal regulations. Nevertheless, you should be aware that state and local administrative regulations also may apply to certain issues, and direct your research inquiries to the appropriate sources. Access to state administrative regulations, especially the regulations of states other than the one in which the researcher resides, can be problematic. However, the online database systems have been adding state administrative regulations to their coverage. As of 1996, WESTLAW provided full text access to the administrative codes for 22 states, while LEXIS provided full text access to 14.

Administrative regulations are binding on agency administrators and on parties whose activities are within their scope. They are enforceable in the courts—assuming, of course, that they conform to statutory and constitutional law, which are superior forms of authority. Administrative regulations are important in some criminal justice contexts, although you should not be surprised if they have no application to many issues.

Printed Sources

When a federal agency first issues a regulation or changes an existing one, this information is promptly published in the **Federal Register**, which is published Monday through Friday except for national holidays. It therefore organizes material chronologically rather than by subject matter. Such an organization would present significant problems for legal researchers if the Federal Register were the only source of federal administrative regulations. Fortunately, it is not.

The **Code of Federal Regulations (C.F.R.)** is a multivolume collection of federal agencies' administrative regulations. The C.F.R. organizes these regulations by subject matter and publishes them in appropriate "titles." For example, title 27 of the C.F.R. covers "Alcohol, Tobacco Products and Firearms"; title 28 pertains to "Judicial Administration"; and title 42 includes regulations governing matters of "Public Health." The C.F.R. titles are updated annually, when new volumes are issued to replace the prior year's set. The titles are replaced on a staggered basis: the first set of new volumes (i.e., titles 1-16) becomes available near the first of the year, while the last set (i.e., titles 42-50) may not be replaced until October of the same year. Regulations that are first published in the Federal Register later are incorporated into the appropriate title of the C.F.R. The C.F.R. contains current regulations. If a regulation is published in the Federal Register and subsequently is revised by a later regulation, only the revised version will appear in the C.F.R. Therefore, researchers conducting historical research must consult issues of the Federal Register if they want to locate the original text of the regulation.

Since the C.F.R. titles are only published annually, it is important to be able to determine whether federal regulations have changed between publication dates. One way to investigate for changes would be to comb the Federal Register each day. Indeed, the Federal Register routinely includes a section, "List of CFR Parts Affected," which identifies changes made in federal regulations and the impact those changes have in the existing C.F.R. There is at least a partial alternative to such a tedious process: At the end of each month, a separate pamphlet of C.F.R. is printed, called **"LSA—List of the CFR Sections Affected" (LSA)**. LSA collects all changes in federal administrative regulations and identifies the different titles and specific sections of the C.F.R. that have been affected. By using the monthly pamphlets in combination with the daily issues of the Federal Register, you can get the most timely information about changes and additions to the body of federal administrative regulations. Another way to search for changes is to check the index to the Federal Register that is issued monthly and cumulates throughout the course of the year. The index includes references by subject and by issuing agency. In addition, Congressional Information Services publishes a commercial index, the **Federal Register Index**, which provides detailed subject access. The Federal Register Index is also published monthly with semiannual cumulations.

The Code of Federal Regulations has its own index, which appears in a separate volume in the C.F.R. collection. The **C.F.R. Index** is arranged by subject, with specific subtopics listed under each major subject heading. Regulations are cited according to the title and section of C.F.R. in which they appear. A substantially similar subject index to C.F.R. is published by Lawyers Cooperative Publishing Company. It is the U.S. Code Service, Lawyers' Edition, **Index and Finding Aids to Code of Federal Regulations**. Since 1978, Congressional Information Services also has published a detailed subject index to the code, **Index to the Code of Federal Regulations**. This index is issued annually, with quarterly supplements.

With the assistance of our hypothetical cases, we can illustrate how C.F.R. is used. We should not expect agencies' administrative regulations to be relevant to all criminal justice issues. For instance, there is virtually no chance that administrative regulations will have anything to do with whether an individual who has tested positive for HIV, and who bites and spits on another person, can be prosecuted for attempted murder or assault with a deadly weapon (Case 2). Nor is it likely that the Federal Communications Commission or any other agency has issued regulations addressing testimony by closed-circuit television in judicial proceedings (Case 1). On the other hand, there is some possibility that a prison bureau or department of corrections regulation governs the cell assignments of smoking and nonsmoking prisoners (Case 3). Although Case 3 involves an aggrieved prisoner in a state penal institution (which would suggest that the particular state prison regulations are applicable, if any exist), we will invoke the privilege that attaches to all hypothetical fact situations and temporarily change the setting to a federal prison so we can investigate whether federal regulations speak to the issue at hand.

We begin our perusal with the C.F.R. Index. As we have done when using other subject indexes, we start with topics that generally describe the issue we are examining. Terms such as **"Prison," "Prisoner," "Smoke," "Tobacco," "Health,"** and others logically might relate to an administrative agency or regulation pertaining to the issue in Case 3. On checking these topics, the most promising references we find appear under **"Prisoners"** and **"Prisons Bureau."** Each of these subjects includes a subtopic, **"Institutional management, inmates,"** and under that subtopic another subdivision: **"Grooming, nondiscrimination, smoking, family planning, organizations, contributions, manuscripts, polygraph tests, and pre-trial inmates**, 28 CFR 551." The 1995 C.F.R. Index listings under **"Prisons Bureau"** are reproduced in Exhibit 3–1. Had we taken these same steps in the U.S.

Exhibit 3–1

Prisons Bureau	**CFR Index**

Traffic in contraband articles in Federal penal and correctional institutions, 28 CFR 6

Prisons Bureau

Admission of inmate to institution, 28 CFR 522

Classification of inmates, 28 CFR 524

Community programs, 28 CFR 570

Computation of sentence, 28 CFR 523

General management and administration

    Acceptance of donations, 28 CFR 504

    Central Office, regional offices, institutions, and staff training centers, 28 CFR 503

    Costs of incarceration fee, 28 CFR 505

    General definitions, 28 CFR 500

    General management policy, 28 CFR 511

    Records access, 28 CFR 513

    Scope of rules, 28 CFR 501

Institutional management, inmates

    Administrative remedy, 28 CFR 542

    Contact with persons in community, 28 CFR 540

Custody, 28 CFR 552

Discipline and special housing units, 28 CFR 541

Education, 28 CFR 544

Food service, 28 CFR 547

Grooming, nondiscrimination, smoking, family planning, organizations, contributions, manuscripts, polygraph tests, and pre-trial inmates, 28 CFR 551

Inmate property, 28 CFR 553

Legal matters, 28 CFR 543

Medical services, 28 CFR 549

Prison drug program, 28 CFR 550

Religious programs, 28 CFR 548

Work and compensation, 28 CFR 545

Parole, 28 CFR 572

Release from custody, 28 CFR 571

Research, 28 CFR 512

Transfers of inmate, 28 CFR 527

Source: Reprinted from CFR.

Code Service, Lawyers' Edition, Index and Finding Aids to Code of Federal Regulations, we would have been referred in precisely the same fashion to title 28 of the Code of Federal Regulations, section 551, or 28 C.F.R. 551 (1995).

It may be useful to become familiar with the general organization of the C.F.R. regulations before we turn to section 551 by consulting the broad outline at the start of the chapter containing these rules. We discover that the title 28 regulations pertain to the U.S. Bureau of Prisons, which is a subdivision of the Department of Justice. Section 551 of this title covers "**Miscellaneous**" matters dealing with "**Institutional Management**" in the federal prison system. (See Exhibit 3–2.) We next refer directly to section 551 and encounter a much more detailed outline of the specific miscellaneous federal Bureau of Prison regulations governing institutional management. We learn that **Subpart N** (sections 551.160 through 551.164) concerns "**Smoking/No Smoking Areas**." (See Exhibit 3–3.)

Finally, we turn to the sections that address smoking. (See Exhibit 3–4.) Section 551.163(c) of the regulations directs that: "To the maximum extent practicable nonsmoking inmates shall be housed in nonsmoking living quarters." While this regulation certainly does not determine whether the plaintiff in Case 3, Deborah Miller, has been subjected to cruel and unusual punishment by being incarcerated with a heavily smoking cellmate, it may be of some use in assessing

Exhibit 3–2

CHAPTER V—BUREAU OF PRISONS, DEPARTMENT OF JUSTICE

SUBCHAPTER A—GENERAL MANAGEMENT AND ADMINISTRATION

Part		*Page*
500	General definitions	393
501	Scope of rules	393
503	Bureau of Prisons central office, regional offices, institutions, and staff training centers	393
504	Acceptance of donations	396
505	Costs of incarceration fee	396
511	General management policy	398
512	Research	400
513	Access to records	404

SUBCHAPTER B—INMATE ADMISSION, CLASSIFICATION, AND TRANSFER

522	Admission to institution	406
523	Computation of sentence	408
524	Classification of inmates	412
527	Transfers	422

SUBCHAPTER C—INSTITUTIONAL MANAGEMENT

540	Contact with persons in the community	426
541	Inmate discipline and special housing units	443
542	Administrative remedy	476
543	Legal matters	478
544	Education	481
545	Work and compensation	490
547	Food service	495
548	Religious programs	496
549	Medical services	497
550	Drug programs	504
551	Miscellaneous	511
552	Custody	521

Source: Reprinted from CFR.

the reasonableness of the warden's nonresponsiveness to her request to be housed with a nonsmoker. As such, it may not be wholly irrelevant to the appropriate resolution of the issue. Having found this regulation, we would next check the cumulative LSA, as well as necessary issues of the Federal Register, to determine whether any changes have been enacted or whether additional relevant regulations have been adopted since the 1995 issue of the C.F.R. was published.

Exhibit 3–3

Bureau of Prisons, Justice **Pt. 551**

PART 551—MISCELLANEOUS

Subpart A—Grooming

Sec.
551.1 Policy.
551.2 Mustaches and beards.
551.3 Hairpieces.
551.4 Hair length.
551.5 Restrictions and exceptions.
551.6 Personal hygiene.
551.7 Bathing and clothing.

Subpart B—Marriages of Inmates

551.10 Purpose and scope.
551.11 Authority to approve a marriage.
551.12 Eligibility to marry.
551.13 Application to marry.
551.14 Special circumstances.
551.15 Furloughs.
551.16 Marriage ceremony in the institution.

Subpart C—Birth Control, Pregnancy, Child Placement, and Abortion

551.20 Purpose and scope.
551.21 Birth control.
551.22 Pregnancy.
551.23 Abortion.
551.24 Child placement.

Subpart D—Inmate Organizations

551.30 Purpose and Scope.
551.31 Approval.
551.32 Dues.
551.33 Meetings.
551.34 Fund-raising projects.
551.35 Special activities.
551.36 Accountability for funds.

Subpart E—Inmate Contributions

551.50 Policy.

Subpart F—Volunteer Community Service Projects

551.60 Volunteer community service projects.

Subpart G—Administering of Polygraph Test

551.70 Purpose and scope.
551.71 Procedures.

Subpart H—Inmate Manuscripts

551.80 Definition.
551.81 Manuscript preparation.
551.82 Mailing inmate manuscripts.
551.83 Limitations on an inmate's accumulation of manuscript material.

Subpart I—Non–Discrimination Toward Inmates

551.90 Policy.

Subpart J—Pretrial Inmates

551.100 Purpose and scope.
551.101 Definitions.
551.102 Commitment prior to arraignment.
551.103 Procedure for admission.
551.104 Housing.
551.105 Custody.
551.106 Institutional employment.
551.107 Pretrial inmate reviews.
551.108 Performance pay.
551.109 Community activities.
551.110 Religious programs.
551.111 Marriage.
551.112 Education.
551.113 Counseling.
551.114 Medical, psychiatric and psychological.
551.115 Recreation.
551.116 Discipline.
551.117 Access to legal resources.
551.118 Property.

continues

Exhibit 3–3   continued

551.119   Release of funds and property of pretrial inmates.

551.120   Visiting.

Subparts K–L—(Reserved)

Subpart M—Victim and/or Witness Notification

551.150   Purpose and scope.

551.151   Definitions.

551.152   Procedures.

551.153   Canceling the notification request.

Subpart N—Smoking/No Smoking Areas

551.160   Purpose and scope.

551.161   Definitions.

551.162   Designated no smoking areas.

Source: Reprinted from CFR.

551.163   Designated smoking areas.

551.164   Notice of smoking areas.

AUTHORITY: 5 U.S.C. 301; 18 U.S.C. 1512, 3621, 3622, 3624, 4001, 4005, 4042, 4081, 4082 (Repealed in part as to offenses committed on or after November 1, 1987), 4161–4166 (Repealed as to offenses committed on or after November 1, 1987), 5006–5024 (Repealed October 12, 1984 as to offenses committed after that date), 5039; 28 U.S.C. 509, 510; Pub. L. 99–500 (sec. 209); 28 CFR 0.95–0.99; Attorney General's August 6, 1991 Guidelines for Victim and Witness Assistance.

SOURCE: 44 FR 38252, June 29, 1979, unless otherwise noted.

Computer-Assisted Legal Research Strategies

Both **LEXIS®** and **WESTLAW** provide full-text access to the Federal Register and the C.F.R. The federal government has made the text of the Federal Register available through its **GPO ACCESS** program on the Internet. The GPO ACCESS is retrospective to 1994 and can be reached through the following uniform resource locator (URL), **http://www.gpo.gov**. The online databases can be most effective in searching either for very current regulations or for regulations dealing with particular factual information not indexed in the printed references.

The printed sources worked well for our hypothetical case involving tobacco smoke in prison living quarters. Below we demonstrate how to search the same topic in the C.F.R. through the LEXIS service. We will be using the FREESTYLE™ feature, which is the LEXIS equivalent of **natural language** searching on WESTLAW. Through the FREESTYLE™ feature we do not have to know how to use proximity operators or how to restrict our search to a particular field or segment of the document; we merely have to enter the key words that describe our topic and follow a few simple procedures to hone our search strategy.

We begin by picking out the key terms that describe our topic. In this example, the key terms include **"second hand,"** **"smoke,"** **"tobacco,"** **"prison,"** **"inmates,"** and **"health."** Our next step is to enter the appropriate database, or file, in the LEXIS service. In Chapter 2 we explained the two-step procedure that is used to access a particular database or file on LEXIS. First, select the appropriate "library"—in this instance **GENFED** (which contains various files including federal legal information). Once in the appropriate library, select the file to search. In this

Exhibit 3–4

Bureau of Prisons, Justice § 551.64

Subpart N—Smoking/No Smoking Areas

SOURCE; 59 FR 34742, July 6, 1994, unless otherwise noted.

§ 551.160 Purpose and scope.

To advance towards becoming a clean air environment and to protect the health and safety of staff and inmates, the Bureau of Prisons will restrict areas and circumstances where smoking is permitted within its institutions and offices.

§ 551.161 Definitions.

For purpose of this subpart, smoking is defined as carrying or inhaling a lighted cigar, cigarette, pipe or other lighted tobacco products.

§ 551.162 Designated no smoking areas.

All areas of Bureau of Prisons facilities and vehicles are no smoking areas unless specifically designated as a smoking area by the Warden as set forth in § 551.163.

§ 551.163 Designated smoking areas.

(a) At all medical referral centers, including housing units, and at minimum security institu-

tions, including satellite camps and intensive confinement centers, the Warden shall identify "smoking areas", ordinarily outside of all buildings and away from all entrances so as not to expose others to second-hand smoke.

(b) At all low, medium, high, and administrative institutions other than medical referral centers, the Warden shall identify outdoor smoking areas and may, but is not required to, designate a limited number of indoor smoking areas where the needs of effective operations so require, especially for those who may be employed in, or restricted to, a nonsmoking area for an extended period of time.

(c) To the maximum extent practicable nonsmoking inmates shall be housed in nonsmoking living quarters.

§ 551.164 Notice of smoking areas.

The Warden shall ensure that smoking areas are clearly identified by the appropriate placement of signs. The *absence* of a sign shall be interpreted as indicating a no smoking area. Appropriate disciplinary action shall be taken for failure to observe smoking restrictions.

Source: Reprinted from CFR.

---

case the file is "CFR," which contains the text of the current C.F.R. Follow the directions at the bottom of the query entry screen and **enter .fr** to switch from the Boolean searching mode to the FREESTYLE searching mode. Next, **enter** the words that describe the search topic, as illustrated in Exhibit 3–5.

The FREESTYLE feature prompts the researcher to search for other concepts related to the search terms, as depicted in Exhibit 3-6. This is very similar to the process of checking the **thesaurus** in **natural language** searching on WESTLAW: Simply follow the directions. For example, concepts related to the term "**smoke**" are displayed on the **Synonym Selection** screen depicted in Exhibit 3–7. Notice that the FREESTYLE feature differs from WESTLAW's **natural language** searching in that the root forms of words are not automatically truncated. Thus, additional terms such as "**smoked**," "**smoker**," and "**smoking**" must be added to the search strategy, as illustrated in Exhibit 3–7.

This process can be repeated to add synonyms for other terms in the search strategy. After this process is complete, return to the query screen and the modi-

Exhibit 3–5

---

second hand tobacco smoke and prison inmates health

Enter your FREESTYLE (TM) Search Description.
Enter phrases in quotation marks.
Example: What are the requirements for a "day care center" license?

Type .bool to exit FREESTYLE and run a Boolean search.

For further explanation, press the H key (for HELP) and then the ENTER key.

Source: CFR, "Tobacco Smoke." Reprinted with the permission of LEXIS-NEXIS, a division of Reed Elsevier Inc. LEXIS and NEXIS are registered trademarks of Reed Elsevier Properties Inc. FREESTYLE, KWIC, SuperKWIC and MEGA are trademarks of Reed Elsevier Properties Inc. SHEPARD'S and SHEPARDIZE are registered trademarks of Shepard's Company, a Partnership.

---

fied search strategy will be depicted. (See Exhibit 3–8.) Notice near the bottom of the screen that other options are available for further refining the search strategy. Those additional options, especially the date restriction, can be helpful in fine-tuning search strategies. At the bottom of the screen, observe that the **Current setting** is "**25**." This indicates that the default option in the FREESTYLE feature retrieves 25 documents. Recall that the default option in **natural language** retrieves 20 documents. The retrieval process used by the FREESTYLE feature operates in a way that is similar to the one used by **natural language**. Documents are retrieved in order of relevancy based on the number of occurrences of a term, with added weight given to terms that appear rarely throughout the entire database. You should be aware that when parentheses are used in a FREESTYLE search, the term appearing immediately in front of the parentheses and all terms within the parentheses are treated as if they had the Boolean operator **or** between them. Phrases within quotation marks are searched for as complete phrases.

There are six elements to the search strategy displayed in Exhibit 3-8:

1. "second hand"
2. tobacco
3. smoke or smoked or smoker or smoking
4. prison or prisoner or "house of reform" or penitentiary
5. inmates or prisoner
6. health

The FREESTYLE feature reads the search query as if the Boolean operator **and** appears between each element of the query. The system searches for documents

Exhibit 3–6

2,3,4,5   TERM SELECTION

Enter numbers for related concepts or synonyms. Example: 1,2,3,6–10

(=1)   Return to Search Options

1   Related concepts for your search description

Search Terms found in thesaurus
– – – – – – – – – – – – – –
2   SMOKE
3   PRISON
4   INMATES
5   HEALTH

For further explanation, press the H key (for HELP) and then the ENTER key.

Source: CFR, "Tobacco Smoke." Reprinted with the permission of LEXIS-NEXIS, a division of Reed Elsevier Inc. LEXIS and NEXIS are registered trademarks of Reed Elsevier Properties Inc. FREESTYLE, KWIC, SuperKWIC and MEGA are trademarks of Reed Elsevier Properties Inc. SHEPARD'S and SHEPARDIZE are registered trademarks of Shepard's Company, a Partnership.

Exhibit 3–7

1,2,3   SYNONYM SELECTION

Synonyms for: SMOKE
Enter synonym numbers to include in search and press ENTER (e.g. 1,2,3–4)

(=1)   Return to Search Options   (=2)   Return to Term Selection

– – – – – – – – – – – – – – – – Term Variations – – – – – – – – – – – – – – – –
    1   smoked   2   smoker   3   smoking
– –
    4   butt   5   cig   6   cigarette
    7   coffin nail   8   fag   9   gasper
   10   pill   11   skag

For further explanation, press the H key (for HELP) and then the ENTER key.

Source: CFR, "Tobacco Smoke." Reprinted with the permission of LEXIS-NEXIS, a division of Reed Elsevier Inc. LEXIS and NEXIS are registered trademarks of Reed Elsevier Properties Inc. FREESTYLE, KWIC, SuperKWIC and MEGA are trademarks of Reed Elsevier Properties Inc. SHEPARD'S and SHEPARDIZE are registered trademarks of Shepard's Company, a Partnership.

Exhibit 3–8

FREESTYLE(TM) SEARCH OPTIONS

Press ENTER to start search.
To use a Search Option, enter an equal sign followed by the number.
Search Description:
    "SECOND HAND" TOBACCO SMOKE (SMOKED SMOKER SMOKING) PRISON (PRISONER
    "HOUSE OF REFORM" PENITENTIARY) INMATES (PRISONER) HEALTH

Press ENTER to start search.
(=1)   Edit Search Description
(=2)   Enter/edit Mandatory Terms
(=3)   Enter/edit Restrictions (e.g., date)
(=4)   Synonyms and Related Concepts
(=5)   Change number of documents Current setting: 25

For further explanation, press the H key (for HELP) and then the ENTER key.

Source: CFR, "Tobacco Smoke." Reprinted with the permission of LEXIS-NEXIS, a division of Reed Elsevier Inc.
LEXIS and NEXIS are registered trademarks of Reed Elsevier Properties Inc. FREESTYLE, KWIC, SuperKWIC and
MEGA are trademarks of Reed Elsevier Properties Inc. SHEPARD'S and SHEPARDIZE are registered trademarks of
Shepard's Company, a Partnership.

---

that contain all six elements, but if there are not 25 documents in the database that contain all six elements, the computer will start dropping elements from the query in order to retrieve 25 documents. After we enter the query, the screen depicted in Exhibit 3-9 is retrieved.

There are several ways of browsing through the retrieved results. The default option is **Key Word in Context (KWIC™)**, which scans the documents and stops at points where one of the search terms appears. The search term is displayed, along with 25 words of text appearing on either side of it. The bibliographic citations to the 25 documents are presented by **entering** the command **.ci**.

Exhibit 3-l0 displays the first two citations. The second citation is a reference to the same part of the C.F.R. that we located by using the printed sources, 28 C.F.R. 551.160. We can retrieve the text of that document by **entering** number "2," but before we do so, note the four commands near the bottom of the screen in Exhibit 3–9. Researchers can use these commands to work more efficiently through the search results.

Exhibit 3–11 displays the first screen of 28 C.F.R. 551.160. At this point, the researcher **enters .fu** to make sure that the full, or complete, text of the document is presented. Then the document can be browsed by using the displayed commands. The relevant part actually appears at 28 C.F.R. 551.163. The documents consist of discrete parts, so the **Next Doc (.ND)** command is used to reach § 551.163, which is displayed in Exhibit 3–12.

Exhibit 3–9

.ci
 Your FREESTYLE search has retrieved the top 25 documents based on statistical ranking. Search terms are listed in order of importance.

PENITENTIARY "SECOND HAND" SMOKER SMOKED PRISONER PRISONER SMOKING INMATES SMOKE PRISON TOBACCO HEALTH "HOUSE OF REFORM"

Press ENTER to view documents in KWIC or use Full, Cite or Segment keys.

(=1) Browse documents in SuperKWIC (.SK)
(=2) Location of search terms in documents (.where)
(=3) Number of documents with search terms (.why)
(=4) Change document order (.sort)

For further explanation, press the H key (for HELP) and then the ENTER key.

Source: CFR, "Tobacco Smoke." Reprinted with the permission of LEXIS-NEXIS, a division of Reed Elsevier Inc. LEXIS and NEXIS are registered trademarks of Reed Elsevier Properties Inc. FREESTYLE, KWIC, SuperKWIC and MEGA are trademarks of Reed Elsevier Properties Inc. SHEPARD'S and SHEPARDIZE are registered trademarks of Shepard's Company, a Partnership.

One of the advantages of doing this search online is that much of the tedious process of ensuring that the regulation has not been amended can be avoided. This search was executed in early July 1996. At that time, LEXIS had updated the C.F.R. through June 19, 1996. Only the intervening two weeks of the Federal Register would have to be checked to make sure that no recent changes had been made to the regulation. This update can be done in LEXIS by changing to the Federal Register database, FEDREG file. You then are given the option of running the existing query in the Federal Register database. (See Exhibit 3–13.) A date restriction can be added by entering =**3**, which displays the screen depicted in Exhibit 3–14. Note the date restriction typed at the upper left portion of the screen, **aft 06/19/96**.

Exhibit 3–15 displays the new query screen with the date restriction added. By entering this query, we are able to determine that no changes had been made in the regulation after June 19, 1996.

Statutes

Much of the law governing the criminal justice system is statutory. Crimes are defined by statute. Pretrial, trial, appellate, and other post-conviction judicial procedures, as well as rules of evidence, typically stem from legislation. Sentencing ranges, parole eligibility, sex offender registration requirements, capital punish-

Exhibit 3–10

To be able to browse preceding or succeeding code sections, enter B. The first page of the document you are currently viewing will be displayed in FULL.

- -

LEVEL 1 - 25 SECTIONS

1. 41 CFR 101-20.105-3, TITLE 41 — PUBLIC CONTRACTS AND PROPERTY MANAGEMENT, SUBTITLE C — FEDERAL PROPERTY MANAGEMENT REGULATIONS SYSTEM, CHAPTER 101 - FEDERAL PROPERTY MANAGEMENT REGULATIONS, SUBCHAPTER D — PUBLIC BUILDINGS AND SPACE, PART 101-20 — MANAGEMENT OF BUILDINGS AND GROUNDS, SUBPART 101-20.1 — BUILDING OPERATIONS, MAINTENANCE, PROTECTION, AND ALTER-ATIONS, 101-20.10 PHYSICAL PROTECTION AND BUILDING SECURITY., 101-20.105-3 Smok-ing., CODE OF FEDERAL REGULATIONS

2. 28 CFR 551.160, TITLE 28 — JUDICIAL ADMINISTRATION, CHAPTER V — BUREAU OF PRISONS, DEPARTMENT OF JUSTICE, SUBCHAPTER C — INSTITUTIONAL MANAGEMENT, PART 551 — MISCELLANEOUS, SUBPART N — SMOKING/NO SMOKING AREAS, 551.160 Pur-pose and Scope., CODE OF FEDERAL REGULATIONS

- -

.MORE	Next Page	.NP	Cite	.CI	Exit FREESTYLE	.BOOL	Print Doc	.PR
.WHERE	Prev Page	.PP	Kwic	.KW	New Search	.NS	Print All	.PA
.WHY	Next Doc	.ND	Full	.FU	Modify	.M	Cmds Off	.COF
.SORT	Prev Doc	.PD	SKWIC	.SK	Chg Library	.CL	Sign Off	.SO

Source: CFR, "Tobacco Smoke." Reprinted with the permission of LEXIS-NEXIS, a division of Reed Elsevier Inc. LEXIS and NEXIS are registered trademarks of Reed Elsevier Properties Inc. FREESTYLE, KWIC, SuperKWIC and MEGA are trademarks of Reed Elsevier Properties Inc. SHEPARD'S and SHEPARDIZE are registered trademarks of Shepard's Company, a Partnership.

ment, and other corrections policies are fixed by statute. The juvenile courts and the rest of the juvenile justice system are creations of the legislature. Civil proce-dures, discovery requirements, civil commitment standards, eligibility for jury service, and countless related matters have statutory origins. It comes as no surprise that Congress and the state legislatures are responsible for many substantive criminal justice policies and the procedures for their administration. Fortunately, statutory materials are relatively easy to locate and convenient to use. They also can be very helpful in directing us to related case law and other references.

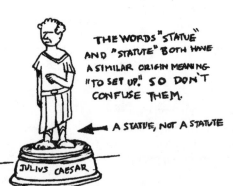

THE WORDS "STATUE" AND "STATUTE" BOTH HAVE A SIMILAR ORIGIN MEANING "TO SET UP." SO DON'T CONFUSE THEM.

◄— A STATUE, NOT A STATUTE

JULIUS CAESAR

Printed Sources: Case 1

Before you check a statute book, you must identify the jurisdiction appropriate for the issue you are re-

Exhibit 3–11

LEVEL 1 — 2 OF 25 SECTIONS

MICHIE'S CODE OF FEDERAL REGULATIONS
Copyright (c) 1996, Michie

*** THIS SECTION IS CURRENT THROUGH THE 6/19/96 ISSUE OF ***
*** THE FEDERAL REGISTER ***

TITLE 28—JUDICIAL ADMINISTRATION
CHAPTER V—BUREAU OF PRISONS, DEPARTMENT OF JUSTICE
SUBCHAPTER C—INSTITUTIONAL MANAGEMENT
PART 551—MISCELLANEOUS
SUBPART N—SMOKING/ NO SMOKING AREAS

.MORE	Next Page	.NP	Cite	.CI	Exit FREESTYLE	.BOOL	Print Doc	.PR
.WHERE	Prev Page	.PP	Kwic	.KW	New Search	.NS	Print All	.PA
.WHY	Next Doc	.ND	Full	.FU	Modify	.M	Cmds Off	.COF
.SORT	Prev Doc	.PD	SKWIC	.SK	Chg Library	.CL	Sign Off	.SO

Source: CFR, "Tobacco Smoke." Reprinted with the permission of LEXIS-NEXIS, a division of Reed Elsevier Inc. LEXIS and NEXIS are registered trademarks of Reed Elsevier Properties Inc. FREESTYLE, KWIC, SuperKWIC and MEGA are trademarks of Reed Elsevier Properties Inc. SHEPARD'S and SHEPARDIZE are registered trademarks of Shepard's Company, a Partnership.

Exhibit 3–12

28 CFR 551.163

551.163 Designated smoking areas.

(a) At all medical referral centers, including housing units, and at minimum security institutions, including satellite camps and intensive confinement centers, the Warden shall identify "smoking areas", ordinarily outside of all buildings and away from all entrances so as not to expose others to second-hand smoke.

(b) At all low, medium, high, and administrative institutions other than medical referral centers, the Warden shall identify outdoor smoking areas and may, but is not required to, designate a limited number of indoor smoking areas where the needs of effective operations so require, especially for those who may be employed in, or restricted to, a nonsmoking area for an extended period of time.

(c) To the maximum extent practicable nonsmoking inmates shall be housed in nonsmoking living quarters.

Next Page	.NP	Next Doc	.ND	Print Doc	.PR		
Prev Page	.PP	Prev Doc	.PD	Print ALL	.PA		
1st Page	.FP	1st Doc	.FD	Cmds Off	.COF		
Focus	.FO	Sel Serv	.SS	View Log	.LOG	Sign Off	.SO

Source: CFR, "Tobacco Smoke." Reprinted with the permission of LEXIS-NEXIS, a division of Reed Elsevier Inc. LEXIS and NEXIS are registered trademarks of Reed Elsevier Properties Inc. FREESTYLE, KWIC, SuperKWIC and MEGA are trademarks of Reed Elsevier Properties Inc. SHEPARD'S and SHEPARDIZE are registered trademarks of Shepard's Company, a Partnership.

Exhibit 3–13

FREESTYLE(TM) SEARCH OPTIONS

This was your last search before changing library/file. Press ENTER to start search or .ns for a new search. Select an option below to edit search.

Search Description:
"SECOND HAND" TOBACCO SMOKE (SMOKED SMOKER SMOKING) PRISON (PRISONER "HOUSE OF REFORM" PENITENTIARY) INMATES (PRISONER) HEALTH

Press ENTER to start search.
(=1)   Edit Search Description
(=2)   Enter/edit Mandatory Terms
(=3)   Enter/edit Restrictions (e.g., date)
(=4)   Synonyms and Related Concepts
(=5)   Change number of documents Current setting: 25

For further explanation, press the H key (for HELP) and then the ENTER key.

Source: Federal Register, "Tobacco Smoke." Reprinted with the permission of LEXIS-NEXIS, a division of Reed Elsevier Inc. LEXIS and NEXIS are registered trademarks of Reed Elsevier Properties Inc. FREESTYLE, KWIC, SuperKWIC and MEGA are trademarks of Reed Elsevier Properties Inc. SHEPARD'S and SHEPARDIZE are registered trademarks of Shepard's Company, a Partnership.

Exhibit 3–14

---

aft 06/19/96

Enter the DATE restriction and press ENTER.   (=0)   Delete restriction.
Example: AFT 05/16/58

(=1) DATE:

(=2) HEADING:

(=3) TEXT:

(=4) NUMBER:

(=5) ADDRESS:

For further explanation, press the H key (for HELP) and then the ENTER key.

Source: Federal Register, "Tobacco Smoke." Reprinted with the permission of LEXIS-NEXIS, a division of Reed Elsevier Inc. LEXIS and NEXIS are registered trademarks of Reed Elsevier Properties Inc. FREESTYLE, KWIC, SuperKWIC and MEGA are trademarks of Reed Elsevier Properties Inc. SHEPARD'S and SHEPARDIZE are registered trademarks of Shepard's Company, a Partnership.

Exhibit 3–15

FREESTYLE (TM) SEARCH OPTIONS

Press ENTER to start search.
To use a Search Option, enter an equal sign followed by the number.
Search Description:
  "SECOND HAND" TOBACCO SMOKE (SMOKED SMOKER SMOKING) PRISON (PRISONER
  "HOUSE OF REFORM" PENITENTIARY) INMATES (PRISONER) HEALTH

Restrictions: DATE (AFT 06/19/96)

Press ENTER to start search.
(=1)   Edit Search Description
(=2)   Enter/edit Mandatory Terms
(=3)   Enter/edit Restrictions (e.g., date)
(=4)   Synonyms and Related Concepts
(=5)   Change number of documents Current setting: 25

For further explanation, press the H key (for HELP) and then the ENTER key.

Source: Federal Register, "Tobacco Smoke." Reprinted with the permission of LEXIS-NEXIS, a division of Reed
Elsevier Inc. LEXIS and NEXIS are registered trademarks of Reed Elsevier Properties Inc. FREESTYLE, KWIC,
SuperKWIC and MEGA are trademarks of Reed Elsevier Properties Inc. SHEPARD'S and SHEPARDIZE are regis-
tered trademarks of Shepard's Company, a Partnership.

---

searching. Obviously, if a homicide occurs in Alabama, you will have little interest in consulting the laws of Michigan to determine whether the offense appears to
be murder or manslaughter, or whether a defense may be available. Each of the 50
states, the federal government, and the District of Columbia have their own collections of statutes, and you must start with the appropriate volumes.

The official reference for federal statutes is the **United States Code (U.S.C.)**.
We mentioned in Chapter 1, and we shall reiterate later in this chapter, that when
dual sets of case reporters exist, the official reporters are not nearly as useful for
legal research as are the unofficial case reports. The same principle applies to
statutory volumes. You will find that both the **United States Code Annotated
(U.S.C.A.)**, which is published by West, and the **United States Code Service
(U.S.C.S.)**, published by Lawyers Cooperative, are far more helpful sources of
federal statutes than is the United States Code.

One reason for this is that both of the unofficial sets of statutes are annotated,
which in this context means that they include brief descriptions or notes of judicial decisions—commonly known as "squibs" or "blurbs"—in which sections of
the statutes have been applied and interpreted. Furthermore, both U.S.C.A. and
U.S.C.S. provide more helpful background information about statutes. They describe legislative history in greater detail than does the U.S. Code. They also cite
additional reference materials that will promote your understanding of the legislation.

All states have their own sets of annotated statutes. Like their federal counterparts, annotated state statutes identify and provide brief synopses of court decisions involving the legislation, describe legislative histories, and sometimes offer other explanatory commentary and list related references. The states' statutory volumes are published by many different companies, so they are not uniform. Nevertheless, all are compiled and used in basically similar ways.

There are several different ways to uncover relevant statutes. The most common beginning point is with the index that accompanies the statutory volumes, which is arranged alphabetically by subject. When you look for specific statutes, such as the murder statute in Alabama, it often helps to think about the general categories that may include your topic. For instance, if you have no luck when you look for "**Murder**" in the index to the Code of Alabama, you should try broader headings such as "**Homicide**" or "**Crimes and Offenses**."

Another way of finding statutes is through using popular-name tables. For example, U.S.C.A. and U.S.C.S. have popular-name tables that cite federal statutory provisions when you look under the name by which the statute commonly is known (e.g., "RICO," the acronym for Racketeer Influenced and Corrupt Organizations, or the Jencks Act). And do not forget that secondary authorities, such as legal encyclopedias, treatises and case books, briefs, A.L.R.s, and law reviews, frequently will direct you to relevant statutes.

We illustrate how research involving state and federal statutory materials can be conducted with the assistance of our three hypothetical case scenarios. For these purposes, we will assume that Case 1, which presents the confrontation issue in John Winston's trial for the sexual abuse of his stepdaughter, arose in state court in New York. We will place Case 2, in which a charge of attempted murder or assault with a deadly weapon may be lodged against Andrew Adams for biting and spitting on Officer Fiegel when Adams knew that he had tested positive for HIV, in the Indiana state courts. In Case 3, Deborah Miller has filed a federal lawsuit alleging that she is being subjected to cruel and unusual punishment by having to breathe the second-hand smoke produced by her prison cellmate, who smokes two or more packs of cigarettes a day. Accordingly, our search for relevant statutory authority in these cases will focus on New York, Indiana, and federal law.

Our research for Case 1 will require us to refer to New York's statutes. Both West and Lawyers Cooperative publish versions of New York statutes. The organization of both sets is similar, and the numbering and the text of the sections of the laws are identical. The two sets differ only in the supplementary material included and the cross-referencing features, which are unique to each publisher. The search process is virtually identical in each set. West's version is called **McKinney's Consolidated Laws of New York Annotated**, and Lawyers Cooperative's set is entitled **New York Consolidated Law Service**. We will use McKinney's in our example.

We begin our research for Case 1 by referring to the General Index volumes of McKinney's, which are arranged alphabetically by subject. We start by identifying search terms describing the issue in the case. When we look first under "Witnesses" and then search for relevant subtopics, we find: "**Television, . . . —child witnesses, closed-circuit use. Infants, generally, this index**." (See Exhibit 3–16.) Similarly, if we initially had consulted "**Television**" in the General Index, we

Exhibit 3–16

WITNESSES—Cont'd
Tampering with—Cont'd
Fourth degree, definition of and classification as
  class A misdemeanor, **PEN 215.10**
Taxes, state board and office of real property
  services, attendance, **RPTL 204**
Telegraphs and telephones, commission's power,
  PUB S 94
Television,
  Cable television, commission, compelling
    attendance of, **PUB S 216**
  Child witnesses, closed-circuit, use. Infants,
    generally, this index

Testamentary trustees, release, claim against
  state, appropriation of trust property,
  testimony by, **SCPA 1508**
Testimonial capacity, **CPL 60.20**
Theatrical syndication financing, investigations
  by attorney general, power to compel
  attendance, examine, etc., **ARTS&CA 23.05**
Town comptroller, proceedings to audit claims,
  TOWN 119
Town courts, single town court, establishing,
  petition form, statement of, **UJCA 106 a**

Source: McKinney's Consol. Laws, Index "Witnesses" p. 733, © West Publishing, used with permission.

would have found the subtopics "**Witnesses, Child witness, closed-circuit use. Infants, generally, this index**," and "**Witnesses, . . . —Infants, closed-circuit television use. Infants, generally, this index**". (See Exhibit 3–17.) Had we begun by looking under the subject, "**Children**," we again would have been referred to "**Infants**." When we consult "**Infants**" in the General Index, under the subtopic "**Witnesses**" and the further subtopic "**Closed-circuit television**," we find a number of discrete subheadings with accompanying citations to specific sections of New York statutes. (See Exhibit 3–18.)

New York's statutes are organized into over 90 different categories, such as Corrections Law, Criminal Procedure Law, Judiciary Law, Penal Law, Vehicle and Traffic Law, and many more. Thus, when you see the subheading, "Defendant remaining in courtroom while child witness testifies from other room, determination to avoid mental or emotional harm, **CPL 65.20**," the reference is to New York's Criminal Procedure Law, section 65.20. Other relevant references are made to sections 65.00, 65.l0, and 65.30 of the Criminal Procedure Law.

We locate the Criminal Procedure Law volumes in McKinney's Consolidated Laws of New York Annotated, and turn to sections 65.00 through 65.30. Section 65.20 of New York's Criminal Procedure Law is a lengthy provision detailing the procedures for using closed-circuit television to present the testimony of a vulnerable child witness and the factors a court should consider when deciding whether to make use of these procedures. We set forth portions of this statute in Exhibit 3–19 and also include the Historical and Statutory Notes that follow the statute, which provide important information about the statute's legislative history. For example, note that subdivisions l0 and 11 of the statute were amended in 1991 and that the statute, which became effective July 24, 1985, has an expiration date of

Exhibit 3–17

TELEVISION—Cont'd

Public service commission. Cable television, generally, ante, this heading

Public Television and Radio, generally, this index

Radio and Television Tubes, generally, this index

Recording, unauthorized, offenses relating to. Records and Recording, generally, this index

Regional college cooperative services boards, plans for use of, **EDUC 498**

Repair ships, personal property, repaired, rebuilt or reconditioned, receipts, **GEN B 597**

School buildings, etc., admission of news media, conditions and uses, **ARTS&CA 61.09**

Sound Recordings, generally, this index

State agencies. Cable television, ante, this heading

Subdivisions, filling fee, offer to sell or lease subdivided lands by means of advertisements, **RPL 337-b**

"Telecommunications equipment" as not including television entertainment, etc., real property tax, general provisions, **RPTL 102**

TELEVISION—Cont'd

Theatrical Employment Agencies, generally, this index

Theatrical Engagement or Employment, generally, this index

Toxic substances, outreach programs, distribution of public service announcements to stations, **PUB HE 4804**

Trust for cultural resources. Cultural Resources, generally, this index

Unauthorized recordings, offenses relating to. Records and Recording, generally, this index

Urban development corporation, regional revolving loan trust fund, exception, financial assistance, **UNCON 6266-a**

Video Tapes, generally, this index

Witnesses,

Child witness, closed-circuit, use. Infants, generally, this index

Infants, closed-circuit television, use. Infants, generally, this index

Source: McKinney's Consol. Laws, Index "Television" p. 70, © West Publishing, used with permission.

November 1, 1996. The Practice Commentaries following the statute, which we reprint in part, are not a part of the legislation; they are provided as an explanatory aid for users.

Other important research aids accompany the statute. A section listing Library References identifies an article in West's legal encyclopedia, C.J.S., that we may wish to consult. (See Exhibit 3–20.) Note that because New York's statutory volumes are published by West, neither Am. Jur. 2d nor other Lawyers Cooperative references are mentioned. The Library References section also provides some case-finding tips under the American Digest System and WESTLAW Research subheadings. These tips will be more understandable when we discuss case law research.

The biggest research boost we get comes from the Notes of Decisions, which you see displayed in Exhibit 3–20. Here, we are provided with a short outline of topics associated with the statute. Outlines are significantly longer and more detailed for some other statutes. The identified topics are discussed in corresponding court decisions, as indicated in the notes. For example, if we are interested in cases

Exhibit 3–18

INFANTS—Cont'd
Witnesses—Cont'd

Child more or less than twelve years old, mental defect, understanding oath, **CPL 60.20**

Closed-circuit television, use, **EXEC 642-a**

A person occupies a position of authority with respect to a child, defined, **CPL 65.00**

Age of child, factor to consider in determining possible mental or emotional harm to child, **CPL 65.20**

Answering papers in motion for order declaring child witness vulnerable, contents, requirements, **CPL 65.20**

Application, grounds, **CPL 65.20**

Chief administrator of courts, monitor, report to chief judge, governor, and legislature, **JUD 216**

Child and defendant in same room during testimony, determination whether such will contribute to mental or emotional harm, **CPL 65.20**

Child witness, defined, **CPL 65.00**

Deadly weapon or dangerous instrument used, factor to consider in determining possible mental or emotional harm to child, **CPL 65.20**

INFANTS—Cont'd
Witnesses—Cont'd

Declaration of vulnerability, **CPL 65.10**

Defendant living with or has access to child, factor in determining possible mental or emotional harm to child, **CPL 65.20**

Defendant remaining in courtroom while child witness testifies from other room, determination to avoid mental or emotional harm, **CPL 65.20**

Defendant's image transmitted to child in testimonial room, **CPL 65.30**

Demand to furnish records, reports, etc., **CPL 65.20**

Discovery upon trial or prior statements and criminal history of witnesses, district attorney to comply with, **CPL 65.20**

Equipment, placed in position limiting child witness from seeing or hearing operator, **CPL 65.30**

Expert testimony that child susceptible to psychological harm if required to testify in open court or in presence of defendant, factor in determining possible mental or emotional harm to child, **CPL 65.20**

Source: McKinney's Consol. Laws, Index "Infants" p. 361, © West Publishing, used with permission.

examining the "**Constitutionality**" of New York Criminal Procedure Law section 65.20, or discussing the "**Right to confrontation**" under the statute, we refer to sections 1 and 2 of the Notes of Decision. *People v. Cintron*, 75 N.Y.2d 249, 552 N.Y.S.2d 68, 551 N.E.2d 561 (1990) surfaces in both sections, and it appears to approve of the procedures for testifying by closed-circuit television that are authorized by the statute.

Recall that the highest court in New York is the New York Court of Appeals. Decisions by that court, such as *People v. Cintron*, are published in three different case reporters: the official New York Reports, 2d Series, West's New York Supplement, 2d Series, and West's regional reporter, North Eastern Reporter, 2d Series. Since it is not safe to rely on the brief excerpts provided in the Notes of Decisions, we should read the full opinion in *People v. Cintron* and then confirm

Exhibit 3–19

§ 65.20   **Closed-circuit television; procedure for application and grounds for determination**
[Eff. until Nov. 1, 1996. See note below.]

1. Prior to the commencement of a criminal proceeding; other than a grand jury proceeding, either party may apply to the court for an order declaring that a child witness is vulnerable.

2. A motion pursuant to subdivision one of this section must be made in writing at least eight days before the commencement of trial or other criminal proceeding upon reasonable notice to the other party and with an opportunity to be heard.

3. The motion papers must state the basis for the motion and must contain sworn allegations of fact which, if true, would support a determination by the court that the child witness is vulnerable. Such allegations may be based upon the personal knowledge of the deponent or upon information and belief, provided that, in the latter event, the sources of such information and the grounds for such belief are stated.

4. The answering papers may admit or deny any of the alleged facts and may, in addition, contain sworn allegations of fact relevant to the motion, including the rights of the defendant, the need to protect the child witness and the integrity of the truth-finding function of the trier of fact.

5. Unless all material facts alleged in support of the motion made pursuant to subdivision one of this section are conceded, the court shall, in addition to examining the papers and hearing oral argument, conduct an appropriate hearing for the purpose of making findings of fact essential to the determination of the motion. Except as provided in subdivision six of this section, it may subpoena or call and examine witnesses, who must either testify under oath or be permitted to give unsworn testimony pursuant to subdivision two of section 60.20 and must authorize the attorneys for the parties to do the same.

6. Notwithstanding any other provision of law, the child witness who is alleged to be vulnerable may not be compelled to testify at such hearing or to submit to any psychological or psychiatric . . .

Historical and Statutory Notes

1991 Amendments. Subd. 10. L.1991, c. 455, § 1, eff. July 19, 1991, required court decision rather than court declaration of possible child witness vulnerability, required hearing outlined in subd. 5 for motion made pursuant to subd. 1, and required fact-finding pursuant to subds. 9 and 11 before determination.

Subd. 11. L.1991, c. 455, § 1, eff. July 19, 1991, substituted reference to deciding vulnerability of child witness for reference to deciding motion made pursuant to subd. 1.

Effective Date; Applicability; Expiration. Section effective July 24, 1985, applicable to proceedings pending on and commenced on or after such date, and to expire Nov. 1, 1996, pursuant to L.1985, c. 505, § 5, as amended, set out as a note under § 65.00.

Practice Commentaries

by Peter Preiser

This section establishes the procedure for obtaining a determination on the issue of whether closed-circuit television testimony will be permitted. The first eight subdivisions prescribe a proceeding initiated by motion with provision for a hearing. The format, though tailored to the situation at hand, basically tracks customary procedure except for the special provisions in subdivisions 6 and 7 designed to shield the child

continues

Exhibit 3–19	continued

from the anomaly of further aggravation of his or her situation brought on by a proceeding initiated to avoid that result.

Subdivision 6 exempts the child from compulsory process in connection with the hearing, but does not bar voluntary participation. Irrespective of whether the child gives evidence at the hearing, his or her prior statements may be received in evidence. But a declaration of vulnerability cannot rest solely upon those statements. Note that while the statute expressly provides that the child's failure to testify shall not be a ground for denial of the motion, there is no expression of legislative guidance regarding the *refusal* to submit to a psychological or psychiatric examination. Sound arguments leading to an . . .

Source: McKinney's, CPL 65.20, © West Publishing, used with permission.

that this case is still reliable authority. (We discuss these procedures when we consider research techniques for judicial decisions.)

We are not yet finished with the statute, however. We should anticipate that there are ways to update primary authorities, much as we discovered ways to update most of the secondary legal authorities we reviewed. Our first step is to check the pocket supplement inserted in the back of the statute volume, or the cumulative, freestanding companion supplement that exists for some statutes. The supplement will alert us to revisions made in the statute after the main volume was published; it will advise us whether the statute was repealed during this interim; and it will provide excerpts in the Notes of Decisions from cases that had not been decided when the main volume went to press.

Our examination of the pocket part corresponding to New York Penal Law sections 65.00 through 65.30 produces nothing of significance. The New York statutes also are kept current by an "Interim Update" supplement, which appears midway through the calendar year. We check this supplement, and again note no significant developments. Some state statutes are unlike New York's in that they are published in ringed notebooks that allow supplemental pages to be inserted in sections close to the original statutory provisions instead of in a separate pamphlet or in a pocket supplement at the rear of a volume.

Since **statutory supplements** typically are published only once a year, they may not keep us apprised of very recent changes that have occurred in legislation. We must check two additional sources. Newly enacted statutes are first printed as **"slip laws,"** which usually are made available promptly through an advance session law service. At the end of a legislative session, slip laws are collected in volumes of **"session laws."** The new statutes normally are published in chronological order in the session laws, and they are indexed by subject matter.

We use **McKinney's Session Laws of New York**, which is published by West, to check for recent changes in the New York statute we have been investigating. The most current volume available for our inspection was published in December 1995. We examine the cumulative table of consolidated laws affected by the leg-

Exhibit 3–20

Library References

American Digest System

Reception of evidence at trial; taking oral testimony in general, see Criminal Law ☞ 667(1) et seq.

Encyclopedia

▶ Trial and proceedings incidental thereto; taking testimony and evidence in general, see C.J.S. Criminal
Law § 1210.

WESTLAW Research

Criminal law cases: 110k[add key number].

Notes of Decisions

▶ **Constitutionality 1**
Evidence of vulnerability 4
Examining physician 5
Hearing 3
▶ **Right to confrontation 2**

---

1. Constitutionality

 Statute providing procedure for testimony of
child witness in sex case by way of two-way
closed circuit television is not facially invalid.
People v. Cintron, 1990, 75 N.Y.2d 249, 552
N.Y.S.2d 68, 551 N.E.2d 561.

2. Right to confrontation

 Statute authorizing use of two-way television to
present testimony of child witness in sex case provides for minimal curtailment of defendant's confrontation rights and adequately requires individualized showing of necessity for use of the
procedure based on clear and convincing
evidence. People v. Cintron, 1990, 75 N.Y.2d 249,
552 N.Y.S.2d 68, 551 N.E.2d 561.

 Allowing child witnesses in child sexual abuse
case to testify by means of live, two-way, closed-
circuit television did not violate defendant father's

right to confrontation, particularly in light of testimony of social worker who specialized in working
with sexually abused children, which established
that, inter alia, children felt abandoned by both
parents and would be particularly susceptible to
psychological harm if required to testify in open
court or in physical presence of defendant. People
v. Guce, 1990, 164 A.D.2d 946, 560 N.Y.S.2d 53,
appeal denied 76 N.Y.2d 986, 563 N.Y.S.2d 775,
565 N.E.2d 524.

3. Hearing

 Before finding child-victim in sexual abuse
prosecution is a vulnerable child, as required to
permit child witness to testify on closed circuit TV
out of actual presence of accused, trial court is required to conduct testimonial hearing to establish
vulnerability, and must make finding on the record
by clear and convincing evidentiary standard that
severe mental or emotional harm will ensue if
child is compelled to testify in presence of accused
and that this situation results from extraordinary
circumstances; it is not sufficient for trial court
judge simply to observe distress of child or child's
inability to answer questions in presence of accused. People v. Rivera, 1990, 160.

Source: McKinney's, CPL 65.20 Library references, © West Publishing, used with permission.

---

islative session that was not fully covered in the statutory supplement, and find no
changes in sections 65.00 through 65.30 of the Criminal Procedure Law. (See
Exhibit 3–21.) Alternatively, we could consult the cumulative subject index to this
volume of McKinney's Session Laws of New York, referring to topics such as

Exhibit 3–21

CONSOLIDATED LAWS

Criminal Procedure Law	Chap.	Sec.
§ 1.20, nt, added	2	72
		73
		152
§ 1.20, subd. 34, par. (q), amended	2	68
		69
§ 2.10, nt, added	2	72
		73
		152
§ 2.10, subd. 4, par. (a), amended	2	70
		71
§ 2.10, subd. 21, par. f, added	658	1
§ 2.10, subd. 27, amended	457	1
§ 2.10, subd. 64, added	206	1
	462	1
	521	1
§ 2.15, subd. 3, amended	522	1
§ 2.15, subd. 9, amended	177	1
§ 140.10, nt, added	349	7
§ 140.10, nt, amended	17	1
	356	1
§ 140.10, subd. 4, par. (b), subpar. (ii), amended	349	4

Source: McKinney's Session Laws, Table T1-9, © West Publishing, used with permission.

"Children and Minors," "Criminal Procedure," "Infants," "Television," and **"Witnesses."** On doing so, we again find no suggestion that any legislative changes had affected sections 65.00 through 65.30 of New York's Criminal Procedure Law. (See Exhibit 3–22.) Remember that considerable time lags occur between legislative sessions and the publication of annual statutory supplements, so you should routinely check for interim supplements, session laws, and slip laws to confirm that no changes have taken place in the legislation in which you are interested.

CALR for Statutes: Case 1 and WESTLAW

We have repeatedly stressed that **computer-assisted legal research (CALR)** is advantageous because it usually provides more current information than print references, and it allows greater flexibility in choosing the key words or phrases for developing search strategies. Another advantage is that the WESTLAW and LEXIS systems allow the researcher to search for the law in multiple jurisdictions

Exhibit 3–22

INDEX
Chapter numbers refer to 1995 Laws unless otherwise indicated

CRIMINAL PROCEDURE
Aggravated harassment in the second degree, jurisdiction, **Ch. 440**
Judicial memorandum, **p. 2741**
Criminal records, copies sent to office of mental health, **Ch. 181**
Executive memorandum, **p. 2311**
Legislative memorandum, **p. 1974**
Electronic court appearances in certain counties, extension of provisions, **Ch. 124**
Legislative memorandum, **p. 1931**
Extradition, expenses, **Ch. 193**
Legislative memorandum, **p. 1979**
Grand jury proceedings, counselors or other professionals, presence during special witness testimony, **Ch. 91**
Legislative memorandum, **p. 1893**
HIV testing of certain defendants, **Ch. 76**

Legislative memorandum, **p. 1872**
Orders of protection, conditions, **Ch. 483**
Judicial memorandum, **p. 2754**
Legislative memorandum, **p. 2173**
Sentencing,
Community service removing graffiti for any graffiti related offense, **Ch. 536**
Executive memorandum, **p. 2354**
Legislative memorandum, **p. 2235**
Conversion of indeterminate sentences to determinant sentences, **Ch. 547**
Legislative memorandum, **p. 2243**
Death penalty, **Ch. 1**
Executive memorandum, **p. 2283**
Legislative memorandum, **p. 1777**
Reform, **Ch. 3**
Legislative memorandum, **p. 1783**
Victims of Crime, generally, this index

Source: McKinney's Session Laws, Index, © West Publishing, used with permission.

at the same time. This latter feature can save the researcher a great deal of labor, as the only way to search for the same topic in the printed versions of the statutes is to replicate the procedure demonstrated above for all the jurisdictions in which the researcher is interested. We will demonstrate searching multiple databases later. First, we explore the use of **CALR** to search for New York statutes relevant to Case 1.

Both WESTLAW and LEXIS provide the full text of all state statutes. Using WESTLAW for this example, we could simply replicate the search strategy that we used with the print version of McKinney's Consolidated Laws of New York, since the General Index for McKinney's can be searched on WESTLAW in the same manner that it is searched in paper. However, we will demonstrate a different approach here in order to illustrate some of the options that online CALR provides the researcher. We must make some preliminary choices before attempting the actual search.

In Chapter 2 we discussed some of the advantages of **natural language** searching versus **Terms and Connectors** searching on WESTLAW. In choosing between these options, you should not only take into account your command of searching techniques but also consider whether the search terms that will be used take on variable meanings in different contexts. Another decision involves whether to search the annotated or unannotated version of McKinney's, since both are available through WESTLAW. The annotated version contains the Notes of Decisions, whereas the unannotated version does not. At first blush, it may be tempting to search the annotated version because so much valuable information is provided by the Notes of Decisions. However, depending on how the search strategy is constructed, searching the annotated version can adversely affect the precision of the outcome of the search and result in the retrieval of a high percentage of irrelevant documents. We decide to search the annotated version of the statutes but use the "**Terms and Connectors**" search mode in order to restrict the portion of the document in which our search terms must appear.

We previously have discussed the procedure for identifying and accessing the appropriate database in WESTLAW. By following those procedures we find that the identifier for the database containing the annotated version of McKinney's Consolidated Laws of New York is **NY-ST-ANN**. The identifier can be entered from any menu screen. Alternatively, you can follow the menu screens until you access the database and the query entry screen depicted in Exhibit 3–23.

The top of the screen confirms that we are in the **Terms and Connectors** search mode. To switch to **natural language** we simply click the button labeled **Nat Lang** at the lower right of the screen. Other helpful features appear on the **Terms and Connectors** "Query" screen. In the box marked **Connectors**, the basic proximity operators are listed and defined. Each document in the statute database is broken down into specific sections called **fields**. Several fields are displayed in the box marked **Fields**, and using the up and down arrow keys will allow you to see fields in addition to those that appear in Exhibit 3–23. A complete listing of all the fields in any database, as well as an explanation of what they contain and how to incorporate them into a search strategy, is included in the online documentation. To see the online documentation, click the **Scope** button. We will limit our search to the text field, labeled **TE**, which includes the actual text of each statute section. Finally, note the **Expanders** box and the "root expander" or truncation symbol !, which instructs the computer to search for various endings of words with a common root.

We have already typed in our search strategy, "**TE(child! /s witness! /p confront! /p televis!)**". The parentheses following the **TE** indicate that the included search terms *must* be within the text field and must meet the requirements established by the proximity indicators. We can interpret the search strategy displayed on the query screen as follows. In the **text** of the section a word beginning with the root "**child**" must appear in the same sentence as a word beginning with the root "**witness**"; furthermore, in the same paragraph, a word beginning with the root "**confront**" and a word beginning with the root "**televis**" both must appear. Before we enter our strategy, we should remember to click the **Thesaurus** button to check for related terms (since we have described that process in Chapter 2 we will not repeat it here). When we click **enter** we get the result displayed in Exhibit 3–24.

We have retrieved the same document, CPL § 65.20, that we did using the printed sources. Note the cite at the top left of the screen (Exhibit 3–24). Also note

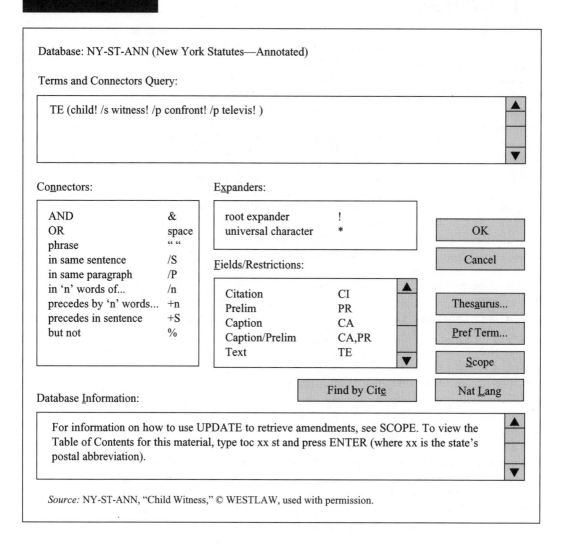

Exhibit 3–23

Database: NY-ST-ANN (New York Statutes—Annotated)

Terms and Connectors Query:

TE (child! /s witness! /p confront! /p televis!)

Connectors:

AND	&
OR	space
phrase	" "
in same sentence	/S
in same paragraph	/P
in 'n' words of...	/n
precedes by 'n' words...	+n
precedes in sentence	+S
but not	%

Expanders:

| root expander | ! |
| universal character | * |

Fields/Restrictions:

Citation	CI
Prelim	PR
Caption	CA
Caption/Prelim	CA,PR
Text	TE

OK

Cancel

Thesaurus...

Pref Term...

Scope

Find by Cite

Nat Lang

Database Information:

For information on how to use UPDATE to retrieve amendments, see SCOPE. To view the Table of Contents for this material, type toc xx st and press ENTER (where xx is the state's postal abbreviation).

Source: NY-ST-ANN, "Child Witness," © WESTLAW, used with permission.

at the top of the screen under **Rank** that this is the first and only document retrieved with our search strategy, "**R 1 of 1**." The small arrows to the left of some lines of information indicate the **hypertext** feature on WESTLAW: By placing the cursor on the arrow and pressing **enter** we can go directly to the text of the document cited to the right of the arrow. When you are researching statutes, it is helpful to examine the beginning of the article that includes the particular section of the statute you have retrieved. By doing so, you will usually find an outline of the several related statutory sections found in the complete article, and some of these may prove to be of interest. We thus retrieve the **Table of Contents** depicted in

Exhibit 3–24

Copr. ©. West 1996 No claim to orig. U.S. govt. works

AUTHORIZED FOR EDUCATIONAL USE ONLY

Citation	Rank (R)	Page (P)	Database	Mode
NY CRIM PRO § 65.20	R 1 of 1	P 1 of 26	NY-ST-ANN	Term
McKinney's CPL § 65.20				

► MCKINNEY'S CONSOLIDATED LAWS OF NEW YORK ANNOTATED
► CRIMINAL PROCEDURE LAW
► CHAPTER 11-A OF THE CONSOLIDATED LAWS
► PART ONE—GENERAL PROVISIONS
► TITLE D—RULES OF EVIDENCE, STANDARDS OF PROOF AND RELATED MATTERS
► ARTICLE 65—USE OF CLOSED-CIRCUIT TELEVISION FOR CERTAIN CHILD WITNESSES
[EFF. UNTIL NOV. 1, 1996. SEE NOTE BELOW.]
Copr. (C) West 1996. All rights reserved.
Current through L. 1996, chs. 1 to 49, and 56 to 60.

► § 65.20 Closed-circuit television; procedure for application and grounds for determination [Eff. until Nov. 1, 1996. See note below.]

1. Prior to the commencement of a criminal proceeding; other than a grand jury proceeding, either party may apply to the court for an order declaring that a child witness is vulnerable.

Source: NY-ST-ANN, "65.20," © WESTLAW, used with permission.

Exhibit 3–25 by clicking on the arrow in front of the reference to "**Article 65,**" shown in Exhibit 3–24.

From the **Table of Contents** screen we can access any of the sections comprising **Article 65** through the use of the hypertext function. We completed this example in July 1996. Observe that in several of our search steps there is a note indicating that Article 65 was due to expire in November 1996. We might question whether the legislature has passed a law extending the life of this article beyond its original expiration date. One of the advantages of searching McKinney's through WESTLAW is that if a section of a statute recently has been amended, a message will appear on the screen indicating that there is an "**update available.**" In this case, no such message appeared. If it had, we could have seen the recent change by entering the **update** command from the screen displaying the text of the statute we are checking.

A second way to determine whether changes have been made in this statute is to check the online version of McKinney's Session Laws. The identifier for McKinney's Session Laws in WESTLAW is **NY-LEGIS**. One helpful feature of WESTLAW (LEXIS also has this advantage) is being able to run a search strategy that already has been constructed in another database. Thus, we could run the search strategy we used in McKinney's Consolidated Laws of New York in **NY-LEGIS** by entering the command **sdb NY-LEGIS**. The command **sdb (search database)** simply tells the computer to run the current search strategy in the database that follows the command.

Since we performed our search in the annotated version of the statutes, we also have retrieved the **Notes of Decisions**, which contain references to court cases. From the screen appearing in Exhibit 3–25, we can return to **CPL 65.20** by using the hypertext feature. We then page through the document until we come to the **Notes of Decisions Index**, as depicted in Exhibit 3–26. Note that there is a reference to the same case, *People v. Cintron*, that we located in the printed set of statutes. The arrow in front of the citation allows us to hypertext directly to the text of *People v. Cintron*. The **update** feature also works with the **Notes of Decisions** section, so an "**update available**" message should appear if recent court cases deal with the retrieved section.

Exhibit 3–25

Copr. © West 1996 No claim to orig. U.S. govt. works

AUTHORIZED FOR EDUCATIONAL USE ONLY

McKinney's NY   Table of Contents   Page 1 of 1

McKinney's NY, CPL, Ch. 11-A, Pt. ONE, T. D, Art. 65

▶ MCKINNEY'S CONSOLIDATED LAWS OF NEW YORK ANNOTATED
▶ CRIMINAL PROCEDURE LAW
▶ CHAPTER 11-A OF THE CONSOLIDATED LAWS
▶ PART ONE—GENERAL PROVISIONS
▶ TITLE D—RULES OF EVIDENCE, STANDARDS OF PROOF AND RELATED MATTERS
ARTICLE 65—USE OF CLOSED-CIRCUIT TELEVISION FOR CERTAIN CHILD WITNESSES
[EFF. UNTIL NOV. 1, 1996. SEE NOTE BELOW.]
Copr. (C) West 1996. All rights reserved.

▶   NY CRIM PRO Ch. 11-A, Pt. ONE, T. D, Art. 65, References and Annotations
▶   § 65.00 Definitions [Eff. until Nov. 1, 1996. See note below.]
▶   § 65.10 Closed-circuit television; general rule; declaration of vulnerability [Eff. until Nov. 1, 1996. See note below.]
▶   § 65.20 Closed-circuit television; procedure for application and grounds for determination [Eff. until Nov. 1, 1996. See note below.]
▶   § 65.30 Closed-circuit television; special testimonial procedures [Eff. until Nov. 1, 1996. See note below.]

Source: NY-ST-ANN, "65-Table of Cont.," © WESTLAW, used with permission.

Exhibit 3–26

Copr. © West 1996 No claim to orig. U.S. govt. works

AUTHORIZED FOR EDUCATIONAL USE ONLY

NY CRIM PRO § 65.20   FOUND DOCUMENT   P 20 OF 26   NY-ST-ANN   Page

ANNOTATIONS (▶ Notes of Decisions Index)

▶ 1. Constitutionality

Statute providing procedure for testimony of child witness in sex case by way of two-way closed circuit television is not facially invalid. People v. Cintron, 1990, ▶ 75 N.Y.2d 249, 552 N.Y.S.2d 68, 551 N.E.2d 561.

Source: NY-ST-ANN, "65.20-Notes of Dec.," © WESTLAW, used with permission.

Print Statutory Sources and Case 2

We substantially replicate the steps used for Case 1 in New York when we consider Case 2 in Indiana. Our task is to determine if the Indiana statutes shed light on whether Andrew Adams' act of biting and spitting on Officer Fiegel when Adams knew he was HIV-positive can constitute attempted murder or assault with a deadly weapon. We begin by examining likely subject headings in the General Index accompanying **Burns Indiana Statutes Annotated**. By referring to topics such as "**Assault**," "**Attempts**," "**Deadly Weapon**," "**Murder**," "**AIDS**," and "**HIV**," we are directed to two major subject headings: "**Criminal Law and Procedure**," and "**Homicide**." We consult those subject headings, and by using the other terms as subheadings, we locate statutory provisions that may be of interest. Specifically, we note the following possibilities: "**Assault and Battery—Conduct constituting aggravated battery § 35-42-2-1.5**," "**Attempt § 35-41-5-1**," "**Definitions—Deadly weapon § 35-41-1-8**," "**HIV—Battery by body waste § 35-42-2-6**," and "**Murder—Conduct constituting § 35-42-1-1**." (See Exhibit 3-27.)

On investigation, we learn that the statute that generally addresses criminal attempts, § 35-41-5-1, promises to be helpful. Subsection (a) of that statute defines the circumstances under which someone attempts to commit a crime, and subsection (b) specifically rules out the defense of impossibility based on the offender's misapprehension of circumstances necessary to commit the intended crime. (See Exhibit 3-28.) In this context, these statutory provisions suggest that even if it is medically impossible to infect another person with HIV by spitting or biting, Andrews may be guilty of attempted murder if he intended to cause Officer Fiegel's death by his actions and if he took a substantial step toward the commission of that crime.

We again refer to the Notes to Decisions to see if we can find cases that may have applied the attempt statute to facts resembling our hypothetical case. We

Exhibit 3–27

1995 INDEX (A–I)

CRIMINAL LAW AND PROCEDURE—
Cont'd
Arrest.
See ARREST.
Arson, §35-43-1-1.
Assault and battery.
 Conduct constituting aggravated battery §35-42-2-1.5.
Conduct constituting battery, §35-42-2-1.
Provocation.
Conduct constituting, §35-42-2-3.
Victim counselors.
General provisions, §§35-37-6-1 to 35-37-6-11.
See VICTIM COUNSELORS.
Assignment of cases, CR. Rule 2.2.
Assisting a criminal, §35-44-3-2.
Associations.
Liability for offenses, §35-41-2-3.
Athletic trainers.
Unlawful practice, §25-5.1-4-2.
Atomic energy.
Radiation.
Violations of provisions, §16-41-35-40.
Attempt, §35-41-5-1.
Attendance of defendants.

CRIMINAL LAW AND PROCEDURE—
Cont'd
Definitions—Cont'd
Animals, §35-46-3-3.
Fighting contests, §35-46-3-4.
Applicability of definitions in chapter, §35-41-1-3.
Bodily injury, §35-41-1-4.
Credit institution, §35-41-1-5.
Crime, §35-41-1-6.
Criminal gangs, §§35-45-9-1, 35-45-9-2.
Criminal intelligence information, §5-2-4-1.

Deadly force, §35-41-1-7.
Deadly weapon, §35-41-1-8.
Deviate sexual conduct, §35-41-1-9.
Dwelling, §35-41-1-10.

See CRIMINAL HISTORY INFORMA-TION.
Juvenile history information, §§5-2-5.1-1 to 5-2-5.1-15.
See JUVENILE HISTORY INFORMA-TION.
HIV.
Battery by body waste, §35-42-2-6.
Home improvement fraud.
General provisions, §§35-43-6-1 to 35-43-6-14.
See HOME IMPROVEMENT FRAUD.

HOMICIDE—Cont'd
Minors.
Death sentence.
Age minimum, §35-50-2-3.
Murder.
Bail, §35-33-8.5-6.
Conduct constituting, §35-42-1-1.
Death sentence, §§35-50-2-3, 35-50-2-9.
Age minimum, §35-50-2-3.
Intestate succession.
Causing suicide, murder or voluntary manslaughter.
Constructive trustee, §29-1-2-12.1.
Penalties, §35-50-2-3.
Death sentence, §§35-50-2-3, 35-50-2-9.
Age minimum, §35-50-2-3.
Previous battery.
Admissible evidence in murder and voluntary manslaughter prosecutions, §35-37-4-14.

Source: The statutes reprinted or quoted verbatim in the following pages are taken from the Burns Indiana Statutes Annotated, Copyright by Michie, a division of Reed Elsevier Inc. and Reed Elsevier Properties Inc., and are reprinted with the permission of Michie. All rights reserved.

Exhibit 3–28

CHAPTER 5
OFFENSES OF GENERAL APPLICABILITY

SECTION.
35-41-5-1. Attempt.
35-41-5-2. Conspiracy.

SECTION.
35-41-5-3. Multiple convictions.

35-41-5-1. Attempt.—(a) A person attempts to commit a crime when, acting with the culpability required for commission of the crime, he engages in conduct that constitutes a substantial step toward commission of the crime. An attempt to commit a crime is a felony or misdemeanor of the same class as the crime attempted. However, an attempt to commit murder is a Class A felony.

(b) It is no defense that, because of a misapprehension of the circumstances, it would have been impossible for the accused person to commit the crime attempted. [IC 35-41-5-1, as added by Acts 1976, P.L. 148, § 1; 1977, P.L. 340, § 22.]

Source: The statutes reprinted or quoted verbatim in the following pages are taken from the Burns Indiana Statutes Annotated, Copyright by Michie, a division of Reed Elsevier Inc. and Reed Elsevier Properties Inc., and are reprinted with the permission of Michie. All rights reserved.

may wish to examine a few topics provided in the outline for relevant Notes to Decisions: for example, "**Attempted murder . . . —Intent . . . —Use of deadly weapon,**" "**Evidence . . . —Sufficiency . . . —Murder,**" and "**Necessary elements . . . —Intent . . . —Murder.**" (See Exhibit 3–29.) As we scan the notes presented, under the heading "**Evidence . . . —Sufficiency . . . —Murder,**" we find a note regarding a case, *State v. Haines*, 545 N.E.2d 834 (Ind. App. 1989), that appears to be similar to our scenario (Exhibit 3–30).

The pocket part to the statutory volume provides no information about changes in the "attempt" statute, nor does it identify additional relevant case law in its Notes to Decisions. We find no revisions in the statute when we check the Indiana sessions laws and slip laws. You recall that we have run across *State v. Haines* previously, in our discussion of both A.L.R. annotations and law review articles in Chapter 2. This is a helpful reminder that secondary legal authorities can be of great help in referring you to primary authorities such as statutes and case law.

CALR and Case 2: LEXIS

Using CALR we now conduct the same search in order to demonstrate how the LEXIS service is used. This particular search topic illustrates a potential peril in performing CALR. We already have pointed out that CALR can be particularly effective in identifying law review articles and court decisions involving the application of general principles of law to particular factual situations. In these instances CALR has an advantage because it allows the researcher to use terms describing the particular case facts as well as general legal principles. This type of

Exhibit 3–29

NOTES TO DECISIONS

ANALYSIS
Abandonment.
Attempted battery.
—Defenses.
—Intent.
Attempted child molesting.
Attempted criminal confinement.
Attempted felony-murder.
Attempted kidnapping.
Attempted manslaughter.
—Involuntary.
Attempted murder.
—Abandonment.
— —Evidence insufficient.
—Defense.
— —Voluntary intoxication.
—Felony murder.
—Informing defendant of charge.
—Intent.
— —Evidence sufficient.
— —Instructions.
— —Use of deadly weapon.
—Lesser included offenses.
— —Criminal recklessness.
—Sentence.
Attempted rape.
Attempted reckless homicide.
Attempted robbery.
Attempted theft.
—Burden of proof.
—Burglary.
—Information.
Attempting to receive stolen property.
Automobile banditry.

Conviction proper.
Double jeopardy.
Evidence.
—Admissibility.
—Agreement.
—Depraved sexual instinct rule.
—Intent to commit felony.
—Sufficiency.
— —Burglary.
— —Murder.
— —Rape.
— —Robbery.
Felony murder.
—Robbery.
Included offenses.
—Not included offenses.
Information
—Sufficient.
Instructions.
—Attempted murder.
—Voluntary intoxication.
Intent.
—Proof.
Necessary elements.
—Controlled substance.
—Intent.
— —Murder.
— —Use of deadly weapon.
—Substantial step toward commission.
— —Found.
— —Present apparent ability.
— —Substantial step found.
Verdict forms.
When attempt occurs.

Source: The statutes reprinted or quoted verbatim in the following pages are taken from the Burns Indiana Statutes Annotated, Copyright by Michie, a division of Reed Elsevier Inc. and Reed Elsevier Properties Inc., and are reprinted with the permission of Michie. All rights reserved.

searching can be problematic when applied to statutes or constitutions because the text of these sources of the law normally will consist only of general principles or rules. This does not mean that you should refrain from using WESTLAW or LEXIS to search for statutes, but it does mean that when you design a search

Exhibit 3–30

— —**Murder.**

Where defendant beat victim about the head and then choked her until she became unconscious after which he stabbed her, evidence was sufficient to show attempted murder. Zickefoose v. State, 270 Ind. 618, 388 N.E.2d 507, 69 Ind. Dec. 70 (1979).

Defendant's knowledge that he had Acquired Immune Deficiency Syndrome (AIDS), and that he unrelentingly and unequivocally sought to kill persons helping him by infecting them with AIDS, provided sufficient evidence for the jury to conclude that he took a substantial step toward the commission of murder. State v. Haines, 545 N.E.2d 834 (Ind. App. 1989).

Source: The statutes reprinted or quoted verbatim in the following pages are taken from the Burns Indiana Statutes Annotated, Copyright by Michie, a division of Reed Elsevier Inc. and Reed Elsevier Properties Inc., and are reprinted with the permission of Michie. All rights reserved.

strategy you must give considerable thought to whether the search terms are likely to appear in the text of the statute or in a court decision interpreting a section of the statute.

When we used CALR to examine the confrontation issue presented in Case 1, we were able to retrieve the relevant New York State statutes even though we limited our search to the actual text of the statute. We are unlikely to be so lucky with our Case 2 hypothetical. The definition and associated elements of what constitutes an "**assault**," "**battery**," or "**attempt to commit murder**" should be included within the text of the statutes. However, the determination of whether a person infected with AIDS or HIV violates that statute by biting another person is almost certain to be found only in a court decision.

For this case we will use the "**Boolean**" search mode and the "**segment**" restriction feature in LEXIS. The **segment** searching feature on LEXIS is the equivalent of using **field** restrictions on WESTLAW. Each document, or statutory section, is broken down into defined segments. Exhibit 3–31 provides an illustration of these segments, which are defined as follows:

STATUS—Contains the date through which this version of the code is current (e.g., "This section is current though the 1996 Supplement (1996 Regular Session)"). Also contains citations to recently enacted legislation (slip laws) that have affected the section.

HEADING—Contains the number and name of the Title, Article, and Chapter (e.g., "Title 35. Criminal Law and Procedure. Article 41. Crimes—General Substantive Provisions. Chapter 5. Offenses of General Applicability").

CITE—Contains the citation to the section (e.g., "Burns Ind. Code Ann. @ 35-41-5-1 (1996)").

Exhibit 3–31

BURNS INDIANA STATUTES ANNOTATED
Copyright (c) 1894–1996 by Michie,
a division of Reed Elsevier Inc. and Reed Elsevier Properties Inc.
All rights reserved.

STATUS — *** THIS SECTION IS CURRENT THROUGH THE 1996 SUPPLEMENT
*** (1996 REGULAR SESSION) ***

HEADING — TITLE 35. CRIMINAL LAW AND PROCEDURE
ARTICLE 41. CRIMES—GENERAL SUBSTANTIVE PROVISIONS
CHAPTER 5. OFFENSES OF GENERAL APPLICABILITY

CITE —————————— Burns Ind. Code Ann. @ 35-41-5-1 (1996)

UNANNO

SECTION — @ 35-41-5-1. Attempt

TEXT

(a) A person attempts to commit a crime when, acting with the culpability required for commission of the crime, he engages in conduct that constitutes a substantial step toward commission of the crime. An attempt to commit a crime is a felony or misdemeanor of the same class as the crime attempted. However, an attempt to commit murder is a Class A felony.

(b) It is no defense that, because of a misapprehension of the circumstances, it would have been impossible for the accused person to commit the crime attempted.

HISTORY—HISTORY: IC 35-41-5-1, as added by Acts 1976, P.L. 148, @ 1; 1977, P.L. 340, @ 22.

NOTES

NOTES:
CROSS REFERENCES. Defenses relating to culpability, IC 35-41-3-1 — IC 35-41-3-10.

INDIANA LAW JOURNAL. The Jury's Role Under the Indiana Constitution, 52 Ind. L.J. 793.

RES GESTAE. Criminal Justice Notes: Attempted Reckless Homicide, 23 Res Gestae 354.

VALPARAISO UNIVERSITY LAW REVIEW. Note, Testimony of Children Via Closed Circuit
Television in Indiana: Face (to Television) to Face Confrontation, 23 Val. U.L. Rev. 455 (1989).

continues

Exhibit 3–31 continued

When Attempt Occurs.

Murder

NOTES
(Cont.)

Defendant's knowledge that he had Acquired Immune Deficiency Syndrome (AIDS), and that he unrelentingly and unequivocally sought to kill persons helping him by infecting them with AIDS, provided sufficient evidence for the jury to conclude that he took a substantial step toward the commission of murder. State v. Haines, 545 N.E.2d 834 (Ind. App. 1989).

CASENOTES

Source: Indiana Statutes Ann., "35-41-5-1". Reprinted with permission of LEXIS-NEXIS, a division of Reed Elsevier Inc. LEXIS and NEXIS are registered trademarks of Reed Elsevier Properties Inc. FREESTYLE, KWIC, SuperKWIC, and MEGA are trademarks of Reed Elsevier Properties Inc. SHEPARD'S and SHEPARDIZE are registered trademarks of Shepard's Company, a Partnership.

SECTION—Section number and name of section (e.g., "@ 35-41-5-1. Attempt").

TEXT—Contains the full text of the section.

HISTORY—Provides citations to the original legislation and subsequent amendments (e.g., "IC 35-41-5-1, as added by Acts 1976, P.L. 148, @ 1; 1977, P.L. 340, @ 22.").

NOTES—References to law review articles, A.L.R., and other secondary sources, as well as squibs of cases interpreting the section.

CASENOTES—Squibs of cases that interpret the section; subset of the NOTES segment.

UNANNO—A group segment that includes all of the segments except for **NOTES** and **CASENOTES**.

The identifier for the Indiana Statutes file on LEXIS is **INCODE**. **INCODE** can be accessed through either the **CODES** or **INDIANA** libraries. The first screen presented after we enter **INCODE** is reproduced in Exhibit 3–32. This screen provides information on related files and also indicates that the researcher can find information on the file's content and coverage by entering the command **incode; .gu**. We will decline the invitation to find this information and proceed directly to our file.

Three concepts figure into the search strategy. The first concept is "**attempt**," which we want to find in association with one of the terms dealing with our second concept, "**assault or murder or battery or homicide or deadly weapon**." These terms should appear in the actual text of statutes dealing with our topic. The search strategy is displayed on the "search request" screen shown in Exhibit 3–33. We have directed that the word "**attempt**" must appear within 10 words of one of

Exhibit 3–32

                              LEXIS

Please ENTER, separated by commas, the NAMES of the files you want to search. You may select as many files as you want, including files that do not appear below, but you must enter them all at one time. To return to the file menu screen, enter .CF or press the CHANGE FILE key.
    DESCRIPTIONS – PAGE 17 of 75 (NEXT PAGE or PREV PAGE for additional files)

NAME	FILE		NAME	FILE

- - - - - - - - - - - - - - - - - - - INDIANA -

NAME	FILE	NAME	FILE
INCODE	Indiana Statutes Annotated	INADMN	Indiana Administrative Code**
INTOC	Table of Contents for IN Code	INRGTR	Tracking of IN Regulations
INALS	Indiana Advance Legis Serv		
INTRCK	Indiana Bill Tracking	INRULE	Indiana Court Rules
INTEXT	Indiana Bill Text		
INBILL	INTRCK & INTEXT files	INARCH	Indiana Statutes Archive

Enter file name; .gu for file content & coverage. Example: incode; .gu

**Neither the State of Indiana, the Legislative Council, the Legislative Services Agency, nor any officer or employee of any of them is liable for any error in or omission from the text of any of these copyrighted materials.

Source: INCODE. Reprinted with the permission of LEXIS-NEXIS, a division of Reed Elsevier Inc. LEXIS and NEXIS are registered trademarks of Reed Elsevier Properties Inc. FREESTYLE, KWIC, SuperKWIC and MEGA are trademarks of Reed Elsevier Properties Inc. SHEPARD'S and SHEPARDIZE are registered trademarks of Shepard's Company, a Partnership.

---

the terms following the **w/10** proximity requirement (refer to Chapter 2 if you need a refresher on LEXIS proximity operators), and that this information must appear in the **text** segment.

    The third concept is "**AIDS** or **HIV**," which we are looking for in the summaries of court decisions interpreting the statute. To do this we first type the name of the segment including the summaries, **CASENOTES** segment, and then include within parentheses the terms we want to search. Notice the **allcaps** command, followed by the terms "**AIDS**" and "**HIV**" within parentheses. The **allcaps** command requires that the immediately following terms must appear in all capital letters as a way of searching for acronyms or initialisms without retrieving the word in other contexts. Also observe that we have searched for the terms "**acquired immunodeficiency**" and "**human immunodeficiency**." Finally, we have used the Boolean operator **and** to link the Text and Casenotes portion of the search strategy. Recall that the **and** operator requires that the documents retrieved must satisfy the search requirements specified on *both* sides of the operator.

Exhibit 3–33

 LEXIS

Text (attempt w/10 assault or battery or murder or homicide or deadly weapon) and Casenotes (allcaps/ aids or hiv) or acquired immunodeficiency or human immunodeficiency)__

Please type your search request then press the ENTER key.
What you enter will be Search Level 1.

Type .fr to enter a FREESTYLE (TM) search.

For further explanation, press the H key (for HELP) and then the ENTER key.

Source: INCODE "Attempt." Reprinted with the permission of LEXIS-NEXIS, a division of Reed Elsevier Inc. LEXIS and NEXIS are registered trademarks of Reed Elsevier Properties Inc. FREESTYLE, KWIC, SuperKWIC and MEGA are trademarks of Reed Elsevier Properties Inc. SHEPARD'S and SHEPARDIZE are registered trademarks of Shepard's Company, a Partnership.

The search results are displayed in Exhibit 3–34, showing that we have retrieved just one document. Of the several options for displaying or browsing our search results we choose the **KWIC** (Keyword in Context) option by entering the command **.kw**. This command will display only the portions of the document where our search terms appear, accompanied by 25 words of text on either side of the terms. This information quickly allows the researcher to determine which documents are relevant.

Exhibit 3–35 displays the fruits of the **.kw** command. The citation for the document appears at the top of the screen, and the terms used to search are highlighted in the body of the text. Preceding the cite to *State v. Haines*, note the "<=175>" that appears toward the bottom of the screen. This is the **LINK**™ or hypertext feature on LEXIS. If we **entered** "=175" we would go directly to the full text of *State v. Haines* without losing our search result. To return to the search result from a document with a LINK marker, **enter** the command **.es**.

Once we have determined that a document is relevant, we can look at the full text of that document by **entering** the command **.fu**. Exhibit 3–36 displays the results of the **.fu** command. One nice feature of online sources is that they are frequently updated. Notice at the middle of the screen the status message indicating that "**This section is current through the 1996 Supplement.**" To confirm that there have been no changes since the issuance of the 1996 Supplement, search the **INALS** (Indiana Advance Legislative Service) file shown as the third item under the "Indiana" heading in Exhibit 3–32. The INALS file contains the recent slip laws for Indiana.

Exhibit 3–34

 LEXIS

TEXT (ATTEMPT W/10 ASSAULT OR BATTERY OR MURDER OR HOMICIDE OR DEADLY WEAPON) AND CASENOTES (ALLCAPS (AIDS OR HIV) OR ACQUIRED IMMUNODEFI-CIENCY OR HUMAN IMMUNODEFICIENCY)

 Your search request has found 1 DOCUMENT through Level 1.
To DISPLAY this DOCUMENT press either the KWIC, FULL, CITE, or SEGMTS key.
To MODIFY your search request, press the M key (for MODFY) and then the ENTER key.
For further explanation, press the H key (for HELP) and then the ENTER key.

Source: INCODE "Attempt" result. Reprinted with the permission of LEXIS-NEXIS, a division of Reed Elsevier Inc. LEXIS and NEXIS are registered trademarks of Reed Elsevier Properties Inc. FREESTYLE, KWIC, SuperKWIC and MEGA are trademarks of Reed Elsevier Properties Inc. SHEPARD'S and SHEPARDIZE are registered trademarks of Shepard's Company, a Partnership.

Exhibit 3–35

 LEXIS

Burns Ind. Code Ann. @ 35-41-5-1 (1996)

commit a crime is a felony or misdemeanor of the same class as the crime attempted. However, an attempt to commit murder is a Class A felony.

(b) It is no defense that, because of a misapprehension of the circumstances it would have been impossible for the accused person to commit the crime attempted.

NOTES:

... marijuana, made his testimony inherently unbelievable. The credibility of this witness was for the jury to determine. <=174> Budd v. State, 499 N.E.2d 1116 (Ind. 1986).
Defendant's knowledge that he had Acquired Immune Deficiency Syndrome (AIDS), and that he unrelentingly and unequivocally sought to kill persons helping him by infecting them with AIDS, provided sufficient evidence for the jury to conclude that he took a substantial step toward the commission of murder. <=175> State v. Haines, 545 N.E.2d 834 (Ind. App. 1989).
— —RAPE.
Where as to one . . .

Source: INCODE "Attempt" display. Reprinted with the permission of LEXIS-NEXIS, a division of Reed Elsevier Inc. LEXIS and NEXIS are registered trademarks of Reed Elsevier Properties Inc. FREESTYLE, KWIC, SuperKWIC and MEGA are trademarks of Reed Elsevier Properties Inc. SHEPARD'S and SHEPARDIZE are registered trademarks of Shepard's Company, a Partnership.

Exhibit 3–36

 LEXIS

To be able to browse preceding or succeeding code sections, enter B. The first page of the document you are currently viewing will be displayed in FULL.

- -

LEVEL 1 - 1 OF 1 DOCUMENT

BURNS INDIANA STATUTES ANNOTATED
Copyright (c) 1894–1996 by Michie,
a division of Reed Elsevier Inc. and Reed Elsevier Properties Inc.
All rights reserved.

*** THIS SECTION IS CURRENT THROUGH THE 1996 SUPPLEMENT
*** (1996 REGULAR SESSION) ***

TITLE 35. CRIMINAL LAW AND PROCEDURE
ARTICLE 41. CRIMES — GENERAL SUBSTANTIVE PROVISIONS
CHAPTER 5. OFFENSES OF GENERAL APPLICABILITY

Burns Ind. Code Ann. @ 35-41-5-1 (1996)

35-41-5-1 Attempt

(a) A person attempts to commit a crime when, acting with the culpability required for commission of the crime, he engages in conduct that constitutes a

Source: INCODE "Attempt" browse. Reprinted with the permission of LEXIS-NEXIS, a division of Reed Elsevier Inc. LEXIS and NEXIS are registered trademarks of Reed Elsevier Properties Inc. FREESTYLE, KWIC, SuperKWIC and MEGA are trademarks of Reed Elsevier Properties Inc. SHEPARD'S and SHEPARDIZE are registered trademarks of Shepard's Company, a Partnership.

We mentioned earlier that CALR allows the researcher to search a topic in the statutes or cases of several jurisdictions at the same time. To illustrate this capability, we can perform this same search in the file containing the statutes for all 50 states, **ALLCDE**. We do this by entering the change library command, **.cl**, after we browse the document displayed in Exhibit 3–36. That command will take us back to the library screen, where we **enter CODES**. The next screen will display the identifiers for all the statute files. At that screen we enter **ALLCDE**.

Exhibit 3–37 depicts the search request screen in **ALLCDE** using our search strategy that was transferred from the previous file. When we press **enter** we retrieve five documents, as indicated in Exhibit 3–38. We can then display the documents to determine if any are relevant. Not surprisingly, the first document is the Indiana statute that we had previously retrieved. Using the **.kw** command, we eventually come to the screen depicted in Exhibit 3–39. This figure refers to a Utah statute regarding attempted assault. Of particular interest is the reference at the bottom of the screen to 13 A.L.R. 5th 628, which is an annotation directly on

Exhibit 3–37

                                 LEXIS

TEXT(ATTEMPT W/10 ASSAULT OR BATTERY OR MURDER OR HOMICIDE OR DEADLY WEAPON) AND CASENOTES(ALLCAPS(AIDS OR HIV) OR ACQUIRED IMMUNODEFICIENCY OR HUMAN IMMUNODEFICIENCY)

The above request is the one you last used before selecting a new library or file. If you now want to use this request again, press the ENTER key.

To edit the above request before you enter it, use the editing keys. Be sure to move the cursor to the end of the request before you enter. For editing instructions, press the E key (for EDIT) and then the ENTER key.

If you do not want to use this search request, press the NEW SEARCH key.

For further explanation, press the H key (for HELP) and then the ENTER key.

Source: ALLCDE "Attempt." Reprinted with the permission of LEXIS-NEXIS, a division of Reed Elsevier Inc. LEXIS and NEXIS are registered trademarks of Reed Elsevier Properties Inc. FREESTYLE, KWIC, SuperKWIC and MEGA are trademarks of Reed Elsevier Properties Inc. SHEPARD'S and SHEPARDIZE are registered trademarks of Shepard's Company, a Partnership.

Exhibit 3–38

                                   LEXIS

TEXT(ATTEMPT W/10 ASSAULT OR BATTERY OR MURDER OR HOMICIDE OR DEADLY WEAPON) AND CASENOTES(ALLCAPS(AIDS OR HIV) OR ACQUIRED IMMUNODEFICIENCY OR HUMAN IMMUNODEFICIENCY)

Your search request has found 5 DOCUMENTS though Level 1.
To DISPLAY these DOCUMENTS press either the KWIC, FULL, CITE or SEGMTS key.
To MODIFY your search request, press the M key (for MODFY) and then the ENTER key.

For further explanation, press the H key (for HELP) and then the ENTER key.

Source: INCODE "Attempt" result. Reprinted with the permission of LEXIS-NEXIS, a division of Reed Elsevier Inc. LEXIS and NEXIS are registered trademarks of Reed Elsevier Properties Inc. FREESTYLE, KWIC, SuperKWIC and MEGA are trademarks of Reed Elsevier Properties Inc. SHEPARD'S and SHEPARDIZE are registered trademarks of Shepard's Company, a Partnership.

Exhibit 3–39

LEXIS

Utah Code Ann. @ 76-5-102 (1995)

@ 76-5-102. Assault
(1) Assault is:
(a) an attempt, with unlawful force or violence, to do bodily injury to another;
(b) a threat, accompanied by a show of immediate force or violence, to do bodily injury to another;
or
(c) an act, committed with unlawful force or violence, that . . .

NOTES:
. . . criminal responsibility, <=19> 68 A.L.R.4th 507.
Admissibility of expert opinion stating whether a particular knife was, or could have been, the weapon used in a crime, <=20> 83 A.L.R.4th 660.
Transmission or risk or transmission of human immunodeficiency virus (HIV) or acquired immunodeficiency syndrome (AIDS) as basis for prosecution or sentencing in criminal or military discipline case, <=21> 13 A.L.R.5th 628.

KEY NUMBERS. —Assault and Battery KEY 48.

Source: ALLCDE "Attempt" display. Reprinted with the permission of LEXIS-NEXIS, a division of Reed Elsevier Inc. LEXIS and NEXIS are registered trademarks of Reed Elsevier Properties Inc. FREESTYLE, KWIC, SuperKWIC and MEGA are trademarks of Reed Elsevier Properties Inc. SHEPARD'S and SHEPARDIZE are registered trademarks of Shepard's Company, a Partnership.

---

point. You will recall that we uncovered the citation to this same annotation through methods described in Chapter 2. We could use the LINK command in LEXIS to go directly to the text of that annotation.

Federal Statutes and Shepardizing: Case 3

We next consider how to find and make use of federal statutes by revisiting Case 3 and Deborah Miller's federal lawsuit against Warden Myers for ignoring her request to be separated from her heavily smoking state prison cellmate, Felicia Liggett. We already have identified the United States Code as the official source for federal statutes, but it is not a very helpful source for legal research purposes because it lacks the case notes and ancillary references that are included in West's United States Code Annotated and Lawyers Cooperative's United States Code Service. You should plan to work with either U.S.C.A. or U.S.C.S., or both, when your research involves federal statutes.

The statute under which Deborah Miller's civil suit has been brought is widely used by people who claim that their federal constitutional rights have been violated by someone acting "under color of" state law. In fact, this statute so commonly serves as the basis for federal lawsuits that you might already know its citation. If you do not, consult the General Index of either U.S.C.A. or U.S.C.S. When we examine U.S.C.A.'s General Index under the heading "**Civil Rights**," and the subtopic "**Prisoners, . . . Civil action for deprivation of rights**," we are referred to 42 U.S.C.A. § 1983, or title 42 of the United States Code Annotated, section 1983. (See Exhibit 3–40.)

The statutory volumes are arranged sequentially by title and section. The opening pages of all volumes of U.S.C.A. identify the 50 different "titles" under which the United States Code is organized. (See Exhibit 3–41.) Note that title 42 pertains to "The Public Health and Welfare." Many other statutes relevant to criminal justice are collected in titles 18 (Crimes and Criminal Procedure), 21 (Food and Drugs), and 28 (Judiciary and Judicial Procedure).

Exhibit 3–40

CIVIL RIGHTS—Cont'd

Organizations, community development banking, studies and reports, examination and audit, consultation, **12 § 4716**

Overseas citizens, voting rights of. Absent Voters, generally, this index

Paperwork Reduction Act, inapplicability to authority of Federal agencies to enforce civil rights laws, **44 § 3518**

Partial invalidity of provisions of law governing, **42 § 2000h-6**

Participating lawfully in speech or peaceful assembly,

Defined, **18 § 245**

Penalties for willful intimidation, interference, etc., with, **18 § 245**

Penalties. Fines and penalties, generally, ante, this heading

Peonage abolished, **42 § 1994**

Persons unable to initiate and maintain proceedings,

Public education, desegregation, **42 § 2000c-6**

Public facilities, desegregation, **42 § 2000b**

Physical obstruction, defined, freedom of access to clinic entrances, reproductive health services, **18 § 248**

CIVIL RIGHTS—Cont'd

Political subdivision,

Defined, **42 § 1973l**

Facilities or institutions for mentally ill, prisoners, etc. Institutionalized persons, generally, ante, this heading

Voting rights, generally, post, this heading

Poll taxes, prohibition, enforced payment, etc., **42 § 1973h**

President, election of. Voting rights, generally, post, this heading

Pretrial detention facilities, State facilities, etc. Institutionalized persons, generally, ante, this heading

Preventive relief, injunction, voting rights, **42 § 1971**

Prisoners,

See, also, Institutionalized persons, generally, ante, this heading

Civil action for deprivation of rights, **42 § 1983**

Prisons, State facilities, etc. Institutionalized persons, generally, ante, this heading

Source: U.S.C.A., Index "Civil Rights," © West Publishing, used with permission.

Exhibit 3–41

TITLES OF UNITED STATES CODE AND
UNITED STATES CODE ANNOTATED

1. General Provisions.
2. The Congress.
3. The President.
4. Flag and Seal, Seat of Government, and the States.
5. Government Organization and Employees.
6. Surety Bonds (*See Title 31, Money and Finance*).
7. Agriculture.
8. Aliens and Nationality.
9. Arbitration.
10. Armed Forces.
11. Bankruptcy.
12. Banks and Banking.
13. Census.
14. Coast Guard.
15. Commerce and Trade.
16. Conservation.
17. Copyrights.
18. Crimes and Criminal Procedure.
19. Customs Duties.
20. Education.
21. Food and Drugs.
22. Foreign Relations and Intercourse.
23. Highways.
24. Hospitals and Asylums.
25. Indians.
26. Internal Revenue Code.
27. Intoxicating Liquors.
28. Judiciary and Judicial Procedure.
29. Labor.
30. Mineral Lands and Mining.
31. Money and Finance.
32. National Guard.
33. Navigation and Navigable Waters.
34. Navy (*See Title 10, Armed Forces*).
35. Patents.
36. Patriotic Societies and Observances.
37. Pay and Allowances of the Uniformed Services.
38. Veterans' Benefits.
39. Postal Service.
40. Public Buildings, Property, and Works.
41. Public Contracts.
42. The Public Health and Welfare. ◀
43. Public Lands.
44. Public Printing and Documents.
45. Railroads.
46. Shipping.
47. Telegraphs, Telephones, and Radiotelegraphs.
48. Territories and Insular Possessions.
49. Transportation.
50. War and National Defense.

Source: U.S.C.A., Titles, © West Publishing, used with permission.

The United States Code Annotated and United States Code Service use the same titles and sections as the United States Code, so it is convenient to look up a statute in any of these series. When we consult title 42 U.S.C.A. § 1983, we find the statute, along with much additional useful information. (See Exhibit 3–42.) We initially are provided with references bearing on the legislative history, including relevant reports from the House of Representatives and the Senate, and other information about the origins of the statute. We can assemble evidence about different legislators' reasons for voting for or against the statute and the law's intended meaning (the "legislative intent") by referring to the Historical and

Exhibit 3–42

CH. 21 GENERALLY   42 § 1983

§ 1983. Civil action for deprivation of rights

Every person who, under color of any statute, ordinance, regulation, custom, or usage, of any State or Territory or the District of Columbia, subjects, or causes to be subjected, any citizen of the United States or other person within the jurisdiction thereof to the deprivation of any rights, privileges, or immunities secured by the Constitution and laws, shall be liable to the party injured in an action at law, suit in equity, or other proper proceeding for redress. For the purposes of this section, any Act of Congress applicable exclusively to the District of Columbia shall be considered to be a statute of the District of Columbia.

(R.S. § 1979; Pub.L. 96–170, § 1, Dec. 29, 1979, 93 Stat. 1284.)

HISTORICAL AND STATUTORY NOTES

Revision Notes and Legislative Reports

1979 Acts. House Report No. 96–548, see 1979 U.S. Code Cong. and Adm. News, p. 2609.

Codifications

R.S. § 1979 is from Act Apr.. 20, 1871, c. 22, § 1, 17 Stat. 13.

Section was formerly classified to section 43 of Title 8, Aliens and Nationality.

Amendments

1979 Amendments. Pub.L. 96–170 added "or the District of Columbia" following "Territory,"

and provisions relating to Acts of Congress applicable solely to the District of Columbia.

Effective Dates

1979 Acts. Amendment by Pub.L. 96–170 applicable with respect to any deprivation of rights, privileges, or immunities secured by the Constitution and laws occurring after Dec. 29, 1979, see section 3 of Pub.L. 96–170, set out as a note under section 1343 of Title 28, Judiciary and Judicial Procedure.

Source: U.S.C.A., "1983," © West Publishing, used with permission.

Statutory Notes following the statute. There, immediately underneath the statute, you will find several cryptic references in parentheses: "R.S. § 1979; Pub.L. 96-170, § 1, Dec. 29, 1979, 93 Stat. 1284."

The "R.S." abbreviation refers to **Revised Statutes**, in which federal legislation was published through 1873. Most of 42 U.S.C.A. § 1983 can be traced to an act dating back to April 20, 1871, R.S. § 1979, which also is reported in **Statutes at Large**, the official collection of federal session laws. Note the citation 17 Stat. 13 in connection with R.S. § 1979 under the "Codifications" heading, and 93 Stat. 1284, which appears in parentheses just under 42 U.S.C.A. § 1983. The "**Stat.**" means Statutes at Large. We also can piece together that the 1871 legislation was amended by Public Law (Pub. L.) 96-170 (Public Law 96-170 was enacted by the 96th Congress, and this provision was the 170th law passed during that congressional session), on December 29, 1979. For further information about the amendment we could read House Report No. 96-548, which we would find reprinted in the massive collection, **United States Code Congressional and Administrative News**. Specifically, we would locate that report in 1979 U.S. Code Cong. Adm. News 2609.

Both the U.S.C.A. (West) and U.S.C.S. (Lawyers Cooperative) references will refer us to a host of secondary legal authorities relevant to the statute, and they both offer notes about cases that have applied and interpreted the legislation. We have seen these features when we used annotated state statutes while researching Cases 1 and 2. This time, we will provide an example from U.S.C.S. We set forth the statute, 42 U.S.C.S. § 1983, and its History; Ancillary Laws and Directives, to show you how similar this service is to U.S.C.A. (See Exhibit 3–43.) We skip over several pages of secondary authorities, consisting of citations to Am. Jur. 2d articles, A.L.R. annotations, treatises, and law review articles discussing different aspects of this law, and present a portion of the Interpretive Notes and Decisions following the statute. The general topic outline reveals a wealth of notes covering "Prison Conditions and Proceedings," including subsection IV.B.5., which pertains to "Living Conditions (notes 638–647)." The more specific outline reveals that note 646 corresponds to "Tobacco smoke." (See Exhibit 3–44.) Another way to find pertinent case notes is to use the Index to Interpretive Notes and Decisions in this same volume of U.S.C.S. When we look there under "Prisons and prisoners," we find the subtopic "tobacco smoke," which again refers us to note 646. (See Exhibit 3–45.)

Most statutes will not even begin to approach the number of case notes that 42 U.S.C.S. § 1983 has. We must turn past several hundreds of pages to find the case descriptions collected under note 646. (See Exhibit 3–46.) Oddly, the U.S. Supreme Court decision that we ran across during our perusal of secondary authorities, *Helling v. McKinney*, 509 U.S. 25, 113 S. Ct. 2475, 125 L. Ed. 2d 22 (1993), is not directly reported. It is only identified as *McKinney v. Anderson*, 959 F.2d 853 (9th Cir. 1992), with the notation that the case was affirmed by the Supreme Court, and that following the remand another opinion was issued by the federal Court of Appeals (5 F.3d 365 (9th Cir. 1993)).

The case excerpts presented in this section of the statute book do not provide a very clear picture of the law governing claims of the type that Deborah Miller has made in our hypothetical situation. This should serve as a reminder not to rely on references of this nature as final authorities. You always should examine the annotated cases in their entirety and check other sources to make sure you have a complete and accurate rendition of the law. Remember to check the pocket part of the statutory volume, and other updating materials. In this instance, we are provided with no additional significant information about the statute or interpretive cases.

We will mention, but not fully describe, one additional process that is helpful for statutory legal research. This process is known as "**Shepardizing**,™" which takes its name from the **Shepard's Citators** that are available for statutes, constitutions, judicial decisions, federal administrative materials, law reviews, and a few other sources. (We discuss Shepardizing a case in much greater detail when we take up case law research.) Once you get the hang of Shepardizing a case, you can transfer the same techniques to Shepardizing legislation. In brief, Shepardizing a statute allows you to determine whether any sections of the law

Exhibit 3–43

THE CODE OF THE LAWS OF THE UNITED STATES OF AMERICA
TITLE 42—THE PUBLIC HEALTH AND WELFARE
CHAPTER 21. CIVIL RIGHTS

§ 1983. Civil action for deprivation of rights

Every person who, under color of any statute, ordinance, regulation, custom, or usage, of any State or Territory or the District of Columbia, subjects, or causes to be subjected, any citizen of the United States or other person within the jurisdiction thereof to the deprivation of any rights, privileges, or immunities secured by the Constitution and laws, shall be liable to the party injured in an action at law, suit in equity, or other proper proceeding for redress. For the purposes of this section, any Act of Congress applicable exclusively to the District of Columbia shall be considered to be a statute of the District of Columbia. (R. S. § 1979; Dec. 29, 1979, P. L. 96–170, § 1, 93 Stat. 1284.)

HISTORY; ANCILLARY LAWS AND DIRECTIVES

Explanatory notes:

This section formerly appeared as U.S.C. § 43.

R.S. § 1979 was derived from Act Apr. 20, 1871, ch 22, § 1, 17 Stat. 13.

Amendments.

1979. Act Dec. 29, 1979 inserted "or the District of Columbia" and "For the purposes of this section, any Act of Congress applicable exclusively to the District of Columbia shall be considered to be a statute of the District of Columbia."

Other provisions:

Application of amendments made by Act Dec. 29, 1979. Act Dec. 29, 1979, P.L. 96–170, § 3, 93 Stat. 1284, which appears as 28 USCS § 1343 note, provided that the amendments made to this section by such Act are applicable with respect to any deprivation of rights, privileges, or immunities secured by the Constitution and laws occurring after enactment on Dec. 29, 1979.

INTERPRETIVE NOTES AND DECISIONS

IV. RIGHTS COGNIZABLE; APPLICABILITY
TO PARTICULAR SITUATIONS
  A. In General (notes 551–576)
  B. Prison Conditions and Proceedings
    1. In General (notes 577–580)
    2. Court Access (notes 581–594)
    3. Discipline and Disciplinary
       Hearings
      a. Hearing, Due Process
        (1). In General (notes 595–609)
        (2). Prior to Particular Sanctions
           (notes 610–617)
      b. Propriety of Particular Discipline
        (1). Solitary Confinement or
           Punitive Isolation (notes
           618–626)

        (2). Other Discipline (notes 627–
           631)
    4. Employment (notes 632–637)
    5. Living Conditions (notes 638–647) ◀
    6. Mail Privileges and Restrictions (notes
       648–655)
    7. Medical Care
      a. In General (notes 656–658)
      b. Particular Circumstances (notes
         659–684)
    8. Personal Property Rights (notes
       685–689)
    9. Physical, Mental or Other Abuse (notes
       690–706)
   10. Place of Confinement (notes 707–710)
   11. Religious Rights (notes 711–720)

Source: U.S.C.S., "1983." Reprinted with the permission of LEXIS-NEXIS, a division of Reed Elsevier Inc. LEXIS and NEXIS are registered trademarks of Reed Elsevier Properties Inc. FREESTYLE, KWIC, SuperKWIC and MEGA are trademarks of Reed Elsevier Properties Inc. SHEPARD'S and SHEPARDIZE are registered trademarks of Shepard's Company, a Partnership.

Exhibit 3–44

---

42 USCS § 1983   **Public Health and Welfare**

637. Work release program
          5. Living Conditions
638. Generally
639. Food
640. Heat, ventilation and exposure
641. Overcrowding

642. Privacy
643. Recreation and exercise
644. Safety hazards
645. Sanitation and hygiene
646. Tobacco smoke ◀
647. Other particular conditions

Source: U.S.C.S., "1983" Outline. Reprinted with the permission of LEXIS-NEXIS, a division of Reed Elsevier Inc. LEXIS and NEXIS are registered trademarks of Reed Elsevier Properties Inc. FREESTYLE, KWIC, SuperKWIC and MEGA are trademarks of Reed Elsevier Properties Inc. SHEPARD'S and SHEPARDIZE are registered trademarks of Shepard's Company, a Partnership.

Exhibit 3–45

---

42 USCS § 1983   **Public Health and Welfare**

**INDEX TO INTERPRETIVE NOTES AND
    DECISIONS**—Cont'd
Prisons and prisoners—Cont'd
—sanitation and hygiene, n 620, 645
—scope of official responsibility, actions within
    n 369
—search and seizure, n 398, 399, 732, 733, 588
—segregation, n 400
—self-inflicted injury, n 396, 677
—services, religious, n 717–719
—sexual assault by other inmate, n 703
—shackled inmate, n 695
—shootings, n 696
—smoke from tobacco, n 646
—solitary confinement, n 400, 618–626

Prisons and prisoners—Cont'd
—standing to sue, n 1351, 1352
—state post-deprivation remedies, availability
    n 580
—strip searches, n 399, 733
—suicide, n 227, 396, 677
—superintendent as liable for deprivation, n 377
—surgery, n 663
—swearing of witnesses, due process, n 603
—telephone privileges, n 734
—time of notice and hearing, disciplinary matters,
    n 598
—tobacco smoke, n 646 ◀
—transfer of funds, n 735
—transfers n 610–614, 629, 709, 1294

Source: U.S.C.S., "1983" Outline-tobacco smoke. Reprinted with the permission of LEXIS-NEXIS, a division of Reed Elsevier Inc. LEXIS and NEXIS are registered trademarks of Reed Elsevier Properties Inc. FREESTYLE, KWIC, SuperKWIC and MEGA are trademarks of Reed Elsevier Properties Inc. SHEPARD'S and SHEPARDIZE are registered trademarks of Shepard's Company, a Partnership.

---

have been repealed, renumbered, revised, or amended by legislative action; to ascertain whether judicial decisions have invalidated the statute on constitutional grounds; and to identify judicial decisions, A.L.R. annotations, and law review articles in which the statute has been cited. Shepard's thus can be a very important source for confirming the continuing validity of a statute and can pro-

Exhibit 3–46

646. Tobacco smoke

Presence of smoke in prison air did not constitute cruel and unusual punishment in violation of Eighth Amendment for purposes of 42 USCS § 1983 action by asthmatic prisoner. Steading v Thompson (1991, CA7 Ill) 941 F2d 498, cert den (1992, US) 117 L Ed 2d 445, 112 S Ct 1206.

Prisoner's occasional involuntary exposure to smoke in 8 foot cell did not constitute Eighth Amendment violation. Clemmons v Bohannon (1992, CA10 Kan) 956 F2d 1523.

Inmate stated 42 USCS § 1983 claim on basis of Eighth Amendment where he alleged that compelled exposure to levels of environmental tobacco smoke posed unreasonable risk of harm to human health and constituted cruel and unusual punishment. McKinney v Anderson (1992, CA9 Nev) 959 F2d 853, 92 CDOS 2622, 92 Daily Journal DAR 4164, affd, remanded (1993, US) 125 L Ed 2d 22, 113 S Ct 2475, 93 CDOS 4501, 93 Daily Journal DAR 7681, 7 FLW Fed S 452, on remand, remanded (1993, CA9) 5 F3d 365, 93 CDOS 6923, 93 Daily Journal DAR 11818.

Source: U.S.C.S., "1983" Case notes 646. Reprinted with the permission of LEXIS-NEXIS, a division of Reed Elsevier Inc. LEXIS and NEXIS are registered trademarks of Reed Elsevier Properties Inc. FREESTYLE, KWIC, SuperKWIC and MEGA are trademarks of Reed Elsevier Properties Inc. SHEPARD'S and SHEPARDIZE are registered trademarks of Shepard's Company, a Partnership.

vide excellent leads to other authorities in which the statute has been examined or applied.

Different volumes of Shepard's exist for federal statutes and the statutes from the individual states. Thus, we would use the Shepard's Citator for New York statutes, for Indiana statutes, and for federal statutes, respectively, to Shepardize the legislation we have uncovered during research for our three hypothetical cases. We can anticipate that Shepardizing the New York statute governing closed-circuit television testimony by children might be especially productive, in light of the specific connection between this statute and the issue we are researching. Nevertheless, we also would want to Shepardize Indiana's "attempts" statute and 42 U.S.C. § 1983 to confirm that these laws have not been altered or invalidated.

Both WESTLAW and LEXIS provide the capability to Shepardize cases and statutes. Indeed, computer-based Shepardizing has two distinct advantages over the printed sources: First, the computer versions are more current than the print versions; and second, they quickly allow the researcher to find citing cases from a particular jurisdiction, or cases that give particular treatment to the cited document, or that make reference to the subject matter in a specific headnote of a cited case. We will demonstrate these advantages more fully when we discuss Shepardizing case law.

Legislative History and Bill-Tracking

When courts interpret a statute, they may review the legislative history in order to ascertain the intent the legislative body had in enacting the law. Primary sources for reconstructing a legislative history include the record of legislative debate, official reports issued by the legislative body at the time of the statute's passage, and to a lesser extent the record of hearings conducted by the legislature prior to passage of the law. In the course of our discussion we have noted occasional references to legislative history appearing in the series containing state and federal statutes.

The record for the U.S. Congress is published in the **Congressional Record**. Most academic libraries, as well as many large public libraries have paper editions of the Congressional Record, which includes its own index. WESTLAW and LEXIS both provide online access to the full text of the Congressional Record, as do other commercial database vendors of government information such as **WASHINGTON ALERT** and **LEGI-SLATE**. The contents of the Congressional Record for recent years also can be accessed through Congress's Internet site, **Thomas (http://thomas/loc/gov)**.

The House and Senate reports that accompany legislation are published in a series called the **Serial Set**, which can be found in most research libraries. House and Senate reports also are reproduced in West's *United States Code, Congressional and Administrative News*, and in various online services. Printed copies of congressional hearings are distributed to research libraries that are designated as "depository" libraries for government publications. These hearings are indexed in the **Monthly Catalog of Government Publications** and other sources. Some hearings also are available through online vendors, and through the government's own **GPO Access** Internet location, **http://www.gpo.gov**. As a general rule, sources of legislative history for state legislation are not as well organized or accessible as references for federal legislation, although there is variation from state to state.

When primary legislative history sources such as those mentioned above are not available, especially for state legislation, criminal justice researchers may find secondary sources helpful. Law reviews may include articles that provide an overview of the legislative history of either a particular law or several laws concerning the same topic. Researchers also can find secondary information about legislative history in the "public policy" literature, which is loosely defined as political science, legal, economic , or other social science or news literature that takes a public policy perspective. PAIS (Public Affairs Information Service Bulletin), discussed previously in Chapter 2, is an excellent index to this genre of literature. It provides bibliographic references and in some cases short descriptive notes for journal articles, books, government documents (including some congressional hearings), and research reports that deal with public-policy aspects of issues. An excellent source for federal legislative history information is *CIS annual (Congressional Information Service annual)*, published by Congressional Information Service, which provides bibliographic citations and abstracts for all congressional publications.

Criminal justice researchers also may be interested in keeping apprised of the status of pending legislation, a process known as **bill-tracking**. Proposed legislation can have a major impact on people working in criminal justice and on criminal justice issues. Computer systems have greatly facilitated the process of monitoring proposed legislation. WESTLAW and LEXIS both have the capability of tracking federal and state legislative proposals. WASHINGTON ALERT and LEGI-SLATE specialize in monitoring federal legislation and in providing a range of background information about federal legislation. Two weekly periodicals, *CQ Weekly Report* and *National Journal*, provide excellent coverage of

Congress and its actions. *CQ Weekly Report* is published by Congressional Quarterly Inc., the same company that produces WASHINGTON ALERT. *National Journal* is published by National Journal Inc., which also produces LEGI-SLATE.

We will not discuss the specific methods and procedures for researching legislative histories and bill tracking. However, you should be aware that this capability exists, and you should be familiar with the major resources for performing these tasks.

Constitutions: Federal and State

The U.S. Constitution is a remarkable legal document. Ratified in 1788, when horses and buggies were the principal means of transportation in a country comprised of 13 states and populated by only a few million people, the Constitution still survives today, after being amended only 27 times, in a country of hundreds of millions, with technological complexities that the framers could not possibly have imagined. The U.S. Constitution endures as the oldest written national government charter in the world. It necessarily was written in broad and general language, to make it capable of evolving with the changing demands of society. Its grand provisions guarantee due process of law, the right to be free from unreasonable searches and seizures, protections against cruel and unusual punishments, and they enshrine other such lofty, fundamental principles.

State constitutions tend to be lengthier, more detailed, and more frequently revised and amended than the federal Constitution, although many of their provisions mirror or resemble the general guarantees and restrictions found in the U.S. Constitution. As we have discussed, state constitutions can offer greater rights to people than the federal Constitution promises, but they cannot cut back on protections. We will have a chance to illustrate this principle shortly. The present point is that both federal and state constitutional provisions speak in broad and general terms, consistent with their fundamental purposes. In this respect they typically differ from administrative regulations and statutes, which tend to be written with greater specificity.

One implication of this difference for legal research is that when you refer to a provision of the U.S. Constitution or a state constitution, you should not expect immediate enlightenment by reading the text of the document. Still, constitutions can be a handy starting point for legal research because, like the statutes we have considered, they are published in annotated editions. Annotated constitutions are used precisely in the same way that annotated statutes are used. The starting place for using printed sources is a general subject index that will refer you to the constitutional provisions most closely associated with the issue you are investigating. When you look up that provision, you will find references to related sources, an outline of notes to judicial decisions in which the provision has been discussed, and then brief notes describing those decisions with the accompanying case citations. You should expect to find either a pocket supplement or a cumulative supplemental pamphlet to keep the bound volume current.

Not all criminal justice issues have constitutional overtones, although a great many obviously do. We know that the hypothetical fact situation in Case 1 raises a constitutional issue concerning the right to confront accusing witnesses, and that the basis for the complaint in Case 3 is an alleged violation of the constitutional right to be free from cruel and unusual punishments. We can illustrate how to investigate these issues by using constitutional sources as legal research tools.

The United States Constitution: Printed Sources and Case 3

Since Deborah Miller's lawsuit is based on an alleged violation of her federal constitutional rights, we will confine our search to the U.S. Constitution. The federal Constitution can be found in the same volumes as federal statutes. The official version is reprinted in the United States Code, but far more helpful annotated versions are included in West's United States Code Annotated and Lawyers Cooperative's United States Code Service. These latter sources should be consulted for legal research purposes.

You often will know enough about the U.S. Constitution to be able to turn directly to the provisions that apply to your research issues. If you do not, consult the General Index in U.S.C.A. or U.S.C.S. for assistance. For example, we can look up **"Cruel and Unusual Punishment"** in the U.S.C.S. General Index, or we could begin with **"Constitution of United States"** in that Index and then refer to the subtopic, **"Cruel and unusual punishment**." Both of these starting points will direct us to the Eighth Amendment of the U.S. Constitution. (See Exhibit 3–47.) We then locate the Eighth Amendment in the appropriate constitutional volume. Note the remarkable brevity of this provision. It is only 16 words long, including the guarantee in which we are interested: "nor cruel and unusual punishments inflicted." (See Exhibit 3–48.)

The advantage of an annotated constitution is that it supplies information that greatly exceeds the unadorned words of the constitution itself. Exhibit 3–48 displays some of the other research aids for the Eighth Amendment that are included in U.S.C.S. We have excised several additional pages of references, including other A.L.R. annotations, treatises, and law review articles that appear in the **"Research Guide"** section. We eventually come to the **"Interpretive Notes and Decisions"** that will help us find cases applying the Eighth Amendment in different contexts. (See Exhibit 3–49.)

The logical beginning point is the general outline. One of the subtopics in the outline is **"Imprisonment or Confinement,"** and a subdivision of that is **"Conditions of Confinement . . . —Living Conditions . . . —Exposure to tobacco smoke**." This subdivision is aligned with note 119. Several cases are collected under note 119 that address whether a prisoner's involuntary exposure to tobacco smoke can amount to cruel and unusual punishment under the Eighth Amendment, including the Supreme Court case we have run across so frequently, *Helling v. McKinney*, 509 U.S. 25, 113 S. Ct. 2475, 125 L. Ed. 2d 22 (1993). (See Exhibit 3–50.) We round out our use of U.S.C.S.'s annotated Eighth Amendment by checking the supplement to this volume for later cases.

Searching Case 3 Using CALR

Case 3 provides us with an opportunity to demonstrate an alternative approach using CALR. Deborah Miller's case is predicated on two related legal claims: first, that the exposure to tobacco smoke violates her civil rights; and, second, that

Exhibit 3–47

UNITED STATES CODE SERVICE   **GENERAL INDEX**

CRUDE OIL WINDFALL PROFIT TAX ACT OF 1980—Cont'd

Interest for past periods resulting from amendments relating to cost recovery oil, prohibition, 26 § 4996 note

Low-income assistance, subaccount established in Windfall Profit Tax Account, 26 § 4986 note

Net revenues of the windfall profit tax, defined, 26 § 4986 note

President of U.S.
    allocation of net revenues, 26 § 4986 note
    report to Congress on tax, 26 § 4986 note

Report to Congress on revenues from tax and their disposition, 26 § 4986 note

Treasury, Windfall Profit Tax Account established in as separate account, 26 § 4986 note

CRUEL AND UNUSUAL PUNISHMENT

Constitution of United States, cruel and unusual punishment not to be inflicted, US Const 8th Amend

Military Justice Code, 10 § 855, 893

Virgin Islands, 48 § 1561

CONSTITUTION OF UNITED STATES— Cont'd

Crime—Cont'd
    extradition of fugitives from justice, US Const Art 4 sec 2 cl 2
    felony, infra
    fugitives, infra
    grand jury presentment or indictment necessity of, US Const 5th Amend
    involuntary servitude, punishment of crime as exception to prohibition of US Const 13th Amend
    jury trial, right, US Const Art 3 sec 2 cl 1; 6th Amend
    place of trial of crime not committed within any state, power of Congress to prescribe, US Const Art 3 sec 2 cl 3
    self-incrimination, privilege against, US Const 5th Amend
    speedy and public trial, right to, US Const 6th Amend
    suffrage, right of, may be denied to persons convicted of crime, US Const 14th Amend sec 2
    treason, infra
    witnesses, confronting and obtaining, US Const 6th Amend
Cruel and unusual punishment, prohibited, US Const 8th Amend

Source: U.S.C.S., Index "Cruel." Reprinted with the permission of LEXIS-NEXIS, a division of Reed Elsevier Inc. LEXIS and NEXIS are registered trademarks of Reed Elsevier Properties Inc. FREESTYLE, KWIC, SuperKWIC and MEGA are trademarks of Reed Elsevier Properties Inc. SHEPARD'S and SHEPARDIZE are registered trademarks of Shepard's Company, a Partnership.

---

the exposure constitutes "cruel and unusual punishment," which is prohibited by the Eighth Amendment to the federal Constitution. The particular factual circumstances of this case, involving a prisoner's exposure to tobacco smoke, are unlikely to be addressed in the text of a federal civil rights statute, and they certainly will not be addressed in the text of the Eighth Amendment. These circumstances are much more likely to be discussed in the squibs of court cases that interpret the statutes and/or the Eighth Amendment. In the printed sources we first had to locate the text of the appropriate statutory section or constitution and then proceed through the rather involved process of trying to locate cases in the **Notes** or **Notes of Decisions** sections following the text. CALR allows us to reverse the search

Exhibit 3–48

AMENDMENT 8

Bail–Punishment.

Excessive bail shall not be required, nor excessive fines imposed, nor cruel and unusual punishments inflicted.

CROSS REFERENCES

Federal statutory provisions as to release and bail, generally, 18 USCS §§ 3141 et seq.

States as prohibited from denying due process or equal protection of laws, generally, USCS Constitution, Amendment 14.

RESEARCH GUIDE

Federal Procedure L Ed:

9A Fed Proc L Ed, Criminal Procedure §§ 22:1658 et seq.

Am Jur:

8 Am Jur 2d, Bail and Recognizance §§ 1 et seq.

16A Am Jur 2d, Constitutional Law §§ 447, 448, 450.

21 Am Jur 2d, Criminal Law §§ 408, 421, 539, 615, 625–629.

21A Am Jur 2d, Criminal Law § 631.

Forms:

7 Fed Procedural Forms L Ed, Criminal Procedure §§ 20:107 et seq.

Annotations:

Comment Note.—What provisions of the Federal Constitution's Bill of Rights are applicable to the states. 18 L Ed 2d 1388, 23 L Ed 2d 985.

Procedural requirements under Federal Constitution in juvenile delinquency proceedings—federal cases. 25 L Ed 2d 950.

Considerations affecting grant, continuance, reduction, or revocation of bail by individual justice of Supreme Court. 30 L Ed 2d 952.

Federal constitutional guaranty against cruel and unusual punishment. 33 L Ed 2d 932.

Implication of private right of action from provision of United States Constitution. 64 L Ed 2d 872.

Denial to incarcerated persons of contact visits as violation of federal constitutional rights. 82 L Ed 2d 1006.

Supreme Court's views on constitutionality of death penalty and procedures under which it is imposed or carried out. 90 L Ed 2d 1001.

Supreme Court's construction and application of provision of Federal Constitution's Eighth Amendment that excessive bail shall not be required. 95 L Ed 2d 1010.

Supreme Court's construction and application of excessive fines clause of Federal Constitution's Eighth Amendment. 106 L Ed 2d 729.

Supreme Court's views as to validity, under due process clause of Federal

Source: U.S.C.S., "8th amend." Reprinted with the permission of LEXIS-NEXIS, a division of Reed Elsevier Inc. LEXIS and NEXIS are registered trademarks of Reed Elsevier Properties Inc. FREESTYLE, KWIC, SuperKWIC and MEGA are trademarks of Reed Elsevier Properties Inc. SHEPARD'S and SHEPARDIZE are registered trademarks of Shepard's Company, a Partnership.

Exhibit 3–49

INTERPRETIVE NOTES AND DECISIONS

I. IN GENERAL
1. Generally
II. BAIL
A. In General
2. Generally
3. Applicability to states
4. Right to bail and prohibition against excessive bail distinguished
5. Validity of particular statutes regulating bail
6. Bail pending appeal
7. Habeas corpus relief
8. Civil action
9. Miscellaneous
B. Factors Considered in Setting Bail
10. Assuring appearance of accused
11. Ability to pay
12. Gravity of offense
III. FINES
13. Generally
14. Applicability to particular proceedings
15. Ability to pay
16. Amounts
17. Imprisonment for nonpayment
18. Fine and imprisonment
19. Miscellaneous
IV. PUNISHMENT
A. In General
20. Generally

---

2. Imprisonment or Confinement
a. In General
62. Generally

63. Aggregation of sentences
64. Consecutive sentences
65. Disparate sentences among codefendants
66. Sentence as habitual criminal or recidivist

---

c. Conditions of Confinement
####### (1). In General
102. Generally
103. Standing
104. Pretrial detainment, generally
####### (2). Disciplinary Measures
105. Generally
106. Segregated confinement
107. —Administrative segregation
108. —Maximum security
109. —Solitary confinement; punitive isolation
110. —Use of force
111. —Beatings
112. —Stun gun
113. —Other particular circumstances
114. —Miscellaneous
####### (3). Living Conditions
115. Generally
116. Cell space; overcrowding
117. —Double celling
118. Exposure to asbestos
119. Exposure to tobacco smoke ◀
120. Ventilation, lighting, temperature, sanitation and hygiene
121. Miscellaneous

Source: U.S.C.S., "8th amend," Interpretive. Reprinted with the permission of LEXIS-NEXIS, a division of Reed Elsevier Inc. LEXIS and NEXIS are registered trademarks of Reed Elsevier Properties Inc. FREESTYLE, KWIC, SuperKWIC and MEGA are trademarks of Reed Elsevier Properties Inc. SHEPARD'S and SHEPARDIZE are registered trademarks of Shepard's Company, a Partnership.

process. We first search in the squibs of court decisions for cases dealing with the issue of tobacco smoke in a prison setting and then limit our search results to documents dealing with either civil rights or the Eighth Amendment.

This search can be done on either WESTLAW or LEXIS. We will demonstrate it using the LEXIS service. We will use segment searching in the Boolean search-

Exhibit 3–50

119. Exposure to tobacco smoke

Health risk allegedly posed by prison personnel's exposure of inmate to environmental tobacco smoke is proper ground for claim for relief under 8th Amendment. Helling v McKinney (1993, US) 125 L Ed 2d 22, 113 S Ct 2475, 93 CDOS 4501, 93 Daily Journal DAR 7681, 7 FLW Fed S 452, on remand, remanded (1993, CA9) 5 F3d 365, 93 CDOS 6923, 93 Daily Journal DAR 11818.

Allegation of involuntary exposure to environmental tobacco smoke states cause of action under Eighth Amendment for unreasonable risk of harm to health whether or not inmate has pre-existing condition sensitive to smoke. McKinney v Anderson (1991, CA9 Nev) 924 F2d 1500, 91 CDOS 1003, 91 Daily Journal DAR 1476, vacated, remanded (1991, US) 116 L Ed 2d 236, 112 S Ct 291 and reinstated, remanded on other grounds (1992, CA9 Nev) 959 F2d 853, 92 CDOS 2622, 92 Daily Journal DAR 4164, affd, remanded (1993, US) 125 L Ed 2d 22, 113 S Ct 2475, 93 CDOS 4501, 93 Daily Journal DAR 7681, 7 FLW Fed S 452, on remand, remanded (1993, CA9) 5 F3d 365, 93 CDOS 6923, 93 Daily Journal DAR 11818.

In deciding whether to allow smoking in prison, officials considered effects on both smokers and nonsmokers, and decision to resolve conflict in favor of smokers cannot be viewed as "punishment" for purpose of Eighth Amendment. Steading v Thompson (1991, CA7 Ill) 941 F2d 498, cert den (1992, US) 117 L Ed 2d 445, 112 S Ct 1206.

Source: U.S.C.S., "8th amend" squib. Reprinted with the permission of LEXIS-NEXIS, a division of Reed Elsevier Inc. LEXIS and NEXIS are registered trademarks of Reed Elsevier Properties Inc. FREESTYLE, KWIC, SuperKWIC and MEGA are trademarks of Reed Elsevier Properties Inc. SHEPARD'S and SHEPARDIZE are registered trademarks of Shepard's Company, a Partnership.

ing mode. The file we will be searching is **USCODE**, which is the online equivalent of the United States Code Service. **USCODE** is accessible through the **CODES** library.

Exhibit 3–51 depicts the search request screen with the search strategy already typed: "**Casenotes(smok! w/15 prison! or inmate* or imprisonment)**". Having used LEXIS in Case 2, you should be able to interpret most of this search strategy. The **Casenotes** segment limits our search results to documents where the terms inside of the parentheses appear in the squibs of court cases that follow the text of a statute or constitutional provision. In each squib, a word beginning with the root "**smok**" must appear within 15 words of a word beginning with the root "**prison**," or the word "**inmate**," or "**imprisonment**." The asterisk after "**inmate**" will retrieve "**inmates**" as well as "**inmate**."

Exhibit 3–52 indicates that we have retrieved 14 documents that meet the requirements of our search strategy. In practice we would probably start browsing through these documents, but for the sake of example, we will take an additional step. We can modify our query by **entering** the command **.m**. Exhibit 3–53 displays the screen with our modification already typed. Notice that we have to begin our modification with a Boolean operator, in this instance **and**. Our modification **and "Heading(civil rights or (constitution and amendment 8))**," requires that the term "**civil rights**," or the term "**constitution**," accompanied by the term "**amendment 8**," appear in the **HEADING** segment. (Recall that the **Heading** segment includes the name of the title, article, or chapter of each section of the code.)

Exhibit 3–51

                                    LEXIS

Casenotes (smok! w/15 prison! or inmate* or imprisonment)

Please type your search request then press the ENTER key.
What you enter will be Search Level 1.

Type .fr to enter a FREESTYLE (TM) search.

For further explanation, press the H key (for HELP) and then the ENTER key.

Source: USCODE, "Smoke." Reprinted with the permission of LEXIS-NEXIS, a division of Reed Elsevier Inc.
LEXIS and NEXIS are registered trademarks of Reed Elsevier Properties Inc. FREESTYLE, KWIC, SuperKWIC and
MEGA are trademarks of Reed Elsevier Properties Inc. SHEPARD'S and SHEPARDIZE are registered trademarks of
Shepard's Company, a Partnership.

Exhibit 3–52

                                    LEXIS

.m
CASENOTES (SMOK! W/15 PRISON! OR INMATE* OR IMPRISONMENT)

Your search request has found 14 DOCUMENTS through Level 1.
To DISPLAY these DOCUMENTS press either the KWIC, FULL, CITE or SEGMTS key.
To MODIFY your search request, press the M key (for MODFY) and then the ENTER key.

For further explanation, press the H key (for HELP) and then the ENTER key.

Source: USCODE, "Smoke" modified. Reprinted with the permission of LEXIS-NEXIS, a division of Reed Elsevier
Inc. LEXIS and NEXIS are registered trademarks of Reed Elsevier Properties Inc. FREESTYLE, KWIC, SuperKWIC
and MEGA are trademarks of Reed Elsevier Properties Inc. SHEPARD'S and SHEPARDIZE are registered trademarks
of Shepard's Company, a Partnership.

Exhibit 3–53

LEXIS

and Heading (civil rights or (constitution and amendment 8))

Your search request is:
    CASENOTES (SMOK! W/15 PRISON! OR INMATE* OR IMPRISONMENT)

Number of DOCUMENTS found with your search request through:
    LEVEL 1 . . . 14

Please enter the modification to your search request (Level 2).
REMEMBER to start your modification with a CONNECTOR.

For further explanation, press the H key (for HELP) and then the ENTER key.

Source: USCODE, "Smoke" result. Reprinted with the permission of LEXIS-NEXIS, a division of Reed Elsevier Inc. LEXIS and NEXIS are registered trademarks of Reed Elsevier Properties Inc. FREESTYLE, KWIC, SuperKWIC and MEGA are trademarks of Reed Elsevier Properties Inc. SHEPARD'S and SHEPARDIZE are registered trademarks of Shepard's Company, a Partnership.

Exhibit 3–54 indicates that four documents remain from our original 14 after the modification. The citations to those documents are displayed in Exhibits 3–55 and 3–56. Notice that three of the cites are to the same section of the civil rights law, 42 U.S.C.S. § 1983. Section 1983 contains so many case notes that it was divided into five sections. We have retrieved the three sections that contain squibs that include our search terms. The fourth citation is a reference to the Eighth Amendment of the U.S. Constitution.

We can browse the documents by using the **.kw** command, and quickly find case references that appear promising. Exhibit 3–57 displays a case squib interpreting 42 U.S.C.S. § 1983, and Exhibit 3–58 displays the squib to the now-familiar *Helling v. McKinney* as an interpretation of the Eighth Amendment.

Case 1: Printed Constitutional Sources

We can tackle the confrontation issue presented in Case 1 from the perspective of both the federal Constitution and a state constitution. This time we will assume that we already know that the U.S. Constitution's Sixth Amendment confrontation clause is implicated. If we did not, we could easily learn that much by looking up "**Confrontation**" in the General Index of U.S.C.A. or U.S.C.S. For this example, we will refer to U.S.C.A. We must select a specific jurisdiction to illustrate how a state constitution's provisions might figure into our analysis. Let us assume again for the purposes of this example that John Winston's sexual assault trial is taking place in Indiana.

Exhibit 3–54

             LEXIS

-

AND HEADING (CIVIL RIGHTS OR (CONSTITUTION AND AMENDMENT 8))

Your search request has found 4 DOCUMENTS through Level 2.
To DISPLAY these DOCUMENTS press either the KWIC, FULL, CITE or SEGMTS key.
To MODIFY your search request, press the M key (for MODFY) and then the ENTER key.

For further explanation, press the H key (for HELP) and then the ENTER key.

Source: USCODE, "Smoke" modified. Reprinted with the permission of LEXIS-NEXIS, a division of Reed Elsevier Inc. LEXIS and NEXIS are registered trademarks of Reed Elsevier Properties Inc. FREESTYLE, KWIC, SuperKWIC and MEGA are trademarks of Reed Elsevier Properties Inc. SHEPARD'S and SHEPARDIZE are registered trademarks of Shepard's Company, a Partnership.

Exhibit 3–55

             LEXIS

To be able to browse preceding or succeeding code sections, enter B. The first page of the document you are currently viewing will be displayed in FULL. NOTE: This ability only applies when the =B= prompt is displayed in the cite.

- -
LEVEL 2 — 4 DOCUMENTS

1. =B= 42 USCS @ 1983 (1996), TITLE 42. THE PUBLIC HEALTH AND WELFARE, CHAPTER 21.
CIVIL RIGHTS, GENERALLY, THE CASE NOTES SEGMENT OF THIS DOCUMENT HAS BEEN
SPLIT INTO 5 DOCUMENTS. THIS IS PART 2. USE THE BROWSE FEATURE TO REVIEW THE
OTHER PART(S)., 1983. Civil action for deprivation of rights, United States Code Service
2. =B= 42 USCS @ 1983 (1996), TITLE 42. THE PUBLIC HEALTH AND WELFARE, CHAPTER 21.
CIVIL RIGHTS, GENERALLY, THE CASE NOTES SEGMENT OF THIS DOCUMENT HAS BEEN
SPLIT INTO 5 DOCUMENTS. THIS IS PART 3. USE THE BROWSE FEATURE TO REVIEW THE
OTHER PART(S)., 1983. Civil action for deprivation of rights, United States Code Service
3. =B= 42 USCS @ 1983 (1996). TITLE 42. THE PUBLIC HEALTH AND WELFARE, CHAPTER 21.
CIVIL RIGHTS, GENERALLY, THE CASE NOTES SEGMENT OF THIS DOCUMENT HAS BEEN
SPLIT INTO 5 DOCUMENTS. THIS IS PART 5. USE THE BROWSE FEATURE TO REVIEW THE
OTHER PART(S)., @ 1983. Civil action for deprivation of rights, United States Code Service

Source: USCODE, "Smoke" citations. Reprinted with the permission of LEXIS-NEXIS, a division of Reed Elsevier Inc. LEXIS and NEXIS are registered trademarks of Reed Elsevier Properties Inc. FREESTYLE, KWIC, SuperKWIC and MEGA are trademarks of Reed Elsevier Properties Inc. SHEPARD'S and SHEPARDIZE are registered trademarks of Shepard's Company, a Partnership.

Exhibit 3–56

	LEXIS	

LEVEL 2 — 4 DOCUMENTS

4. =B= USCS Const. Amend. 8 (1996), CONSTITUTION OF THE UNITED STATES OF AMERICA AMENDMENTS, AMENDMENT 8, Bail—Punishment., United States Code Service

Source: USCODE, "Smoke" citations. Reprinted with the permission of LEXIS-NEXIS, a division of Reed Elsevier Inc. LEXIS and NEXIS are registered trademarks of Reed Elsevier Properties Inc. FREESTYLE, KWIC, SuperKWIC and MEGA are trademarks of Reed Elsevier Properties Inc. SHEPARD'S and SHEPARDIZE are registered trademarks of Shepard's Company, a Partnership.

Exhibit 3–57

	LEXIS	

42 USCS @ 1983 (1996)

NOTES:

. . . CA6 Ohio) 972 F2d 712, reh den (1992, CA6) <=1437> 1992 US App LEXIS 23375.

Prison superintendent was entitled to qualified immunity in <=1438> 42 USCS @ 1983 action by inmate alleging that he was involuntarily exposed to environmental tobacco smoke in violation of Eighth Amendment, where at time superintendent assigned inmate to cell with smoking cellmate, there was no clearly established constitutional right to be free from exposure to environmental tobacco smoke. <=1439> Murphy v Dowd (1992, CA8 Mo) 975 F2d 435, cert den (1993,US) <=1440> 122 L Ed 2d 698, 113 S Ct 1310.

Evidence was insufficient to sustain finding that . . .

NOTES:

. . . 1019, 100 L Ed 2d 221, 108 S Ct 1760.

Tobacco company did not become state actor for purposes of <=2412> 42 USCS @ 1983 action by asthmatic inmate challenging presence of smoke in air of prison, on basis that company sold its products to government. <=2413> Steading v Thompson (1991, CA7 Ill) 941 F2d 498, cert den (1992, US) <=2414> 117 L Ed 2d 445, 112 S Ct . . .

NOTES:

. . . provision of toiletries and clothing changes cannot be deemed

Source: USCODE, "Smoke" browse. Reprinted with the permission of LEXIS-NEXIS, a division of Reed Elsevier Inc. LEXIS and NEXIS are registered trademarks of Reed Elsevier Properties Inc. FREESTYLE, KWIC, SuperKWIC and MEGA are trademarks of Reed Elsevier Properties Inc. SHEPARD'S and SHEPARDIZE are registered trademarks of Shepard's Company, a Partnership.

Exhibit 3–58

                        LEXIS

USCS Const. Amend. 8 (1996)

inmate was permitted to smoke cigarettes occasionally, to take showers, and to consult with his attorney concerning postconviction proceedings for which his temporary transfer was initially made. <=1906> Gladson v Rice (1988, CA8 Iowa) 862 F2d 711.
  Law was clearly . . .

  . . . by permitting asbestos to fall from ceiling without notice, information, warning, or protection. <=2062> Nasim v Warden, Md. House of Correction (1995, CA4 Md) 42 F3d 1472.

119. Exposure to tobacco smoke
  Health risk allegedly posed by prison personnel's exposure of inmate to environmental tobacco smoke is proper ground for claim for relief under 8th Amendment. <=2063> Helling v McKinney (1993, US) 125 L Ed 2d 22, 113 S Ct 2475, 93 CDOS 4501, . . .

  . . . 7 FLW Fed S 452, on remand, remanded (1993, CA9) <=2064> 5 F3d 365, 93 CDOS 6923, 93 Daily Journal DAR 11818.
  Allegation of involuntary exposure to environmental tobacco smoke states cause of action under Eighth Amendment for unreasonable risk of harm to health whether or not inmate has pre-existing condition sensitive to smoke. <=2065> McKinney v Anderson (1991, CA9 Nev) 924 F2d 1500, 91 CDOS 1003, 91

Source: USCODE, "Smoke" Helling. Reprinted with the permission of LEXIS-NEXIS, a division of Reed Elsevier Inc. LEXIS and NEXIS are registered trademarks of Reed Elsevier Properties Inc. FREESTYLE, KWIC, SuperKWIC and MEGA are trademarks of Reed Elsevier Properties Inc. SHEPARD'S and SHEPARDIZE are registered trademarks of Shepard's Company, a Partnership.

We first look up the Sixth Amendment in the appropriate "**Constitution**" volume of U.S.C.A. The operative section of this Amendment is: "In all criminal prosecutions, the accused shall enjoy the right . . . to be confronted with the witnesses against him." (See Exhibit 3–59.) Since this is an annotated version of the Constitution, we are provided with **Historical Notes and Cross References**, and then several different sources related to the Sixth Amendment under the Library References section. But our most helpful research aid is likely to be the Notes of Decisions outline and the casenotes and citations we find through this service. The broad topic outline for the Notes of Decisions reflects that confrontation clause issues are covered under Roman numerals XVIII through XXIII. (See Exhibit 3–60.) The more detailed outlines will help narrow our inquiry to the specific issue we are researching. A promising note appears under section XXI, "**Confrontation with Witnesses—Admissibility of Prosecution Evidence; Hearsay.**" The specific subtopic and note that seem to pertain to our issue are: "**Videotaping of . . . Sexually abused child testimony 1394.**" (See Exhibit 3–61.)

The hardbound volume of the U.S.C.A. that contains the Sixth Amendment was published in 1987, so the cases discussed in the accompanying notes will be rela-

Exhibit 3–59

AMENDMENT VI—JURY TRIAL FOR CRIMES, AND PROCEDURAL RIGHTS

In all criminal prosecutions, the accused shall enjoy the right to a speedy and public trial, by an impartial jury of the State and district wherein the crime shall have been committed, which district shall have been previously ascertained by law, and to be informed of the nature and cause of the accusation; to be confronted with the witnesses against him; to have compulsory process for obtaining witnesses in his favor, and to have the Assistance of Counsel for his defense.

HISTORICAL NOTES

Proposal and Ratification

The first ten amendments to the Constitution were proposed to the Legislatures of the several States by the First Congress on September 25, 1789, and were ratified on December 15, 1791. For the States which ratified these amendments, and the dates of ratification, see Historical notes under Amendment I.

CROSS REFERENCES

Jury trial in criminal case, see section 2, clause 3, of Article 3.

LIBRARY REFERENCES

Forms

Arraignment, see West's Federal Forms § 7251 et seq.

Assignment of counsel, see West's Federal Forms § 7811 et seq.

Subpoenas, see West's Federal Forms § 7391 et seq.

Trial by jury or by court, see West's Federal Forms § 7451 et seq.

Trial jurors, see West's Federal Forms § 7471 et seq.

Law Reviews

Fair play: Evidence favorable to an accused and effective assistance of counsel. Barbara Allen Babcock, 34 Stan.L.Rev. 1133 (1982).

Ineffective assistance of counsel. Joel Jay Finer, 58 Cornell L.Rev. 1077 (1973).

Jones v Barnes, the Sixth, and the Fourteenth Amendments: Whose appeal is it, anyway? Karen A Krisher, 47 Ohio St.L.J. 179 (1986).

State constitutions and the protection of individual rights. William J. Brennan, 90 Harvard L.Rev. 489 (1977).

Texts and Treatises

Absolute right to conviction by unanimous verdict, see Wright & Miller, Federal Practice and Procedure: Civil § 2492.

Admissibility of irrelevant evidence, see Wright & Graham, Federal Practice Procedure: Evidence § 5202.

Admissible or inadmissible evidence, see Wright & Graham, Federal Practice and Procedure: Evidence § 5191 et seq.

Admission of incriminating extrajudicial statements, see Wright, Federal Practice and Procedure: Criminal 2d § 224.

Appointment of interpreters, see Wright, Federal Practice and Procedure: Criminal 2d § 451 et seq.

Assistance of counsel, scope and elements, see LaFave and Israel, Criminal Procedure § 11.1 et seq.

Compulsion of admission of certain evidence, see Wright & Graham, Federal Practice and Procedure: Evidence § 5387.

Source: U.S.C.A., "Amend VI," © West Publishing, used with permission.

Exhibit 3–60

JURY TRIAL FOR CRIMES   **AMEND. 6**

NOTES OF DECISIONS

I.	GENERALLY 1–50
II.	SPEEDY TRIAL—GENERALLY 51–100
III.	DELAYS PROTECTED AGAINST 101–150
IV.	CONSIDERATIONS GOVERNING VIOLATION GENERALLY 151–210
V.	PREJUDICE FROM DELAY 211–260
VI.	REASONS FOR DELAY 261–340
VII.	WAIVER 341–380
VIII.	PUBLIC TRIAL 381–460
IX.	JURY TRIAL—GENERALLY 461–530
X.	PROCEEDINGS IN WHICH AVAILABLE 531–570
XI.	OFFENSES TRIABLE BY JURY 571–620
XII.	IMPARTIALITY OF JURY GENERALLY 621–690
XIII.	COMPOSITION OR SELECTION OF JURY 691–800
XIV.	PUBLICITY SURROUNDING TRIAL 801–830
XV.	WAIVER 831–880
XVI.	STATE AND DISTRICT OF TRIAL 881–940
XVII.	INFORMATION RESPECTING ACCUSATION 941–1030
XVIII.	CONFRONTATION WITH WITNESSES—GENERALLY 1031–1140
XIX.	PRESENCE OF DEFENDANT 1141–1190
XX.	DISCOVERY, DISCLOSURE, OR INSPECTION 1191–1230
XXI.	ADMISSIBILITY OF PROSECUTION EVIDENCE; HEARSAY 1231–1420
XXII.	CROSS-EXAMINATION BY DEFENSE 1421–1560
XXIII.	WAIVER OF RIGHT 1561–1620
XXIV.	COMPULSORY PROCESS 1621–1730
XXV.	ASSISTANCE OF COUNSEL—GENERALLY 1731–1790
XXVI.	PROCEEDINGS TO WHICH AMENDMENT APPLIES 1791–1860
XXVII.	STAGES AT WHICH REQUIRED 1861–2020
XXVIII.	CHOICE OF COUNSEL 2021–2060
XXIX.	APPOINTMENT OF COUNSEL 2061–2100
XXX.	EFFECTIVE ASSISTANCE 2101–2360
XXXI.	CONFLICT OF INTEREST 2361–2460
XXXII.	WAIVER 2461–2540
XXXIII.	SELF-REPRESENTATION 2541–2581

Source: U.S.C.A., "Amend VI" Notes-table, © West Publishing, used with permission.

Exhibit 3–61

Source: U.S.C.A., "Amend VI" table-sub, © West Publishing, used with permission.

tively dated. We first check the cases excerpted under note 1394 in the bound volume, but we can expect to have better luck in the paperbound supplementary pamphlet, which collects more recent case notes. We turn to note 1394 that accompanies the Sixth Amendment in that supplementary pamphlet. Beginning with the Supreme Court's decision in *Maryland v. Craig*, we are provided with several pages of notes and citations to both federal and state cases that may be of interest to our problem. (See Exhibit 3–62.)

We eventually will read *Maryland v. Craig* in one of the case reporters for U.S. Supreme Court decisions. Based on what we have learned from different secondary authorities and notes of decisions, it appears that *Craig* stands for the proposition that the televised presentation of testimony by a child who is the alleged victim of a sexual assault and who is outside of the defendant's physical presence does not invariably violate the accused's Sixth Amendment right to confront accusing witnesses. However, even if the federal Constitution permits such a procedure, it is possible that the state constitution in the jurisdiction where the defendant is being tried prohibits it. We must check for this possibility by consulting the Indiana Constitution, since we are assuming that John Winston's trial is taking place in a state court in Indiana.

The Indiana Constitution is found with Indiana's statutory collection. We can investigate whether confrontation rights are guaranteed under the state constitution by consulting the General Index in the annotated Indiana Constitution. Subjects are defined very generally in this index. Our most promising strategy is to begin with "Criminal Law and Procedure" and review the accompanying subheadings. The index of the Indiana Constitution, which covers the state Bill of

Exhibit 3–62

1394.—Videotaping of testimony

Child assault victim's testimony at trial of child abuse defendant through use of one-way closed circuit television procedure authorized by Maryland child witness protection statute did not impinge upon the truth seeking nor symbolic purposes of the confrontation clause; procedure required that child witness be competent to testify and testify under oath, defendant retained full opportunity for contemporaneous cross-examination, and judge, jury and defendant were able to view witness' demeanor and body by video monitor. Maryland v. Craig, U.S.Md.1990, 110 S.Ct. 3157, 497 U.S. 836, 111 L.Ed.2d 666, on remand 588 A.2d 328, 322 Md. 418.

Violation of child sexual molestation defendant's confrontation rights, when trial court allowed jury to view videotaped interview between child and police in which child accused defendant of molesting her and demonstrated what had allegedly occurred using anatomically correct dolls, was not mere harmless error, without regard to sufficiency of other nontainted evidence; tre-

mendous impact and emotional appeal of videotape was such that, despite medical and other evidence that child had been sexually abused, Court of Appeals could not say, on federal habeas review, that error in admitting videotape did not have substantial and injurious effect or influence in determining jury's verdict, particularly given evidence of prior fabrications of sexual abuse by child. Offor v. Scott, C.A.5 (Tex.) 1995, 72 F.3d 30.

State trial court's findings about possibility of harm to five-year-old victim were insufficient to avoid violation of defendant's confrontation clause rights by virtue of victim testifying before television camera outside of courtroom and outside presence of defendant in prosecution for attempted capital sexual battery; no one at trial appeared to have considered defendant's confrontation clause rights and there was nothing to indicate that victim was afraid of defendant or that testifying by closed circuit television would enhance protection she needed. Cumbie v. . . . .

Source: U.S.C.A., "Amend VI" squib, © West Publishing, used with permission.

Rights, lists "**Right to jury trial**," "**Rights of accused**," and "**Witnesses—Right of accused to compulsory process**." (See Exhibit 3–63.) Although we have not located a direct reference to confrontation rights, we would be remiss if we did not examine the Bill of Rights protections of the state constitution, specifically Article 1, § 13. We find in Article 1 § 13 a guarantee that is like the Sixth Amendment's right of confrontation but that is worded in interestingly different terms: "In all criminal prosecutions, the accused shall have the right . . . to meet the witnesses face to face." (See Exhibit 3–64.) If this provision is applied literally, it very well may confer rights on criminal defendants that exceed the Sixth Amendment's protections as construed in *Maryland v. Craig*. We can investigate this possibility by using the Notes to Decisions to look for cases that have applied this provision. We quickly spot the heading "**Confrontation of witnesses**" and the subtopic "**Statements of children**," in the subject listings. (See Exhibit 3–65.)

Instead of numbered notes, which were used with the constitutional provisions in U.S.C.S. and U.S.C.A., an alphabetical arrangement of topics and subtopics is used for the "Notes to Decisions" in the annotated Indiana Constitution. The pocket supplement, under "**Confrontation of Witnesses**" and the subheading

Exhibit 3–63

CONSTITUTION OF INDIANA

CRIMINAL LAW AND PROCEDURE—Cont'd
Bill of rights—Cont'd
  Jury.
    Right to jury trial, Const. Ind., Art. 1, §13.
  Prisoners.
    Unnecessary rigor prohibited, Const. Ind.,
    Art. 1, §15.
  Reformation as basis of penal code, Const.
  Ind., Art. 1, §18.

Rights of accused, Const. Ind., Art. 1, §13.
Self-incrimination, Const. Ind., Art. 1, §14.
Treason, Const. Ind., Art. 1, §§28, 29.
Witnesses.
  Right of accused to compulsory process,
  Const. Ind., Art. 1, §13.
  Self-incrimination, Const. Ind., Art. 1, §14.
Prosecutions, Const. Ind., Art. 7, §18.

Source: The statutes reprinted or quoted verbatim in the following pages are taken from the Burns Indiana Statutes Annotated, Copyright by Michie, a division of Reed Elsevier Inc. and Reed Elsevier Properties Inc., and are reprinted with the permission of Michie. All rights reserved.

Exhibit 3–64

Art. 1, § 13   CONSTITUTION OF INDIANA

§ 13. Rights of accused.—In all criminal prosecutions, the accused shall have the right to a public trial, by an impartial jury, in the county in which the offense shall have been committed; to be heard by himself and counsel; to demand the nature and cause of the accusation against him, and to have a copy thereof; to meet the witnesses face to face, and to have compulsory process for obtaining witnesses in his favor.

Source: The statutes reprinted or quoted verbatim in the following pages are taken from the Burns Indiana Statutes Annotated, Copyright by Michie, a division of Reed Elsevier Inc. and Reed Elsevier Properties Inc., and are reprinted with the permission of Michie. All rights reserved.

"**Statements of Children,**" proves to have a more promising collection of case notes than does the main volume. We display this section of the supplement in Exhibit 3–66.

The different case summaries appear to be somewhat in conflict, or at least ambiguous, regarding the law that governs our issue. This should serve as a reminder about how important it is to read case decisions in full, rather than to rely on the brief notes that accompany an annotated constitution or statute. For instance, the first case presented, *Brady v. State*, 575 N.E.2d 981 (Ind. 1991), appears to hold that the statute authorizing the use of videotaped testimony of a child witness is consistent with the Sixth Amendment to the U.S. Constitution but that it violates an accused's right under the Indiana Constitution to "meet the witness face to face." However, the very next case that is noted, *Hart v. State*, 578 N.E.2d 336 (Ind. 1991), apparently holds that the Indiana Constitution is not offended by the practice of presenting videotaped testimony. It is not clear from either of these

Exhibit 3–65

NOTES TO DECISIONS

ANALYSIS

In general.
Alibi witness.
Applicability of section.
—Traffic infractions.
Appointment of experts.
Confrontation of informer.
Confrontation of witnesses.
—Compulsory process.
—Habitual offender proceedings.
—Hearsay evidence.
—Limitation on examination.
—Presence at deposition.
—Statements of children.
Contempt proceedings.
Counsel.
—As witness.
—Change of counsel.
——From private to public.
—Dismissal on morning of trial.
—Inadequate representation.
——Found.
——Not found.
—Informing of right.
—Joint representation.
—Out-of-state attorney.
—Preparation for trial.
—Right to be heard.
——Competency.

—Supplied by court.
—Waiver.
—Withdrawal.
Course of the trial.
Criminal proceedings.
Definiteness of indictment.
Errors.
—Cumulative effect.
Essential elements of crime.
—Necessity to plead.
Grand juries.
Guilty plea.
Impartial judge.
Impartial jury.
—Burden of proof.
—Consumption of alcohol.
—Defendant in prison garb.
—Habitual offender statute.
—Poll as to publicity.
—Replay of trial tapes.
—Voir dire.
Indefiniteness in criminal pleading.
Informant.
—Materiality.
Instructions.
—Objection.
Instructions outside issues.
Jury selection.

Source: The statutes reprinted or quoted verbatim in the following pages are taken from the Burns Indiana Statutes Annotated, Copyright by Michie, a division of Reed Elsevier Inc. and Reed Elsevier Properties Inc., and are reprinted with the permission of Michie. All rights reserved.

case excerpts whether the accused and the witness met face-to-face during the videotaping of the child's testimony. Thus, while we have found some potentially interesting information by using the annotated Indiana Constitution, we will use what we have learned to stimulate additional research involving the cited cases, rather than treating the information as the end of our research efforts.

CALR and Case 1

We can safely assume that we are not going to find references to the particular fact situation in Case 1 in the text of the constitution. Rather, the factual circumstances are likely to be found in the court decision squibs that follow the text. In

Exhibit 3–66

Art. 1, § 13   CONSTITUTION OF INDIANA

Confrontation of Witnesses. (Cont'd)
—Statements of Children.

IC 35-37-4-8, which authorizes the use of videotaped testimony of child witnesses at trial, satisfies the requirements of the federal confrontation right guaranteed by the sixth amendment but violates the "face-to-face" requirement of this section. Brady v. State, 575 N.E.2d 981 (Ind. 1991).

The statutory authorization in IC 35-37-4-8 permitting a child's testimony to be taken before trial and videotaped for use at trial does not offend the Indiana constitutional face-to-face requirement. Hart v. State, 578 N.E.2d 336 (Ind. 1991).

Failure to apply the face-to-face requirement of the Indiana Constitution did not constitute fundamental error, where the videotaped testimony of a child-molestation victim was consistent with traditional judicial fact-finding procedures, and the witness's bodily movements and facial expression provided the jury viewing the videotape a full and adequate opportunity to assess demeanor and credibility. Hart v. State, 578 N.E.2d 336 (Ind. 1991).

IC 35-37-4-6, which authorizes the use of videotaped testimony of child witnesses at trial, is valid under the Indiana Constitution. Casselman v. State, 582 N.E.2d 432 (Ind. App. 1991).

Out-of-court statements made by three-year-old boy regarding molestation by defendant were sufficiently reliable to be admissible, and the use of this statement was not a violation of the right of confrontation guaranteed by Art. I, § 13 of the Indiana Constitution where a meeting occurred between defendant and the boy at the pre-trial hearing, which was recorded and transcribed, during which the child repeated the accusations. Arndt v. State, 642 N.E.2d 224 (Ind. 1994).

Source: The statutes reprinted or quoted verbatim in the following pages are taken from the Burns Indiana Statutes Annotated, Copyright by Michie, a division of Reed Elsevier Inc. and Reed Elsevier Properties Inc., and are reprinted with the permission of Michie. All rights reserved.

---

this example we are only interested in investigating sections of the federal and state constitutions relevant to the particular confrontation issue.

We will use WESTLAW for this example and start our search in the **USCA** database. **USCA** is the online equivalent of United States Code Annotated. It includes the text and annotations to the U.S. Constitution as well as the text and annotations of federal statutory law. Since WESTLAW includes the index to **USCA**, we could simply replicate the process used with the printed versions. Such a technique would not provide a very good example of CALR, nor would it be much fun, so we will employ a strategy similar to the one we used with LEXIS in the previous example. Whereas documents in LEXIS (remember that each section of a statute is considered a document) are broken down into **segments**, in WESTLAW the documents are similarly divided into portions called **fields**. We will not list and describe all the fields in WESTLAW's statutory databases here. Instead, we confine our discussion to the fields that are important to our current search topic.

The **text (TE)** field in WESTLAW contains the same information as the text segments in LEXIS, i.e., the actual text of the document. The **prelim (PR)** field on WESTLAW contains the same information as the **heading** segment in LEXIS—that is, the names of the title, article, and chapter. The **annos** field in WESTLAW includes the squibs of court decisions and is comparable to the

casenotes segment in LEXIS. For reasons that will soon become apparent, it is important to note that only the **text** information is identical in both systems.

We rely on the **Terms and Connectors** search mode in WESTLAW, which allows us to use field restrictions as well as proximity requirements and Boolean operators. We have typed the search strategy depicted in Exhibit 3–67, "**PR (constitution & amendment) & ANNOS(child! /p witness! testimony /p confron! / p televis!)**". The computer thus searches for the word "**constitution**" and the word "**amendment**" in the **prelim** field. In the **annotation** field it searches for a word beginning with the root "**child**" in the same paragraph as either a word beginning

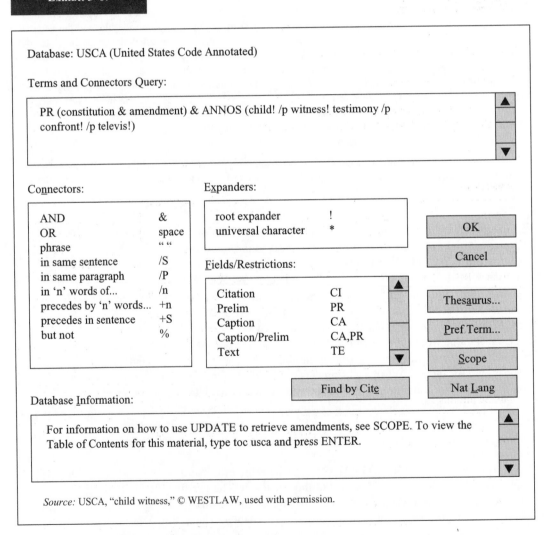

Exhibit 3–67

Database: USCA (United States Code Annotated)

Terms and Connectors Query:

PR (constitution & amendment) & ANNOS (child! /p witness! testimony /p confront! /p televis!)

Connectors:

AND	&
OR	space
phrase	" "
in same sentence	/S
in same paragraph	/P
in 'n' words of...	/n
precedes by 'n' words...	+n
precedes in sentence	+S
but not	%

Expanders:

root expander	!
universal character	*

Fields/Restrictions:

Citation	CI
Prelim	PR
Caption	CA
Caption/Prelim	CA,PR
Text	TE

OK

Cancel

Thesaurus...

Pref Term...

Scope

Find by Cite

Nat Lang

Database Information:

For information on how to use UPDATE to retrieve amendments, see SCOPE. To view the Table of Contents for this material, type toc usca and press ENTER.

Source: USCA, "child witness," © WESTLAW, used with permission.

with the root "**witness**" or the word "**testimony**." This combination must be in the same paragraph with a word beginning with the root "**confront**" and in the same paragraph with a word beginning with the root "**televis**."

Before we enter our search strategy, we want to check the **Thesaurus** to determine if there are additional terms that should be added. We have previously discussed this process on WESTLAW, so we need not retrace it. We pick out the following related terms to add to our search strategy: "**minor**" as a related term for "**child!**"; "**face**" as a related term for "**confront!**"; and "**T.V.**" as a related term for "**televis!**". The edited search query appears in Exhibit 3–68.

Exhibit 3–68

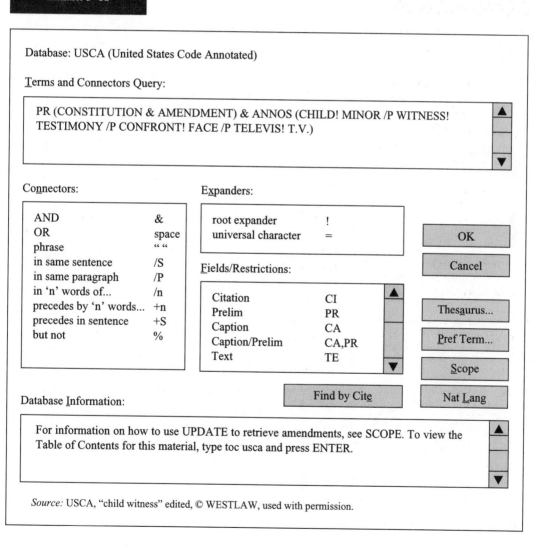

Source: USCA, "child witness" edited, © WESTLAW, used with permission.

The first page of the search result is displayed in Exhibit 3–69. Notice that the complete text of the Sixth Amendment appears on the first page (screen) of the document, but that the document itself is 4983 pages long. Most of the remaining pages are consumed with squibs of court decisions interpreting the Sixth Amendment. One of the advantages of performing this search on WESTLAW or LEXIS is that the computer systems provide us with quick access to the squibs containing our search terms. Before we check those squibs, we want to draw your attention to the fact that the Sixth Amendment appears as **Amendment VI** in the **prelim** field. The use of the Roman numeral **VI** instead of **6** is one difference in format between the United States Code Annotated and the United States Code Service. This seemingly small difference can be a major problem in CALR. The computer reads only what is entered and does not translate terms. We could have included the Roman numeral **VI** in our search strategy, but sometimes it is important not to add more information than you actually need.

We now can examine the squibs of court decisions interpreting the Sixth Amendment. In **Terms and Connectors** searching, the default browsing mode is **term**, as indicated in the top right of Exhibit 3–69. Recall that in the **term** mode we will only be looking at the initial screen of a document and those portions of the document in which the search terms appear. In our example, we only have to hit the **enter** key twice to arrive at the screen depicted in Exhibit 3–70. This squib

---

Exhibit 3–69

Copr. © West 1996 No claim to orig. U.S. govt. works

AUTHORIZED FOR EDUCATIONAL USE ONLY

Citation	Rank (R)	Page (P)	Database	Mode
USCA CONST Amend. VI	R 1 OF 3	P 1 OF 4983	USCA	Term
U.S.C.A. Const. Amend. VI				

▶ UNITED STATES CODE ANNOTATED
▶ CONSTITUTION OF THE UNITED STATES
▶ AMENDMENT VI—JURY TRIAL FOR CRIMES, AND PROCEDURAL RIGHTS
Copr. (C) West 1996. All rights reserved.
Current through P.L. 104–133, approved 4–25–96

▶ Amendment VI. Jury trials for crimes, and procedural rights

In all criminal prosecutions, the accused shall enjoy the right to a speedy and public trial, by an impartial jury of the State and district wherein the crime shall have been committed, which district shall have been previously ascertained by law, and to be informed of the nature and cause of the accusation; to be confronted with the witnesses against him; to have compulsory process for obtaining witnesses in his favor, and to have the Assistance of Counsel for his defence.

Source: USCA, "child witness" result, © WESTLAW, used with permission.

Exhibit 3–70

Copr. © West 1996 No claim to orig. U.S. govt. works
AUTHORIZED FOR EDUCATIONAL USE ONLY

USCA CONST Amend. VI   R 1 OF 3   P 3956 OF 4983   USCA   Term

ANNOTATIONS ▶ Subdivision Index XXI

▶ 1394, —— Videotaping of testimony

Child assault victim's testimony at trial of child abuse defendant through use of one-way closed circuit television procedure authorized by Maryland child witness protection statute did not impinge upon the truth seeking nor symbolic purposes of the confrontation clause; procedure required that child witness be competent to testify and testify under oath, defendant retained full opportunity for contemporaneous cross-examination, and judge, jury and defendant were able to view witness' demeanor and body by video monitor. Maryland v. Craig, U.S.Md.1990, ▶ 110 S.Ct. 3157, 497 U.S. 836, 111 L.Ed.2d 666, on remand ▶ 588 A.2d 328, 322 Md. 418.

Violation of child sexual molestation defendant's confrontation rights, when trial court allowed jury to view videotaped interview between child and police in which child accused defendant of molesting her and demonstrated what had allegedly occurred using anatomically correct dolls, was not mere harmless error, without regard to sufficiency of other nontainted evidence; tremendous impact and emotional appeal of videotape was such that, despite medical and other evidence that child had been sexually abused, Court of Appeals could not . . .

Source: USCA, "child witness" squib, © WESTLAW, used with permission.

corresponds to *Maryland v. Craig*. Recall that you can proceed directly to the text of *Maryland v. Craig* by placing the cursor on the arrow preceding the cite and pressing **enter**. To return from the case to your search result, enter **gb** for "**go back**".

The next step is to look for the section in the Indiana Constitution that deals with the same topic. Recall that in WESTLAW you can run the current search strategy in a different database through the use of the command **sdb**, or "search database," followed by the identifier of the new database. This command can be used from any screen in the search results. In our example, we enter "**sdb IN-ST-ANN**". (**IN-ST-ANN** is the identifier for the annotated version of the Indiana Code.)

When we performed this command, we received the message that no documents in the database met the requirements of our search strategy. Before jumping to the conclusion that there is nothing in the Indiana Code that addresses our issue, we should reexamine the search strategy to determine if it needs editing. Observe that the strategy requires that both the word "**constitution**" and the word "**amendment**" must appear in the **prelim** field. We are interested only in sections of the state constitution pertaining to our topic so we want to retain "**constitution**" in the **prelim** field. However, the section of the Indiana Constitution dealing with this particular right does not necessarily have to be in an amendment; it could appear in a section of the original constitution. Therefore, we alter our search strategy by **entering q** to get back our query screen and then deleting "**amendment**." The edited query retrieves the document displayed in Exhibit 3–71.

This search retrieves just one document, but it is the section of the Indiana Constitution dealing with rights of the accused in criminal prosecutions. The text of this provision is displayed on page 1, but the document is 1210 pages long, indicating that a huge number of squibs of court decisions are provided that interpret this section of the constitution. Before we refer to the squibs, notice that neither the word "**confront**" nor the word "**confrontation**" appears in the text of this section; instead, the phrase "**face to face**" is used. Our search strategy did not search the text field, but if we had, we would not have retrieved this document if we had not included related terms for "**confront**." This example should reinforce the importance of including related terms in your strategy.

We are in **term** mode. By twice pressing the **enter** key we arrive at the screen displayed in Exhibit 3–72. We find the same squibs we located through our print-based search, with the added feature that we can use the arrows in front of the cites to hypertext to the cases.

Exhibit 3–71

Copr. © West 1996 No claim to orig. U.S. govt. works

AUTHORIZED FOR EDUCATIONAL USE ONLY

Citation	Rank (R)	Page (P)	Database	Mode
IN CONST Art. 1, §13	R 1 OF 1	P 1 OF 1210	IN-ST-ANN	Term
Const. Art. 1, § 13				

▶ WEST'S ANNOTATED INDIANA CODE
▶ CONSTITUTION OF THE STATE OF INDIANA [ANNOTATED]
▶ ARTICLE 1. BILL OF RIGHTS.
Copr. (C) West 1996. All rights reserved.
Current with amendments received through 7–1–96

▶ § 13 Rights of accused in criminal prosecutions

Section 13. In all criminal prosecutions, the accused shall have the right to a public trial, by an impartial jury, in the county in which the offense shall have been committed; to be heard by himself and counsel; to demand the nature and cause of the accusation against him, and to have a copy thereof; to meet the witnesses face to face, and to have compulsory process for obtaining witnesses in his favor.

PROPOSED AMENDMENT

< Senate Enrolled Joint Resolution No. 8 [P.L.177–1994] of the 1994

Source: IN-ST-ANN "child witness" result, © WESTLAW, used with permission.

Exhibit 3–72

Copr. © West 1996 No claim to orig. U.S. govt. works
AUTHORIZED FOR EDUCATIONAL USE ONLY
IN CONST Art. 1, § 13   R1 OF 1   P 1137 OF 1210   IN-ST-ANN   Term
ANNOTATIONS ▶ Subdivision Index VII

Indiana Constitution cannot be abridged by judicial or legislative action, Brady v. State, 1991, ▶ 575 N.E.2d 981.

Indiana constitutional guarantee to criminal defendants of right to meet witnesses face to face is in nature of privilege which concerns individual defendant and bears only upon the procedure at trial. Brady v. State, 1991, ▶ 575 N.E.2d 981.

Victim's mere nervousness is not sufficient to constitute particularized finding of need for child to avoid confrontation with accused that would be sufficient to outweigh defendant's constitutionally protected right to be physically confronted by witness against him. Casada v. State, App. 1 Dist. 1989, ▶ 544 N.E.2d 189, transfer denied.

Confrontation clause under this article requires face to face confrontation, while federal clause mandates only general right to be confronted with witnesses, so that rights guaranteed by this article and Federal Constitution may differ to some degree. Miller v. State, 1987, ▶ 517 N.E.2d 64.

Source: IN-ST-ANN "child witness" squib, © WESTLAW, used with permission.

Case Law

We have repeatedly emphasized the importance of finding judicial decisions in the course of legal research efforts. This goal should be especially clear now that we have considered other forms of primary legal authority, which are so heavily dependent on interpretive case law to give them meaning. Locating case law on point for the issues you are investigating is both the most challenging and the most rewarding aspect of legal research. In this section we describe how to find judicial decisions that shed light on the issues you are exploring and, once having found those cases, how to confirm their continuing validity and how to use them as a springboard for finding still other cases that may be of interest.

At the risk of some redundancy, we remind you not to forget the numerous ways we already have covered for finding relevant judicial decisions. Virtually all of the secondary and other primary authorities we have discussed can direct you to case law pertaining to an issue. Once you find even one case dealing with the point that is of interest, you will be amazed at how quickly your search for related cases can snowball. Thus, you should never hesitate to start your hunt for judicial decisions by using A.L.R. annotations, law review articles, annotated statutes and constitutions, or other forms of authority. We now focus our attention on the techniques and materials that allow you to find case law directly.

Case Digests, Descriptive Word Indexes, and West's Key Number System

The first steps toward finding case law are taken using case digests and the accompanying **Descriptive Word Indexes**. We will focus on digests and indexes arranged according to West Publishing Company's **key number system**. This remarkable system is what makes case law research possible in printed sources. We describe the general operation of West's key number system and then give examples of its use as we continue to research the issues in the three hypothetical cases we have been examining.

The editors at West Publishing Company have developed a detailed scheme for classifying the legal issues addressed in judicial decisions. They have identified seven **Main Divisions of Law** (Persons, Property, Contracts, Torts, Crimes, Remedies, and Government), which in turn are subdivided into well over 400 discrete **Digest Topics**. The individual digest topics, which are listed in the front of all West case digests, include headings such as Assault and Battery, Civil Rights, Constitutional Law, Criminal Law, Drugs and Narcotics, Habeas Corpus, Homicide, Prisons, Searches and Seizures, Weapons, and hundreds more related to other legal issues. Each of the 400-plus topics is further divided into what may amount to thousands of unique subtopics. Each subtopic is assigned its own number. A digest topic (for example, Assault and Battery) and the number assigned to one of its subtopics combine to make a West **key number**—for example, Assault and Battery ☞ 56.

Each unique topic and key number corresponds to a principle of law announced in a judicial decision. As the editors at West prepare judicial decisions for publication in West case reporters, they systematically scour a court's lead opinion and identify the different legal principles that are discussed. Dissenting and concurring opinions are not subjected to this process. The editors then superimpose the de-

tailed key number classification scheme on the legal principles they have extracted from the court opinion. The published decisions of all courts—from the lowest state courts
to the U.S. Supreme Court—are processed in this fashion.
The same topic and key number are assigned to the same
principle of law no matter what court decides a case.
Through this process, West is able to identify all cases in-

A SQUIB?

volving the same principle of law and collect them under the same topic and key
number, such as Assault and Battery ⌖ 56.

As they cull judicial opinions in this fashion, West's editors extract or create a
brief summary of the court's discussion of the identified legal principles. Each of
these brief case excerpts is then transferred to the appropriate **case digest**. A case
digest collects and displays the brief summaries from decisions, sometimes referred to as "**squibs**" or "**blurbs**", under the appropriate topic and key number,
along with the case name and citation. West publishes several different case digests, which cover the various court systems or jurisdictions.

The most inclusive set is the **American Digest System**, which covers all reported state and federal court decisions. When you use these volumes you almost
always will be working with the **Decennial Digests** and the **General Digests**,
although a **Century Edition** also exists. The Century Edition of the American
Digest System presents case squibs under topic and key numbers for cases decided between 1658 and 1896. Thereafter, West began publishing the Decennial
Digests, which include squibs under topic and key numbers for cases decided
during ten-year intervals. The First Decennial Digest covers decisions from the
decade 1897-1906; the Second Decennial Digest covers 1907-1916, and so on. The
last decade covered in full is the Eighth Decennial Digest, which corresponds to
the 1966-1976 period.

The West editors then decided that ten years was too long a period to cover
before publishing the collected cases. Although West retains the name Decennial
Digest, a new set of digests is now published every five years. Thus, the Ninth
Decennial Digest, Part 1 collects notes of decisions from 1976 through 1981, and
the Ninth Decennial Digest, Part 2 covers 1981 through 1986. As this book is
being written, the latest complete set is the Tenth Decennial Digest, Part 1, which
covers cases decided from 1986 through 1991. Part 2 of the Tenth Decennial Digest, covering cases from the 1991 to 1996 period, is presently being compiled
but the set remains incomplete as of this writing.

Legal researchers always welcome the dawning of a new five-year cycle, with
an up-to-date Part 1 or Part 2 of a Decennial Digest. The reason for this concerns
what must be done during the years between the publications of each five years'
worth of cases. The supplementation of each five-year collection involves West's
General Digests, which are used as companions to the Decennial Digest.

The modern Decennial Digests, with their Part 1 and Part 2 format, bring together in one book all cases decided over a five-year period that involve the same
principle of law. For example, all cases decided between 1986 and 1991 that involve the legal principle associated with Assault and Battery ⌖ 56 can be conveniently located by referring to a single volume of the Tenth Decennial Digest,
Part 1; cases decided in the 1991 through 1996 interval are included in the Tenth

Decennial Digest, Part 2 under the same topic and key number. After consulting these cases in the most recently published Decennial Digest, the researcher will have to check the General Digest, a series of individual digests that are published roughly 14 times each year, that include case squibs under the relevant topic and key numbers.

As we write this, the researcher would have to check 10 individual volumes of the General Digest, Ninth Series, under Assault and Battery ⚷ 56 in order to find squibs from cases decided after 1996, or the period not fully covered by the Tenth Decennial Digest, Part 2. This task is made slightly less onerous through the use of the **Table of Key Numbers**, which can be found in every tenth volume of the General Digest series. The researcher can look up a particular key number in the table and determine which volumes include squibs for that key number. For example, a researcher consulting the Table of Key Numbers appearing in volume 10 of the General Digest and looking under key number Assault and Battery ⚷ 56, may find that volumes 1, 3, 4, 5, 6, 7, 8, and 9 contain squibs for that key number. When it becomes available, the researcher can check the Table of Key Numbers in volume 20 for references to squibs for Assault and Battery ⚷ 56 appearing in volumes 11 through 20, and proceed in a similar manner through the rest of the volumes in the General Digest series.

Because the Tenth Decennial Digest, Part 2 has not yet been fully compiled, the task becomes much more tedious for some topics. We will have to consult all General Digests published after the Tenth Decennial Digest, Part 1 that include Assault and Battery ⚷ 56 to find the case squibs from 1991 through the present that deal with the principle of law associated with that topic and key number. Whenever you use the Decennial Digests, you should always anticipate having to refer to the accompanying General Digests in order to keep abreast of cases decided after the Decennial Digest volumes were published.

As we have indicated, the Decennial and General Digests include squibs from *all* published state and federal court decisions, and they accordingly are the most comprehensive of West's digests. There may be times when you do not want or need all of the information presented in the Decennial and General Digests. For example, you may wish to confine your search to a particular state court system, or to the federal courts, or to the U.S. Supreme Court. West publishes other case digests that enable you to focus your search for cases more narrowly.

West's case digests are published for all states except Delaware, Nevada, and Utah. Case squibs from North Dakota and South Dakota are combined in the Dakotas Digest, and the Virginia and West Virginia Digest serves both of those states. Decisions from the District of Columbia Court of Appeals are collected in the Maryland Digest. State digests cover both state court decisions and the decisions of federal court cases originating in the focal state. A researcher frequently will be interested primarily in case law from a particular state, so state digests often come in handy.

West also publishes digests that correspond to four of its seven regional reporters, which helps explain how researchers interested in cases from Delaware, Nevada, and Utah can make do. These regional digests are the Atlantic, North Western, Pacific, and South Eastern Digests. Unlike the state digests, the **regional digests** present squibs only from state court decisions; they do not include federal court decisions.

West's **Federal Practice Digest** covers U.S. district court, U.S. court of appeals, and U.S. Supreme Court decisions. The Federal Practice Digest 4th is the most recent set in this series. Case digests that cover earlier federal court decisions begin with the Federal Digest and then progress through the Modern Federal Practice Digest, and the Federal Practice Digest 2d and Federal Practice Digest 3d. Although U.S. Supreme Court decisions are digested in the Federal Practice Digest series, West also publishes the **United States Supreme Court Digest**, which includes only Supreme Court decisions.

The West key number system is used in all of the West case digests, and it remains uniform throughout. Thus, if you find that a digest topic and key number correspond to the issue you are researching in the New York Digest 4th, you can safely assume that the same topic and key number will help you find related federal cases in the Federal Practice Digest 4th and related state cases from West's Pacific region in the Pacific Digest. Lawyers Cooperative publications do not use the West key number system. Since only West digests cover state and lower federal court decisions, the West system is absolutely indispensable for case law research. For this reason, we focus on the West key-number system, and the digests and Descriptive Word Indexes that make use of it.

There are different ways to find the topic and key number most appropriate for researching an issue. One way is by using the Descriptive Word Index. Each of the West case digests has its own Descriptive Word Index, although they are all structured similarly. We have urged that some combination of logic, common sense, ingenuity, and persistence is necessary when you consult subject indexes for other types of legal authorities and you are striving to find terms that describe the issues you are investigating. We renew this advice as you prepare to examine the Descriptive Word Indexes that accompany case digests, although more systematic strategies for identifying key words and terms describing your issue are available.

West recommends that search terms can be identified by analyzing a problem according to the following elements: (1) the *parties* involved in the case; (2) the *places* where the facts arose, and the *objects* and *things* involved; (3) the *basis* of the case or the case *issue*; and (4) the *relief sought*. Lawyers Cooperative recommends breaking a research problem down into similar parts: (1) the *things* involved in a case; (2) the *acts* relevant to the case; (3) *persons* involved in the case; and (4) *places* where the facts arose. You should try these strategies if you find them useful to parse issues as you employ Descriptive Word Indexes for case research.

Descriptive Word Indexes help link an abstract definition of the problem you are investigating to the topic and key number assigned to that issue under West's classification system. We illustrate how this process works by attempting to locate judicial decisions related to the issues in our hypothetical cases. We start with Case 2, in which Andrew Adams, knowing that he was infected with the HIV, has bitten and spit on Officer Fiegel. The prosecutor is interested in determining whether Adams can be charged with either attempted murder or assault with a deadly weapon.

Let us assume that, at least as far as we know, such an occurrence is unprecedented in the state in which this case arose, and we would like to learn whether these issues have been considered in other jurisdictions. We thus approach this search from the broadest possible jurisdictional perspective. If we are interested in case law from any court system, state or federal, throughout the country, that addresses these issues, we must select the case digests that capture this wide range of decisions. West's Decennial Digest and the supplemental General Digests are the appropriate starting points.

Terms that help describe the issue presented in Case 2 include "**AIDS**," "**assault**," "**attempt**," "**deadly weapon**," "**HIV**," and "**murder**," as well as "**bite**," "**spit**," and "**teeth**." Because the case involves a criminal prosecution, we also can check "**criminal law**," which is a convenient catchall category you should keep in mind as a potentially useful search term for criminal issues. Descriptive Word Indexes make heavy use of cross-referencing, so whichever term we look up first should not make a difference; we arrive at the same destination, no matter where we start or what path we take to get there. We start by looking up the terms we have chosen in the Descriptive Word Index accompanying the Tenth Decennial Digest, Part l, as that is the most recent compilation of the Decennial Digests available at the time of this writing.

Sometimes a Descriptive Word Index will point unambiguously to a search topic and key number; sometimes it will not. We have the misfortune of being in the latter category for this issue. The terms we have chosen do not lead precisely to a solution to our question, but they will have to do. Some of the terms are complete washouts; others offer at least a glimmer of hope. For example, we find two subtopics that may be of interest under "**Assault and Battery**," each of which refers us elsewhere: "**DANGEROUS or deadly weapons, see this index Dangerous and Deadly Weapons**," and "**HOMICIDE—Assault with intent to kill, see this index Homicide**." (See Exhibit 3–73.) If we start differently, and look up "attempt" in the Descriptive Word Index, we find the subcategory "**Homicide**," and the further subtopic, "**Evidence . . . —Weight and Sufficiency. Homic 256**." (See Exhibit 3–74.) This is our first concrete lead to a topic (homicide) and key number (256) in the Digest. Had we started with "**deadly weapon**" in the Descriptive Word Index, we would have been referred to "**Dangerous and Deadly Weapons**." Under that heading we see the subtopic "**ASSAULT with. Assault 56, 78, 92, 96(8)**." (See Exhibit 3–75.) "**Homicide**" has numerous subtopics, some of which may be interesting, including "**ASSAULT with intent to kill . . . — Evidence. Homic 176, 257**," "**ATTEMPTS, see this index Attempt**," and "**EVIDENCE—Ability . . . Execute intent in assault with intent to kill. Homic 257(2)**." (See Exhibit 3–76.)

The Descriptive Word Index has given us some leads but no direct hits as we move to the Tenth Decennial Digest, Part l. A sensible strategy at this point is to try to find out a bit more about the two digest topics to which we have been referred, "**Homicide**" and "**Assault**." We can do this by going to the Digest volumes that contain those terms and looking at the topic outlines at the beginning of the respective sections. We first go to the broad topic outline for "**Homicide**." (See Exhibit 3–77.) Recall that the Descriptive Word Index has referred us to Homicide ☞ 256, 257, and 257(2), among other key numbers, so we will want to pay particular attention to these sections. We turn to the detailed outline and learn that Homicide ☞ 256 concerns evidence of "**Attempt . . . to kill**," and thus corresponds conceptually to the issue we are investigating. (See Exhibit 3–78.) We pursue a similar inquiry with the lead we have received regarding "**Assault**." We

Exhibit 3–73

38–10th D Pt 1—90

ASSAULT AND BATTERY—Cont'd
BODILY harm, assault with intent to do great
  bodily harm. **Assault 55, 78, 92(2), 96(9)**
  Conviction of lesser grade or degree of offense
    than that charged. **Ind & Inf 189(4)**
BROKEN bottles—
  Dangerous weapons. **Assault 56**
BURDEN of proof. **Assault 82**
  Civil liability. **Assault 26**
  Instructions to jury. **Assault 43(4)**
CARRYING weapon—
  Probative value versus prejudicial effect. **Evid
    146**
CHARACTER of parties, evidence of. **Assault 29,
  85**
CIVIL liability—
  Insane defendants—
    Judgment. **Mental H 512**
CIVIL rights violation. **Civil R 113**
COMPLAINT in civil action. **Assault 24(1)**
CONDITIONS precedent to actions. **Assault 19**
CONDUCT of trial in criminal prosecution. **As-
  sault 94**
CONDUCT regardless of life. **Assault 48**
CONSENT as defense. **Assault 11, 65**
  Informed consent, medical procedures—
    Assault 2
    Phys 15(8, 15)
CONSPIRACY, evidence. **Consp 47(8)**
CONVICTS—
  Employees' liability for injuries in prisoner's
    assault on another prisoner. **Convicts 2**
  Liability—Prison employees. **Convicts 2**
COSTS. **Assault 46**
CRIMINAL responsibility. **Assault 47–100**
  Certainty in statutes. **Crim Law 13.1(7)**
CULPABLE mental state. **Assault 49**
DAMAGES. **Assault 36–40**
  Evidence of damages. **Assault 33, 34**
  Instructions. **Assault 43(5)**

ASSAULT AND BATTERY—Cont'd
  Passengers, damages for assault on. **Carr 319**
DANGEROUS or deadly weapons, see this index ◀
  Dangerous and Deadly Weapons
DECLARATION in civil actions. **Assault 24(1)**
DEFENSE of another. **Assault 14, 68**

EVIDENCE—Cont'd
  Civil actions. **Assault 25–35**
  Criminal prosecutions. **Assault 81–92**
    Other offenses, see this index **Other
      Offenses**
  Harmless error in criminal prosecutions. **Crim
    Law 1169.1(3)**
  Notice of intent to use—
    Victim's promiscuity. **Assault 85**
  Presumptions, generally, post
EXCESSIVE force in doing lawful act. **Assault 7**
EXEMPLARY damages. **Assault 39**
EXTENT of injury, evidence of. **Assault 32, 90**
FEDERAL control of railroad, liability for assault
  arising during Federal control. **R R 5¹/₂(11)**
FELLOW servants. **Emp Liab 114**
  Evidence. **Emp Liab 209**
FINE. **Assault 100**
FORMER jeopardy, see this index **Former Jeop-
  ardy**
GROUNDS and conditions precedent to actions.
  Assault 19
GUN as "deadly weapon". **Assault 54**
HANDS—
  Whether dangerous instruments or deadly
    weapons. **Assault 56**
HOMICIDE—
  Assault with intent to kill, see this index ◀
    Homicide
  Commission of or attempt to commit assault.
    Homic 63

Source: Dec. Digest, DWI "Homicide," © West Publishing, used with permission.

Exhibit 3–74

ATTEMPT
ARSON. **Arson 13**
  Indictment or information. **Arson 24**
ASSAULT and battery. **Assault 61**
  Indictment or information. **Assault 79**
BURGLARY. **Burg 11**
  Evidence. **Burg 41(10)**
  Indictment or information. **Burg 26**
CONVICTION of attempt to commit on charge of
  commission of offense. **Ind & Inf 190**
CRIME in general. **Crim Law 44**
DRUG offense. **Drugs & N 73**
ESCAPE. **Escape 5¹/₂**
  Relevancy of evidence. **Crim Law 351(9)**
FACTUAL impossibility. **Mil Jus 838**
FALSE pretenses. **False Pret 21**
FORGERY. **Forg 19**

HOMICIDE. **Homic 25**
  Admissibility of evidence of previous
    attempts. **Homic 167(8)**
  Evidence. **Homic 167(8), 256**
    Weight and sufficiency. **Homic 256**
  Indictment or information. **Homic 140**
  Manslaughter. **Homic 31**
  Other offense, homicide in attempt to commit.
    Homic 61–67
  Previous attempts as showing first-degree
    murder. **Homic 253(7)**
  Subsequent attempts by accused,
    incriminating or exculpatory
    circumstances. **Homic 174(4)**
INCLUDED offense instructions. **Crim Law**
  795(2.85)

Source: Dec. Digest, DWI "Attempt," © West Publishing, used with permission.

Exhibit 3–75

DAMS AND RESERVOIRS—Cont'd
  STORAGE of water—
    Irrigation or other agricultural purposes.
      Waters 243
  TAX valuation. **Tax 348.1(6)**
  WARNINGS. **Ship 11**
  WATER supply—
    Domestic and municipal purposes. **Waters**
    193

DANCE HALLS
  NUISANCE. **Nuis 3(9)**

REGULATION, licenses, and taxes. **Theaters**
  3.50

DANGEROUS AND DEADLY WEAPONS
  ASSAULT with. **Assault 56, 78 92, 96(8)**
    Intent to kill—
      Indictment or information. **Homic**
      141(9)
  FIST or hand. **Assault 56**
  PIPE—
    Aggravated robbery. **Rob 11**
  WEAPONS in general, see this index **Weapons**

Source: Dec. Digest, DWI "Dangerous," © West Publishing, used with permission.

consult the broad topic outline and the more detailed outline at the start of the topic, "**Assault and Battery**," in the Tenth Decennial Digest Part 1. (See Exhibits 3–79 and 3–80.) Note that Assault & Battery ☞ 56 corresponds to "**Assault with dangerous or deadly weapon**," which should be of interest as we continue to explore this issue.

Exhibit 3–76

HOMICIDE—Cont'd
APPEARANCES doctrine—
  Self-defense. **Homic 116(2)**
ARREST, homicide in making or resisting arrest,
  see this index **Arrest**
ASSAULT with intent to kill. **Homic 84–100**
  Automobiles. **Autos 345**
  Defense of—
    Another. **Homic 97**
    Habitation. **Homic 98**
    Property. **Homic 99**
  Evidence. **Homic 176, 257**
    Deadly character of assault. **Homic 257(8)**
  Indictment or information. **Homic 141**
    Conviction of lesser grade or degree than
      that charged. **Ind & Inf 189(3)**
    Duplicity in indictment or information. **Ind
      & Inf 125(29)**
    Following language of statute. **Ind & Inf
      110(17)**
  Instructions to jury, post

Means or instrument used, post
Nature of assault with intent to kill. **Homic 84,
  89**
ATROCITY, circumstances of as sufficient to es-
  tablish first-degree murder. **Homic 253(4)**
ATTEMPTS, see this index **Attempt**
AUTHORITY or duty, homicide in exercise of—
  Evidence. **Homic 184, 242**
  Instructions. **Homic 298**
  Justification. **Homic 103–106**

ESCAPE, homicide in connection with, see this
  index **Escape**
EVIDENCE—
  Ability. **Homic 168**
    Execute intent in assault with intent to kill.
      Homic 257(2)
  Accident. **Homic 248**
  Admissibility in general. **Homic 153–199**

Source: Dec. Digest, DWI "Assault," © West Publishing, used with permission.

Our next step is to turn to the pages in the Tenth Decennial Digest, Part l associated with the topic and key numbers we have identified for the issues we are investigating. Since Homicide ⌐⊸ 256 applies generally to "attempts, threats, or solicitation to kill," we can anticipate being referred to many cases that have nothing to do with the issue we are investigating. Indeed, we find this to be true for the first several case squibs under this key number. Note that the squibs begin with the federal court of highest jurisdiction. After the federal cases have run their course, state cases are presented, in alphabetical order of the states. (See Exhibit 3–81.) If we are persistent, our efforts are rewarded by finding a case decided by the Indiana Court of Appeals (a case we have encountered before) that involves an attempted murder charge against a defendant infected with the AIDS virus who bit and spat on another person: *State v. Haines*, 545 N.E.2d 834 (Ind. App. 1989). (See Exhibit 3–82.) Further investigation is likely to produce additional relevant cases.

Remember that the Tenth Decennial Digest, Part l only includes federal and state court cases decided during the 1986-1991 interval. To find more recent cases, we must check each and every one of the General Digests accompanying the Decennial Digest, under Homicide ⌐⊸ 256 because the Tenth Decennial Digest,

Exhibit 3–77

HOMICIDE

Analysis

    I. **THE HOMICIDE,** ☞ **1–6.**

   II. **MURDER,** ☞ **7–30(3).**

  III. **MANSLAUGHTER,** ☞ **31–83.**

  IV. **ASSAULT WITH INTENT TO KILL,** ☞ **84–100.**

   V. **EXCUSABLE OR JUSTIFIABLE HOMICIDE,** ☞ **101–126.**

  VI. **INDICTMENT AND INFORMATION,** ☞ **127–142(10).**

 VII. **EVIDENCE,** ☞ **143–257(8).**

     (A) PRESUMPTIONS AND BURDEN OF PROOF, ☞ 143–152.

     (B) ADMISSIBILITY IN GENERAL, ☞ 153–199.

     (C) DYING DECLARATIONS, ☞ 200–221.

     (D) PROCEEDINGS AT INQUEST, ☞ 222–227.

     (E) WEIGHT AND SUFFICIENCY, ☞ 228–257.

VIII. **TRIAL,** ☞ **258–315.**

     (A) CONDUCT IN GENERAL, ☞ 258–267.

     (B) QUESTIONS FOR JURY, ☞ 268–282½.

     (C) INSTRUCTIONS, ☞ 284–311.

     (D) VERDICT, ☞ 312–315.

  IX. **NEW TRIAL,** ☞ **316–321.**

   X. **APPEAL AND ERROR,** ☞ **322–349.**

Source: Dec. Digest, Homicide, © West Publishing, used with permission.

Part 2, has not been completed through this topic at the time of this writing. Once again, we can expect to be referred to many irrelevant cases, and most of the General Digests may include no cases that are of interest. From time to time we do get lucky, however, as with volume 54 of West's General Digest (1995). Here, under Homicide ☞ 256, we find a squib pertaining to *Weeks v. Scott,* 55 F.3d 1059 (5th Cir. 1995). (See Exhibit 3–83.)

We next consult the digests under Assault & Battery ☞ 56, which provides case squibs pertaining to "assault with dangerous or deadly weapon." (We reproduce here only the initial page of the Tenth Decennial Digest, Part l corresponding to this topic and key number.) The search process is strictly analogous to what we have just completed for Homicide ☞ 256. (See Exhibit 3–84.) We continue our search through the Decennial Digest and the accompanying General Digests under Assault & Battery ☞ 56 with the hope that we find one or more cases involving facts similar to Case 2.

Exhibit 3–78

HOMICIDE

VII. EVIDENCE.—Cont'd

246. Defense of habitation.
247. Defense of property.
248. Accident or misfortune.
249. Principals and accessories.
250. Degree of homicide in general.
251. Degree of murder.
252. —In general.
253. —First degree.
    (1). In general.
    (2). Sufficiency of circumstantial evidence in general.
    (3). Circumstances of cool blood, deliberation, and premeditation.
    (4). Circumstances of atrocity, cruelty, malignity and depravity.
    (5). Nature of means or instrument used.
    (6). Commission of or attempt to commit other offense.
    (7). Threats, preparations, and previous attempts.
    (8). Provocation or other extenuating circumstances.

254. —Second and lesser degrees.
255. Degree of manslaughter.
    (1). In general.
    (2). Voluntary or involuntary.
    (3). Degrees.
256. Attempt, threats, or solicitation to kill. ◀
257. Assault with intent to kill.
    (1). In general.
    (2). Ability to execute intent.
    (3). Showing as to grade of offense had death ensued.
    (4). Sufficiency of circumstantial evidence in general.
    (5). Conclusiveness of circumstances in general.
    (6). False and improbable statements accompanied by other evidence.
    (7). Threats accompanied by other evidence.
    (8). Deadly character of assault.

Source: Dec. Digest, Homicide-attempt, © West Publishing, used with permission.

Exhibit 3–79

ASSAULT AND BATTERY

Analysis

I. **CIVIL LIABILITY,** ⌐ **1–46.**
  (A) ACTS CONSTITUTING ASSAULT OR BATTERY AND LIABILITY THEREFOR, ⌐ 1–18.
  (B) ACTIONS, ⌐ 19–46.

II. **CRIMINAL RESPONSIBILITY,** ⌐ **47–100.**
  (A) OFFENSES, ⌐ 47–71.
  (B) PROSECUTION AND PUNISHMENT, ⌐ 72–100.

Source: Dec. Digest, Assault, © West Publishing, used with permission.

Exhibit 3–80

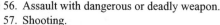

II. CRIMINAL RESPONSIBILITY.

(A) OFFENSES.

 47. Nature and elements of criminal assault.
48. —In general.
49. —Intent and malice.
50. —Ability to execute intent.
51. —Overt act in general.
52. —Unlawful act.
53. —Use of weapons.
54. Aggravated assault.
55. Assault with intent to do great bodily harm.
56. Assault with dangerous or deadly weapon.
57. Shooting.
58. Stabbing or other wounding.
59. Indecent assault.
60. Degrees.

Source: Dec. Digest, DWI "Assault-dangerous," © West Publishing, used with permission.

Exhibit 3–81

22 10th D Pt 1—1502

256. Attempt, threats, or solicitation to kill.
Library references
C.J.S. Homicide § 320.

C.A.5 (La.) 1987. Convictions for seven counts of attempted first-degree murder were supported by evidence that defendant fired shots at pursuing police cars in order to avoid arrest.—Procter v. Butler, 831 F.2d 1251, certiorari denied 109 S.Ct. 219, 488 U.S. 888, 102 L.Ed.2d 210.

Convictions on two counts of attempted murder were supported by evidence that defendant's car directly struck police car at roadblock and that police officer standing near police car was visible to defendant, and that defendant hit another police officer at another roadblock while he was in control of vehicle.—Id.

C.A.10 (Utah) 1991. Evidence supported convictions for attempt to kill defendant's former supervi-

sor that was employee of the Internal Revenue Service (IRS) on account of her performance of her official duties and for assault and intimidation of district director of the IRS on account of performance of her official duties, using deadly weapon in commission of such act, although "Molotov cocktail" firebombs thrown on roof of supervisor's residence did relatively little damage; Molotov cocktails were deadly weapons, defendant had made numerous threats against supervisor's life, and finding of fourth shell casing from handgun used to kill defendant's wife on night that firebombs were thrown permitted inference that defendant had handgun with him at time of firebombing. 18 U.S.C.A. §§ 111, 1114.—U.S. v. Treff, 924 F.2d 975, certiorari denied 111 S.Ct. 2272, 114 L.Ed.2d 723.

N.D.Ind. 1990. Sufficient evidence in prosecu-

continues

Exhibit 3–81   continued

tion for attempted murder and robbery supported finding that defendant used deadly weapon in manner likely to cause death, despite contention that shooting was accident; shotgun could only have been fired by pulling back on trigger with force in excess of 6 pounds, and victim's testimony indicated that defendant had taken time to level gun at victim.—Beadin v. Clark, 762 F.Supp. 243, affirmed 931 F.2d 58.

S.D.N.Y. 1987. Habeas petitioner's attempted murder conviction was sufficiently supported by testimony of victim, that habeas petitioner and petitioner's accomplice had stabbed him repeatedly with ice picks, though victim fatally injured accomplice and allegedly had motive to lie in order to avoid criminal prosecution. N.Y. McKinney's Penal Law §§ 110.00, 125.25.—Vargas v. Hoke, 664 F.Supp. 808.

E.D.Pa. 1989. Evidence was sufficient to sustain jury finding of guilt as to racketeering act involving conspiracy and attempted murder in RICO prosecution. 18 U.S.C.A. § 1961 et seq.—U.S. v. Scarfo, 711 F.Supp. 1315, affirmed U.S. v. Pungitore, 910 F.2d 1084, rehearing denied, certiorari denied Virgilio v. U.S., 111 S.Ct. 2009, 114 L.Ed.2d 98, 111 S.Ct. 2009, 114 L.Ed.2d 98, Pungitore v. U.S., 111 S.Ct. 2010, two cases, 114 L.Ed.2d 98, Staino v. U.S., 111 S.Ct. 2010, 114 L.Ed.2d 98, Scafidi v. U.S., 111 S.Ct. 2010, 114 L.Ed.2d 98, Grande v. U.S., 111 S.Ct. 2010, 114 L.Ed.2d 98, Ciancaglini v. U.S., 111 S.Ct. 2011, 114 L.Ed.2d 98 and Narducci v. U.S., 111 S.Ct. 2011, 114 L.Ed.2d 98.

Evidence was sufficient to sustain murder and attempted murder predicate acts in RICO prosecution. 18 U.S.C.A. § 1961 et seq.—Id.

Ala. 1987. Finding that attempted murder defendant had stabbed her newborn infant with scissors shortly after unassisted birth did not flow naturally from the evidence that the infant had eight or nine wounds in its neck all roughly parallel to each other, which were each one-fourth to three-fourth inches deep, in light of defendant's testimony that the wounds on infant's neck could have occurred when the defendant was attempting to force the infant out of the birth canal when its

shoulders would not pass, and evidence that the wounds could have been made by her fingernails.—Ex parte Mauricio, 523 So.2d 87.

Mother's explanation that wounds on newborn infant's neck occurred when mother attempted to force infant out of birth canal during unassisted birth was consistent with circumstantial evidence, which thus could not support her conviction of attempted murder.—Id.

Ala.Cr.App. 1990. Evidence was sufficient to support jury verdict of guilty on attempted murder charge, even if there was no proof that defendant actually participated in beating of victim, where there was sufficient proof, including defendant's statement and victim's videotaped statement, that defendant aided and abetted in commission of offense.—Ready v. State, 574 So.2d 894.

Ala.Cr.App. 1990. There was sufficient evidence to support defendant's conviction for attempted murder of woman, notwithstanding conflicting evidence; woman and her husband testified that defendant crouched and aimed a revolver at woman and that defendant fired one shot from revolver into the door of truck in which woman was sitting. Code 1975, §§ 13A–4–2, 13A–6–2.—Barnes v. State, 571 So.2d 372.

Ala.Cr.App. 1990. Evidence, although conflicting, including testimony of victim and investigating police officers, was sufficient to prove that defendant was shooting at victim with intent to kill him and to support defendant's conviction for attempted murder; State was not required to prove that other reasonable explanations did not exist. Code 1975, §§ 13A–4–2, 13A–6–2.—Wheeler v. State, 570 So.2d 876.

Ala.Cr.App. 1990. Finding that defendant who was charged with attempted murder thought that gun was loaded was supported by evidence that sheriff had used gun to fire warning shots which caused defendant to stop and that, shortly before defendant tried to shoot victim, defendant was involved in scuffle with sheriff over gun during which gun discharged seven or eight times.—Dover v. State, 570 So.2d 784.

Ala.Cr.App. 1989. Evidence was sufficient to establish that defendant intended to kill victim, . . .

Source: Dec. Digest, Homicide squib, © West Publishing, used with permission.

Exhibit 3–82

HOMICIDE ⌐ 256

. . . of shooting victim and eyewitnesses.—Toledano v. State, 498 N.E.2d 979.

Ind. 1986. Conviction for voluntary manslaughter and attempted voluntary manslaughter was sufficiently supported by testimony that, a few hours after victim and defendant quarreled, defendant returned to victim's apartment and shot victim and brother.—Bryant v. State, 498 N.E.2d 397.

Ind. 1986. Evidence sufficiently established that defendant convicted of attempted murder had requisite intent to commit murder; testimony included details of defendant's unprovoked infliction of multiple stabbings into victim's chest penetrating his heart and lungs, witnesses observed knife in defendant's hand, and police recovered knives from defendant's person.—Jenkins v. State, 497 N.E.2d 549.

Ind. 1986. Evidence, which indicated that defendant fired his rifle at garage owner after owner confronted him in his garage, was sufficient to support defendant's conviction for attempted voluntary manslaughter.—Mellott v. State, 496 N.E.2d 396, on rehearing 500 N.E.2d 173.

Ind. 1986. Testimony of surviving victim that, during drug sale, defendant shot surviving victim in face and that, as he fell, victim heard two other shots which killed second victim was sufficient to support defendant's convictions of murder and attempted murder, even though bullets in second victim came from two different weapons and an-

other individual was in room at time of shooting.—Allen v. State, 496 N.E.2d 53.

Ind. 1986. Evidence that, as victim proceeded to drive out of parking lot, defendant thrust his pocketknife into open driver's window and slashed victim across neck and shoulder, sustained conviction for attempted murder.—Lamotte v. State, 495 N.E.2d 729.

Ind.App. 1 Dist. 1991. Evidence that defendant fired more than one bullet into the heads of each of his victims was sufficient to support convictions for attempted murder.—Lowrance v. State, 565 N.E.2d 375.

Ind.App. 2 Dist. 1989. Attempted murder convictions were supported by evidence that defendant carried AIDS (Acquired Immune Deficiency Syndrome) virus, was aware of infection, believed it to be fatal, and intended to inflict others with disease by spitting, biting, scratching, and throwing blood, regardless of whether it was actually possible for him to spread virus that way. IC 35–41–5–1, 35–41–5–1(b) (1988 Ed.).—State v. Haines, 545 N.E.2d 834, rehearing denied, transfer denied.

La.App. 1 Cir. 1990. Conviction for attempted first-degree murder of sheriff was sufficiently supported by evidence that defendant fired three shots in direction of sheriff. LSA–R.S. 14:27, subd. A, 14:30, subd. A(2).—State v. Donahue, 572 So.2d 255.

Source: Dec. Digest, Homicide squib, © West Publishing, used with permission.

Now let us turn our attention to Case 1, where the issue is whether, consistent with the defendant's right to confront accusing witnesses, the testimony of a child who allegedly has been sexually assaulted can be presented by closed-circuit television so that the child does not have to testify in the presence of the accused. For illustrative purposes, we will choose a set of digests different from those we used for our research on Case 2. We begin with West's **Federal Practice Digest 4th**, which collects the most recent federal court decisions under the key-number system that we now have seen in operation.

Exhibit 3–83

⌐ **256. Attempt, threats, or solicitation to kill.** C.A.5 (Tex.) 1995. There was constitutionally sufficient evidence that act of spitting on another person by accused infected with human immunodeficiency virus (HIV) tended to cause death to support accused's conviction of attempted murder under Texas law, in light of expert testimony presented by prosecution that HIV could be transmitted through saliva, despite "mountain" of scientific evidence presented by accused that HIV could not be transmitted through saliva. V.T.C.A., Penal Code § 15.01(a).—Weeks v. Scott, 55 F.3d 1059.

Source: Dec. Digest, Homicide-attempt squib, © West Publishing, used with permission.

Exhibit 3–84

ASSAULT & BATTERY ⌐ **56**

⌐ **56. Assault with dangerous or deadly weapon.**
Library references.
C.J.S. Assault and Battery §§ 77–79.
C.A.11 (Ala.) 1991. A desk, overturned onto a person, was a "dangerous weapon" for purposes of the offense of assault with a dangerous weapon. 18 U.S.C.A. § 111.—U.S. v. Gholston, 932 F.2d 904.

C.A.11 (Fla.) 1986. Boat used to ram pursuing customs vessel was "deadly weapon" for purposes of assault with deadly weapon charge, where defendants were traveling at high rates of speed when attempted rammings occurred, and had driver been successful he could have severely damaged customs vessel and posed substantial danger to lives of personnel aboard.—U.S. v. Gualdado, 794 F.2d 1533, rehearing denied 802 F.2d 1399, certiorari denied Fernandez v. U.S., 107 S.Ct. 1327, 479 U.S. 1101, 94 L.Ed.2d 178.

C.A.7 (Ill.) 1989. Functionality of weapon is not prerequisite for prosecution under assaulting official of United States Penitentiary with dangerous weapon statute. 18 U.S.C.A. § 111.—U.S. v. Gometz, 879 F.2d 256, rehearing denied, certiorari denied 110 S.Ct. 752, 493 U.S. 1033, 107 L.Ed.2d 768.

"Zip gun" used by defendant in attempt to shoot prison guard constituted "dangerous weapon," within meaning of assaulting official of United States Penitentiary with "dangerous weapon" statute, even though "zip gun" was inoperable due to defective wiring. 18 U.S.C.A. § 111.—Id.

C.A.8 (Minn.) 1988. Almost any weapon, as used or attempted to be used, may endanger life or inflict great bodily harm; as such, in appropriate circumstances, it may be dangerous and deadly weapon.—U.S. v. Moore, 846 F.2d 1163, rehearing denied.

It is not necessary that object used by defendant actually caused great bodily harm, as long as it has capacity to inflict such harm in way it was used, in order for particular object to be considered deadly and dangerous weapon.—Id.

C.A.6 (Tenn.) 1990. Under statute prohibiting assault with dangerous weapon with intent to cause bodily injury, to be dangerous, weapon need not be gun or knife; object is "dangerous weapon" if it is used in manner likely to cause bodily harm. 18 U.S.C.A. § 113(c).—U.S. v. Gibson, 896 F.2d 206, denial of post-conviction relief affirmed 948 F.2d 1288.

Ala.Cr.App. 1989. For purposes of conviction for assault in second degree, portions of broken bottle inserted into rectum and vagina of victim were "dangerous instruments," and furthermore, such insertion constituted recklessness. Code

continues

Exhibit 3–84   continued

1975, §§ 13A–2–2, 13A–6–21.—Rothchild v. State, 558 So.2d 981.

Ala.Cr.App. 1989. Fists may constitute deadly weapons or dangerous instruments, depending upon the circumstances and manner of their use.—Brock v. State, 555 So.2d 285, appeal after remand 580 So.2d 1390.

Alaska App. 1988. Bare hand could be dangerous instrument used during third-degree assault; to determine whether hand was dangerous instrument, precise manner in which it was used needed to be examined. AS 11.41.220(a)(2), 11.81.900(b)(11, 40, 50).—Konrad v. State, 763 P.2d 1369.

Before hand may be deemed dangerous instrument in third-degree assault, State must present particularized evidence from which reasonable jurors could conclude that hand posed actual and substantial risk of death or serious physical injury, rather than merely hypothetical or abstract risk. AS 11.41.220(a)(2), 11.81.900(b)(11, 40, 50).—Id.

Defendant did not use his bare hands as "dangerous instruments" when he hit his estranged spouse on the head and ribs and did not commit third-degree assault, even though defendant offered to take spouse to doctor; defendant did not inflict serious injury, had never received martial arts training and was not otherwise skilled in using hands to inflict injury; and spouse's bleeding spleen healed without treatment within short period of time. AS 11.41.220(a)(2), 11.81.900(b)(11, 40, 50).—Id.

Cal.App. 2 Dist. 1989. Assault with a deadly weapon is "general intent crime" because it is not necessary to find specific intent to cause particular injury, but only general intent to willfully commit "battery," an act which has the direct, natural and probable consequences, if successfully completed, of causing injury to another; intent to frighten or mere reckless conduct is insufficient.—People v. Brown, 261 Cal. Rptr. 262, 212 C.A.3d 1409, opinion modified, review denied.

Cal.App. 5 Dist. 1991. Defendant's knowledge of probability of success of his or her intended action is not relevant to determination of whether defendant committed assault with deadly weapon. West's Ann.Cal.Penal Code § 245(a)(1).—People v. Craig, 278 Cal.Rptr. 39, 227 C.A.3d 644, review denied.

Cal.App. 5 Dist. 1989. Victim's fear, lack of fear, injury, or lack of injury are not elements which need to proved or disproved to convict defendant of assault with deadly weapon. West's Ann.Cal.Penal Code §§ 240, 245(a)(2).—People v. Griggs, 265 Cal.Rptr. 53, 216 C.A.3d 734, review denied.

The naming of the particular victim is not element of assault with a deadly weapon. West's Ann.Cal.Penal Code §§ 240, 245(a)(2).—Id.

Defendant could be convicted of assault upon the "person of another" with a firearm, even though intended victim was unknown, where defendant was clearly aware that crowd was assembled in direction in which he fired gun. West's Ann.Cal.Penal Code §§ 240, 245(a)(2).—Id.

Cal.Super. 1991. A "deadly weapon or instrument" within meaning of statute proscribing assault with deadly weapon must be an instrument used in such a manner that its use could produce and is likely to produce death or great bodily injury. West's Ann.Cal.Penal Code § 245(a)(1).—People v. Nealis, 283 Cal.Rptr. 376, 232 C.A.3d Supp. 1.

A dog trained to attack humans on command, or one without training that follows such a command, and which is of sufficient size and strength relative to its victim to inflict death or great bodily injury, may be considered a "deadly weapon or instrument" within meaning of statute proscribing assault with a deadly weapon. West's Ann.Cal.Penal Code § 245(a)(1).—Id.

D.C.App. 1990. Stationary bathroom fixtures were not "dangerous weapons" with which defendant could be armed within meaning of mayhem while armed and malicious disfigurement while armed statutes; attached sink, toilet, and bathtub against which defendant alleged hurled his wife were preexisting part of surroundings in which defendant found himself while perpetrating assault and not something which defendant could possess or with which he could arm himself. D.C.Code

continues

Exhibit 3–84   continued

1981, §§ 22–502, 22–506, 22–3202.—Edwards v. U.S., 583 A.2d 661.

D.C.App. 1988. Pencil is capable of causing bodily harm and thus may in some circumstances be "dangerous weapon." D.C.Code 1981, §§ 22–502, 22–506, 22–3202, 22–3202(a).—Wynn v. U.S., 538 A.2d 1139.

In determining whether weapon is dangerous weapon, best evidence of dangerous character is injury actually inflicted by weapon. D.C.Code 1981, §§ 22–502, 22–506, 22–3202, 22–3202(a).—Id.

Elements of "attempted battery" assault with dangerous weapon are attempt or effort, with force or violence, to do injury to person of another, apparent present ability to carry out such attempt or effort, general intent to engage in attempt or effort and dangerous weapon used in perpetration thereof. D.C.Code 1981, § 22–502.—Id.

D.C.App. 1987. To convict defendant of assault on police officer with dangerous weapon, the Government must prove elements of simple assault, use of dangerous weapon by defendant . . .

Source: Dec. Digest, Homicide-attempt squib, © West Publishing, used with permission.

Exhibit 3–85

97 F P D 4th—263

CHILD CARE CENTERS
LICENSE—
  Free exercise of religion. Const Law 84.5(7)
REGULATION—
  Moral character. Asyl 4

CHILD LABOR
See this index—

Infants
Worker's Compensation

CHILD WELFARE
See this index Parental and Child Welfare

CHILDREN
See this index Infants

Source: FPD4th, DWI "Children," © West Publishing, used with permission.

Our first step is to identify search terms we will use in the Federal Practice Digest 4th Descriptive Word Index. It might pay to look in the Descriptive Word Index under "**Children**," "**Confrontation**," "**Television**," "**Testimony**," "**Witnesses**," and perhaps other options, including the general category, "**Criminal Law**." The heading **Children** simply refers us to "**Infants**," which in turn has a subtopic, "**TESTIMONY**," which directs us to "see generally, Witnesses, post." (See Exhibits 3–85 and 3–86.) We do not find "**Confrontation**," but the "**Confronting Witnesses**" heading has a subtopic, "**CHILD molestation case—. . . Infants 17**," which may prove interesting. (See Exhibit 3–87.) Under "**Television and Radio**" we find two possibilities: "**CLOSED circuit, testimony by—Crim**

Exhibit 3–86

INFANTS

INFANTS—Cont'd
CONSORTIUM—
    Children's consortium, loss of. **Parent &
      C 7(1–14)**
    Parents' consortium, loss of. **Parent & C 7.5**
CORPORAL punishment. **Parent & C 2(1)**
CREDIBILITY as witness—
    Expert testimony. **Crim Law 474.3(3)**
CUSTODY and control—
    Habeas corpus—
      Determination and disposition. **Hab Corp
        798**
      Existence and exhaustion of other
        remedies. **Hab Corp 280**
      Grounds for relief. **Hab Corp 531–536**
      Hearing. **Hab Corp 744**
      Jurisdiction. **Hab Corp 614(1,2), 624**
      Mootness and prematurity of application.
        Hab Corp 232
      Petition, sufficiency. **Hab Corp 670(10)**
      Venue and personal jurisdiction. **Hab Corp
        636**
      Weight and sufficiency of evidence. **Hab
        Corp 731**

Interference. **Parent & C 18**
Relinquishment of custody. **Infants 19.4**
Tender years doctrine—
    Infants 19.2(4)
    Parent & C 2(3.2)
Threats—
    Removal of children. **Const Law 274(5)**
FAMILY purpose doctrine, see generally, this in-
    dex **Family Purpose Doctrine**
NEGLECT—
    Failure to support, see generally, Parent and
      child, post
OBVIOUS danger—
    Appreciation of. **Neglig 85(4)**
PARENT and child—
    Negligent entrustment. **Parent & C 13(1)**
    Negligent supervision. **Parent & C 13(1)**
PRECONCEPTION negligence—
    Rights of action. **Infants 72(2)**
TENDER years doctrine, see Custody and control,
    ante
TESTIMONY, see generally, Witnesses, post

Source: FPD4th, DWI "Infants," © West Publishing, used with permission.

Exhibit 3–87

CONFRONTING WITNESSES
CHILD molestation case—
    Admin Law 475
    Infants 17
CIVIL actions. **Trial 38**
CRIMINAL proceedings—
    Generally. **Crim Law 662–662.80**
    Military cases. **Mil Jus 1248**

DELINQUENT children—
    Transfer to adult court. **Witn 68.7(3)**
DISBARMENT proceeding. **Atty & C 53(1)**
PAROLE revocation hearing. **Pardon 88**
PARTY to civil actions. **Trial 38**
PRESENTENCE reports. **Crim Law 662.40**
STANDING to challenge evidentiary statute.
    Const Law 42.1(6)

Source: FPD4th, DWI "Confronting," © West Publishing, used with permission.

Exhibit 3–88

TELEVISION AND RADIO
Generally. **Tel 381–448**
ACTIONS. **Tel 418–427**
Charges and rates. **Tel 448**

CLOSED circuit, testimony by—
Crim Law 662.1
Witn 228
COERCION, license or permit. **Tel 394**
COLOR television, license or permit. **Tel 395**
COMMON carrier, status as. **Tel 3**
COMMUNITY antenna television systems. **Tel 449**
COMPETITION and interference, license and regulation. **Tel 403–407**
Administrative proceedings, competitors' remedies. **Tel 415**
Allocation as to states, areas or population served. **Tel 406**
Economic injury. **Tel 404**

Particular cases and problems. **Tel 405–407**
CONCEALMENT, license or permit. **Tel 397**
CONDUCT—
License or permit. **Tel 393**
Persons connected with. **Tel 393**
CONSTITUTIONAL and statutory provisions. **Tel 384**
CONTRACTS. **Tel 442–443**
Control of station. **Tel 443**
Ownership of station. **Tel 443**
CONTROL of station, contracts. **Tel 443**
CONTROVERSIAL questions on program. **Tel 435**
COURTROOM use—
Local rule—
Applicability to view. **Fed Civ Proc 1968**
CRIMINAL prosecutions, presenting trial events.
Crim Law 633(1)
DIRECTIONAL radio and radar, license or regulation. **Tel 396**

Source: FPD4th, DWI "Television," © West Publishing, used with permission.

Exhibit 3–89

TESTIMONY
See this index—
Evidence
Witnesses
TETANUS

Source: FPD4th, DWI "Testimony," © West Publishing, used with permission.

Law 662.1, Witn 228," and "**CRIMINAL prosecutions, presenting trial events. Crim Law 633(1).**" (See Exhibit 3–88.) "**Testimony**" is not a very helpful term. It simply refers us to "**Evidence**" and "**Witnesses**" in the Descriptive Word Index. (See Exhibit 3–89.) On the other hand, there is a bit more promise to "**Witnesses.**" The subtopic "**CHILDREN**" refers us only to "**this index In-**

Exhibit 3–90

WITNESSES
    Credibility and impeachment, see this index
      Credibility and Impeachment of
      Witnesses
CHILDREN, see this index **Infants**
CIVIL rights—

CONFESSION of crime. **Crim Law 517(8)**
CONFIRMING obligation to tell truth. **Witn 318**
CONFRONTATION, see this index **Confront-**
    ing Witnesses

CONGRESSIONAL investigating bodies, see
    this index **Congress**

    Prosecuting attorney, privilege of communica-
      tions to. **Witn 203**
ONE-WAY television. **Witn 228**
OPINION evidence, see this index **Opinion**
    Evidence
PARTIES—
    Competency, see this index **Competency of**
    Witnesses

Source: FPD4th, DWI "Witnesses," © West Publishing, used with permission.

Exhibit 3–91

CRIMINAL LAW—Cont'd
COMPETENCY—
    To be executed. **Crim Law 981(1)**
CONFESSIONS, see this index **Confession of**
    Crime
CONFLICTING evidence, see this index **Con-**
    flicting Evidence
CONFLICTING presumptions. **Crim Law 325**

CONFORMITY to state practice. **Fed Civ Proc**
    56
CONFRONTING witnesses, see this index **Con-**
    fronting Witnesses
CONFUSING evidence. **Crim Law 338(7)**
CONGRESSIONAL hearing testimony—
    Admissibility—

Source: FPD4th, Outline "Criminal," © West Publishing, used with permission.

fants"; and the subtopic "**CONFRONTATION**" directs us to "**this index Con-fronting Witnesses**"; but there also is the subtopic "**ONE-WAY television. Witn 228**." (See Exhibit 3–90.) The best lead under "**Criminal Law**" is the subtopic "**CONFRONTING witnesses**," which simply refers us back to "**this index Con-fronting Witnesses**." (See Exhibit 3–91.)

    If nothing else, this exercise should reassure you that it really does not matter where you begin looking in a Descriptive Word Index. The cross-referencing system eventually will point you in the right direction, no matter how roundabout your route is. We have a few concrete leads about the right direction for our present search, now that we have identified several specific topic and key numbers through the Descriptive Word Index. Our behind-the-scenes investigation, which you are invited to replicate, suggests that Criminal Law ⚷ 662.1 is the best choice to pursue.

We can learn what issue corresponds to this key number by examining the topic outline at the beginning of "**Criminal Law**" in West's Federal Practice Digest 4th. The general outline identifies Criminal Law ⚷ 662.1 as falling within the broad domain of the "Reception of Evidence" at "Trial." (See Exhibit 3–92.) The more detailed outline reveals that the specific topic is "Right of accused to confront witnesses—In general." (See Exhibit 3–93.) This key number promises to include a relatively diverse set of confrontation clause issues, among which we hope to find cases dealing with the presentation of

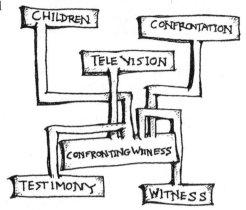

Exhibit 3–92

CRIMINAL LAW

XI. **TIME OF TRIAL AND CONTINUANCE,** ⚷ **573–617.**
   (A) TIME OF TRIAL, ⚷ 573–577.16(11).
      1. DECISIONS PRIOR TO 1967, ⚷ 573–577.
      2. DECISIONS SUBSEQUENT TO 1966, ⚷ 577.1–577.16(11).
   (B) CONTINUANCE, ⚷ 578–617.
XII. **TRIAL,** ⚷ **618–904.**
   (A) PRELIMINARY PROCEEDINGS, ⚷ 618–632.5.
   (B) COURSE AND CONDUCT OF TRIAL IN GENERAL, ⚷ 633–660.
➤   (C) RECEPTION OF EVIDENCE, ⚷ 661–689.
   (D) OBJECTIONS TO EVIDENCE, ⚷ MOTIONS TO STRIKE OUT, AND EXCEPTIONS, ⚷ 690–698.
   (E) ARGUMENTS AND CONDUCT OF COUNSEL, ⚷ 699–730.
   (F) PROVINCE OF COURT AND JURY IN GENERAL, ⚷ 731–768.
   (G) NECESSITY, REQUISITES, AND SUFFICIENCY OF INSTRUCTIONS, ⚷ 769–823.
   (H) REQUESTS FOR INSTRUCTIONS, ⚷ 824–836.
   (I) OBJECTIONS TO INSTRUCTIONS OR REFUSAL THEREOF, AND EXCEPTIONS, ⚷ 838–847.
   (J) CUSTODY, CONDUCT, AND DELIBERATIONS OF JURY, ⚷ 848–868.
   (K) VERDICT, ⚷ 870–894.
   (L) WAIVER AND CORRECTION OF IRREGULARITIES AND ERRORS, ⚷ 895–904.
XIII. **MOTIONS FOR NEW TRIAL AND IN ARREST,** ⚷ **905–976.**
XIV. **JUDGMENT, SENTENCE, AND FINAL COMMITMENT,** ⚷ **977–1003.**

Source: FPD4th, Outline "Criminal," © West Publishing, used with permission.

Exhibit 3–93

CRIMINAL LAW

(C) RECEPTION OF EVIDENCE.

⌾⇒ 661.   Necessity and scope of proof.

➤ 662.   Right of accused to confront witnesses.

➤ 662.1 —In general.

    662.3 —Nature or stage of proceeding.

    662.4 —Failure to produce or disclose witnesses or evidence.

    662.5 —Informants, failure to produce or disclose.

    662.6 —Refusal of codefendants or others to testify.

    662.7 —Cross-examination and impeachment.

    662.8 —Out-of-court statements and hearsay in general.

Source: FPD4th, Outline "Criminal-Trial," © West Publishing, used with permission.

---

televised testimony by a witness who testifies outside of the presence of the defendant.

Our next step is to look at the case squibs under Criminal Law ⌾⇒ 662.1 in this same volume of the Digest. When we do, we find squibs and citations for federal cases addressing a variety of issues related to the right to confront accusing witnesses. Supreme Court cases are presented first, followed by U.S. court of appeals decisions. (See Exhibit 3–94.) Although not pictured, U.S. district court opinions will be noted after court of appeals cases. Among the squibs presented in Exhibit 3–94, none describes the specific issue we are researching, although two Supreme Court decisions, *Coy v. Iowa* and *Pennsylvania v. Ritchie*, appear to be generally relevant to whether literal face-to-face confrontation between prosecution witnesses and a criminal defendant inevitably is required by the Constitution. By now, checking for pocket supplements should be reflexive. When we consult the digest's pocket part we find several squibs describing *Maryland v. Craig*, the Supreme Court case we previously have run across and that appears to be directly on point to the issue we are examining. (See Exhibit 3–95.) Other recent federal court decisions classified under Criminal Law ⌾⇒ 662.1 also are presented in the pocket supplement.

As you recall, the Federal Practice Digest 4th covers only federal court decisions. Now that we have found the topic and key number that seem to fit our research problem, it will be an easy matter to search for court decisions from different jurisdictions. We simply stick with Criminal Law ⌾⇒ 662.1 and make use of different case digests in the West network. For example, by using the Pacific Digest, we can look for cases decided by the several state courts included in West's Pacific region. (See Exhibit 3–96.) If we wanted to focus on New York cases, we look in West's New York Digest 4th under Criminal Law ⌾⇒ 662.1 (See

Exhibit 3–94

CRIMINAL LAW

⌘ 662.1.—In general.

U.S.Del. 1985. The Confrontation Clause [U.S.C.A. Const.Amend. 6] includes no guaranty that every witness called by the prosecution will refrain from giving testimony that is marred by forgetfulness, confusion, or evasion.

Delaware v. Fensterer, 106 S.Ct. 292, 474 U.S. 15, 88 L.Ed.2d 15, on remand 509 A.2d 1106.

U.S.Iowa 1988. Confrontation clause provides criminal defendant right to "confront" face-to-face witnesses giving evidence against him at trial; such confrontation helps to insure integrity of fact-finding process by making it more difficult for witnesses to fabricate testimony. U.S.C.A. Const.Amend. 6.

Coy v. Iowa, 108 S.Ct. 2798, 487 U.S. 1012, 101 L.Ed.2d 857, on remand 433 N.W.2d 714.

Placement of screen between defendant and child sexual assault victims during their testimony at trial violated defendant's right to face-to-face confrontation under confrontation clause. U.S.C.A. Const.Amend. 6.

Coy v. Iowa, 108 S.Ct. 2798, 487 U.S. 1012, 101 L.Ed.2d 857, on remand 433 N.W.2d 714.

U.S.N.Y. 1987. Witness whose testimony is introduced in joint trial with limited instruction that it be used only to assess guilt of one defendant is not considered to be witness against other defendants for purpose of confrontation clause. U.S.C.A. Const.Amend. 6.

Cruz v. New York, 107 S.Ct. 1714, 481 U.S. 186, 95 L.Ed.2d 162, on remand People v. Cruz, 519 N.Y.S.2d 959, 70 N.Y.2d 733, 514 N.E.2d 379.

U.S.Pa. 1987. Confrontation clause provides two types of protection for criminal defendant:

right physically to face those who testify against him, and right to conduct cross-examination. (Per Justice Powell, with the Chief Justice and two Justices concurring and one Justice concurring in result.) U.S.C.A. Const.Amend. 6.

Pennsylvania v. Ritchie, 107 S.Ct. 989, 480 U.S. 39, 94 L.Ed.2d 40.

Confrontation clause is not constitutionally compelled rule of pretrial discovery. (Per Justice Powell, with Chief Justice and two Justices concurring and one Justice concurring in result.) U.S.C.A. Const.Amend. 6

Pennsylvania v. Ritchie, 107 S.Ct. 989, 480 U.S. 39, 94 L.Ed.2d 40.

Normally right to confront one's accusers is satisfied if defense counsel receives wide latitude at trial to question witnesses. (Per Justice Powell, with Chief Justice and two Justices concurring and one Justice concurring in result.) U.S.C.A. Const.Amend. 6.

Pennsylvania v. Ritchie, 107 S.Ct. 989, 480 U.S. 39, 94 L.Ed.2d 40.

C.A.D.C. 1988. Right to confront prosecution witnesses is not absolute and reasonable restrictions may be imposed. U.S.C.A. Const. Amend. 6.

U.S. v. Tarantino, 846 F.2d 1384, 269 U.S. App.D.C. 398, certiorari denied Burns v. U.S., 109 S.Ct. 108, 102 L.Ed.2d 83 and 109 S.Ct. 174, 102 L.Ed.2d 143.

C.A.11 (Ala.) 1985. Sixth Amendment [U.S.C.A. Const.Amend. 6] guarantees only opportunity to confront adverse witnesses; it does not guarantee right to confront witnesses who testified not against but rather in favor of party asserting right.

U.S. v. Andrews, 765 F.2d 1491, rehearing denied 772 F.2d 918, certiorari denied Royster v. U.S., 106 S.Ct. 815, 474 U.S. 1064, 88 L.Ed.2d 789.

continues

Exhibit 3–94   continued

C.A.9 (Ariz.) 1987. Evidence regarding police department's discipline of officer for improperly using police vehicle on off-duty job was only peripherally relevant to officer's credibility as witness; accordingly, trial court's decision to exclude evidence did not deprive defendant of constitutional right to confront witnesses against him. U.S.C.A. Const.Amend. 6.

Pool v. Dowdle, 834 F.2d 777.

Source: Dec. Digest, Homicide-attempt squib, © West Publishing, used with permission.

Exhibit 3–97.) To broaden our search, we follow the same procedures illustrated in connection with our research for Case 2. We first consult the Tenth Decennial Digest, Part 1 under Criminal Law ☞ 662.1 (see Exhibit 3–98) and then examine the succeeding General Digests in order to update the Decennial Digest. (See Exhibit 3–99.)

By this time, you should be relatively comfortable with using Descriptive Word Indexes, case digests, and West's key number system to find judicial decisions relevant to the issues you are researching. The very same process that has given us leads for case law addressing the issues in Cases 1 and 2 should serve us as well for Case 3. You should be able to identify likely search terms (e.g., "**prisoners,**" "**prisons,**" "**cruel and unusual punishment,**" "**smoking,**" "**conditions of confinement,**" "**civil rights**") describing the issue in Case 3, and use appropriate Descriptive Word Indexes and case digests to search for judicial decisions. We recommend that you sharpen your skills by practicing on this issue.

The CALR Approach to Case Law Research

We have seen previously that computer-assisted legal research (CALR) has several advantages over the more traditional printed sources. One key advantage is that information is more up-to-date in using the computer method. CALR also allows the researcher to create his or her own search terms rather than relying solely on subject terms assigned to documents in a particular database. These advantages are particularly telling in performing case law research. WESTLAW and LEXIS generally have the full text of U.S. Supreme Court decisions available on their respective systems on the same day that the decisions are issued. While lower federal court cases and state court decisions take longer to appear in the computer systems, they nevertheless are available considerably sooner than the printed versions. The key word searching capabilities of WESTLAW and LEXIS provide the researcher with many alternative methods of accessing relevant cases that are not possible using the printed sources.

There are three basic ways of finding a case in the printed sources. First, if you know the title of a case you can look up the cite to the case in one of the sources providing **tables of cases**. Second, you may already have the name and the cite to a case from a law review article, encyclopedia article, or some other source. Third, as described above, you can look for cases by using the subject approach provided by the various digest series. With CALR, the researcher has many more options.

Exhibit 3–95

CRIMINAL LAW

⚷ **662.1.—In general.**

U.S.Md. 1990. The central concern of the confrontation clause is to ensure the reliability of the evidence against a criminal defendant by subjecting it to rigorous testing in the context of an adversary proceeding before the trier of fact. U.S.C.A. Const.Amend. 6.—Maryland v. Craig, 110 S.Ct. 3157, 497 U.S. 836, 111 L.Ed.2d 666, on remand Craig v. State, 588 A.2d 328, 322 Md. 418.

A face-to-face confrontation enhances the accuracy of fact-finding by reducing the risk that a witness will wrongfully implicate an innocent person. U.S.C.A. Const.Amend. 6.—Id.

Face-to-face confrontation with witnesses is not an indispensable element of the Sixth Amendment's guarantee of the right to confront one's accusers. U.S.C.A. Const.Amend. 6.—Id.

Child assault victim's testimony at trial of child abuse defendant through use of one-way closed circuit television procedure authorized by Maryland child witness protection statute did not impinge upon the truth seeking nor symbolic purposes of the confrontation clause; procedure required that child witness be competent to testify and testify under oath, defendant retained full opportunity for contemporaneous cross-examination, and judge, jury and defendant were able to view witness' demeanor and body by video monitor. Md.Code, Courts and Judicial Proceedings, § 9–102, U.S.C.A. Const.Amend. 6.—Id.

If the State makes an adequate showing of necessity, the State's interest in protecting child witnesses from the trauma of testifying in a child abuse case is sufficiently important to justify the use of a special procedure permitting a child witness in abuse case to testify at trial in the absence of face-to-face confrontation with the defendant. U.S.C.A. Const.Amends. 6, 14.—Id.

Determination of whether use of procedure permitting a child witness to testify in a child abuse case without face-to-face confrontation with the defendant is justified by the State's interest in protecting witness from the trauma of testifying must be made on a case specific basis; trial court must determine whether use of one-way closed circuit television procedure is necessary to protect welfare of particular child witness, must find that child witness would be traumatized by the presence of the defendant, not by the courtroom generally, and must find that the emotional distress is more than mere nervousness, excitement or reluctance to testify. Md.Code, Courts and Judicial Proceedings, §§ 9–102, 9–102(a)(1)(ii); U.S.C.A. Const.Amend. 6.—Id.

Testimony of child witnesses in child abuse case by one-way closed circuit television would be admissible under the confrontation clause to the extent that a proper finding was made that use of procedure was necessary to protect child witness from trauma; witnesses were under oath, were subject to full cross-examination and could be observed by judge, jury and defendant as they testified. Md.Code, Courts and Judicial Proceedings, § 9–102; U.S.C.A. Const.Amend. 6.—Id.

Observation of child abuse victims' behavior in defendant's presence and consideration of less restrictive alternatives to one-way closed circuit television procedure, although possibly strengthening grounds for use of protective measures, were not categorically prerequisites to use of television testimony procedure as a matter of federal constitutional law. Md.Code, Courts and Judicial Proceedings, § 9–102; U.S.C.A. Const.Amends. 6, 14.—Id.

U.S.Mich. 1991. Michigan's rape-shield statute implicated Sixth Amendment to extent it authorized preclusion of evidence of defendant's own past sexual conduct with victim for defendant's failure to comply with notice-and-hearing requirement. U.S.C.A. Const.Amend. 6; M.C.L.A. § 750.520.—Michigan v. Lucas, 111 S.Ct. 1743, 500 U.S. 145, 114 L.Ed.2d 205, on remand People v. Lucas, 484 N.W.2d 685, 193 Mich.App. 298, appeal after remand 507 N.W.2d 5, 201 Mich.App.

continues

Exhibit 3–95   continued

717, appeal denied 521 N.W.2d 606, 445 Mich. 936, certiorari denied 115 S.Ct. 593, 130 L.Ed.2d 505.

Restrictions on criminal defendant's right to confront adverse witnesses and to present evidence may not be arbitrary or disproportionate to purposes they are designed to serve. U.S.C.A. Const.Amend. 6.—Id.

Legitimate interests served by notice requirement of rape-shield statute can, under some circumstances, justify precluding evidence of prior sexual relationship between rape victim and criminal defendant without running afoul of Sixth Amendment. U.S.C.A. Const.Amend. 6.—Id.

Michigan trial court's exclusion of evidence of defendant's prior sexual relationship with victim for defendant's failure to comply with notice-and-hearing requirement of Michigan's rape-shield statute was not per se violation of Sixth Amendment; without determining whether Michigan's notice requirement, which demanded written motion and offer of proof to be filed within ten days after arraignment, was overly restrictive, whether Michigan's rape-shield statute in fact authorized preclusion, and whether preclusion violated defendant's rights on facts of case, the Sixth Amendment was not so rigid as to prohibit preclusion under any circumstances, without regard to specific circumstances of case. U.S.C.A. Const.Amend. 6; M.C.L.A. § 750.520.—Id.

C.A.D.C. 1996. Confrontation Clause violations are found primarily where defendants have been given no realistic opportunity to allow discriminating appraisal of witness's motives and bias. U.S.C.A. Const.Amend. 6.—U.S. v. Graham, 83 F.3d 1466.

C.A.D.C. 1993. Confrontation clause does not bar judge from imposing reasonable limits on defense counsel's inquiries. U.S.C.A. Const.Amend. 6.—U.S. v. Derr, 990 F.2d 1330, 301 U.S.App.D.C. 60.

C.A.D.C. 1988. U.S. v. Tarantino, 846 F.2d 1384, 269 U.S.App.D.C. 398, certiorari denied Burns v. U.S., 109 S.Ct. 108, 488 U.S. 840, 102 L.Ed.2d 83, certiorari denied 109 S.Ct. 174, 488 U.S. 867, 102 L.Ed.2d 143, appeal after remand U.S. v. Bell, 905 F.2d 458, 284 U.S.App.D.C. 353.

C.A.9 (Ariz.) 1993. Introduction of witness' testimony that search for defendant's tax return was conducted pursuant to court order, and that no record was found, as well as certification of lack of record for individual tax return, did not violate defendant's Sixth Amendment right to confrontation, even absent proof that person who researched tax records was unavailable. U.S.C.A. Const.Amend. 6; Fed.Rules Evid.Rule 803(10), 28 U.S.C.A.—U.S. v. Hutchison, 22 F.3d 846, as amended, and as amended on denial of rehearing and reh. en banc.

Source: FPD4th, squibs-pocket "662.1," © West Publishing, used with permission.

For example, we have already indicated that in doing a subject search the researcher is not limited to the subjects that appear in the printed digests, such as those categories itemized by West's key numbers, but instead can make up his or her own key words or phrases to search throughout the text of the documents in a database. On WESTLAW or LEXIS it is possible to retrieve cases by keying in the name of the particular judge who wrote the decision or the judge who wrote a dissenting or concurring opinion. It is even possible to retrieve cases in which a particular lawyer participated or in which a particular group filed an amicus brief.

The online systems also allow more flexibility in limiting retrieved cases to a particular court or by multiple courts' locations or levels. For example, a strategy could be used to retrieve cases only from the highest courts in California and New

Exhibit 3–96

☞ **662.1** CRIMINAL LAW

. . . given opportunity for effective cross-examination. U.S.C.A. Const.Amend. 6.—Merritt v. People, 842 P.2d 162.

Colo. 1991. Defendant has constitutional right to present evidence on his behalf, and to confront adverse witnesses.—People v. Chard, 808 P.2d 351, certiorari denied Chard v. Colorado, 112 S.Ct. 186, 116 L.Ed.2d 147.

Colo. 1990. Defendant's right of confrontation guaranteeing face-to-face meeting with witnesses appearing before trier of fact is not absolute and must occasionally give way to considerations of public policy and necessities of the case. West's C.R.S.A. § 18-3-413(4); West's C.R.S.A. Const. Art. 2, § 16; U.S.C.A. Const.Amends. 6, 14.—Thomas v. People, 803 P.2d 144, habeas corpus denied 754 F.Supp. 833, affirmed 962 F.2d 1477.

Hawaii App. 1987. The Confrontation Clause reflects a preference for face-to-face confrontation at trial, and a primary interest secured by the clause is the fight of cross-examination.—State v. Rodrigues, 742 P.2d 986, 7 Haw.App. 80.

In *Ohio v. Robert*, 448 U.S. 56, 100 S.Ct. 2531, 65 L.Ed.2d 597 (1980), the Supreme Court held that the Confrontation Clause restricts the range of admissible hearsay in two ways. First, the prosecution must either produce, or demonstrate the unavailability of, the declarant whose statement it wishes to use against the defendant. Second, the hearsay statement must be marked by adequate indicia of reliability. Since the right to confrontation is basically a trial right, the two requirements must be satisfied at the time of trial itself.—Id.

Idaho App. 1992. In determining whether Sixth Amendment rights were violated by exclusion of evidence, court considers whether evidence proffered is relevant and then asks whether other legitimate interests outweighed defendant's interest in presenting evidence. U.S.C.A. Const.Amend. 6.—State v. Peite, 839 P.2d 1223, 122 Idaho 809, review denied.

Kan. 1992. Right to confrontation under Kansas and United States Constitutions includes right of accused to face-to-face confrontation while victim accuser is testifying against accused. U.S.C.A. Const.Amend. 6; K.S.A. Const.Bill of Rights, § 10.—State v. Chisholm, 825 P.2d 147, 250 Kan. 153.

Kan. 1991. Right of confrontation includes right of accused to face-to-face confrontation while accuser is testifying against accused. U.S.C.A. Const.Amend. 6; K.S.A. Const.Bill of Rights, § 10.—State v. Hamons, 805 P.2d 6, 248 Kan. 51.

Mont. 1991. Defendant's right to confront witnesses only precluded application of shield law to prevent defendant from introducing victim's prior accusations or allegations of sexual conduct if such accusations were proven or admitted to be false. U.S.C.A. Const.Amend. 6; Const. Art. 2, § 24; MCA 45–5–511(4).—State v. Van Pelt, 805 P.2d 549, 247 Mont. 99.

Mont. 1990. State v. LaPier, 790 P.2d 983, 242 Mont. 335, denial of habeas corpus affirmed LaPier v. McCormick, 986 F.2d 303.

N.M.App. 1991. Defendant's confrontation clause rights were not violated by precluding defendant from admitting evidence of letter opinion from prior civil litigation between defendant and complainant, where defendant was permitted to . . .

Source: Pac. Digest, squibs "662.1," © West Publishing, used with permission.

Exhibit 3–97

⌐‒‒ **662.1**   **CRIMINAL LAW**

N.Y.Sup. 1988. Use of "two-way" closed circuit television set-up in questioning minor witness, who was alleged victim of rape, sodomy, and sexual abuse by her father, who was the defendant, did not violate defendant's right to confront his accuser; witness was able to view defendant on screen before her and defendant was similarly able to see witness on screen in courtroom, and jury was instructed not to draw any inferences unfavorable to defendant merely because the "two-way" system was being used. U.S.C.A. Const.Amend. 6.

> People v. Rivera, 535 N.Y.S.2d 909, 141 Misc.2d 1031.

N.Y.Sup. 1988. New York statute, which permits child abuse victims to testify to abuse outside of defendant's physical presence, but only upon individualized determination that child is "vulnerable witness," does not violate defendant's Sixth Amendment rights to confront witnesses against him. U.S.C.A. Const.Amend. 6; McKinney's CPL § 65.00 et seq.

> People v. Logan, 535 N.Y.S.2d 322, 141 Misc.2d 790.

N.Y.Sup. 1986. Statute authorizing examination of child victims of sexual abuse by live closed-circuit television does not violate defendant's rights to confront adverse witness, effective assistance of counsel or due process. U.S.C.A. Const.Amends. 5, 6, 14.

> People v. Henderson, 503 N.Y.S.2d 238, 132 Misc.2d 51.

N.Y.Sup. 1986. In prosecution for rape, sodomy and sexual abuse, examination of child witness outside of courtroom did not infringe on defendant's right to confront witnesses against him, though defendant was not present in room where child witness was examined, where defense counsel was present in room, where defendant and jurors were able to observe witness over closed circuit television, and where child witness was aware of the fact she was being watched. U.S.C.A. Const.Amend. 6; McKinney's Const. Art. 1, § 6; McKinney's CPL § 65.00 et seq.

> People v. Algarin, 498 N.Y.S.2d 977, 129 Misc.2d 1016.

Statute providing that child witness, in prosecution for sexual abuse, may testify outside courtroom did not violate defendant's right to confront witnesses against him, even assuming that out-of-court examination to some extent infringed on that right, where statute allowed child to be removed from defendant's presence only upon finding that he or she was likely to otherwise suffer severe mental or emotional harm, where defendant and jurors were able to observe witness' demeanor over closed circuit television, and where compelling State interest in protecting emotional well-being of child sex offense victims more than outweighed such minimal infringement on defendant's rights. U.S.C.A. Const.Amends. 6, 14; McKinney's Const. Art. 1, § 6; McKinney's CPL § 65.00 et seq.

> People v. Algarin, 498 N.Y.S.2d 977, 129 Misc.2d 1016.

Source: NY Digest, squibs "662.1," © West Publishing, used with permission.

York. Date restrictions allow CALR researchers to restrict their results to decisions issued on a particular date, before a date, after a date, or within a range of dates. These examples are just a few of the search enhancements provided by WESTLAW and LEXIS. Of course, what makes CALR even more advantageous is that it allows researchers to combine the various access options in one search strategy. For example, a researcher could search for cases on WESTLAW that

Exhibit 3–98

⌔ **662.1**
CRIMINAL LAW

⌔ **662.1.—In general.**

U.S.Iowa 1988. Confrontation clause provides criminal defendant right to "confront" face-to-face witnesses giving evidence against him at trial; such confrontation helps to insure integrity of fact-finding process by making it more difficult for witnesses to fabricate testimony. U.S.C.A. Const.Amend. 6.—Coy v. Iowa, 108 S.Ct. 2798, 487 U.S. 1012, 101 L.Ed.2d 857, on remand 433 N.W.2d 714.

Placement of screen between defendant and child sexual assault victims during their testimony at trial violated defendant's right to face-to-face confrontation under confrontation clause. U.S.C.A. Const.Amend. 6.—Id.

U.S.Md. 1990. The central concern of the confrontation clause is to ensure the reliability of the evidence against a criminal defendant by subjecting it to rigorous testing in the context of an adversary proceeding before the trier of fact. U.S.C.A. Const.Amend. 6.—Maryland v. Craig, 110 S.Ct. 3157, 111 L.Ed.2d 666, on remand 588 A.2d 328, 322 Md. 418.

A face-to-face confrontation enhances the accuracy of fact-finding by reducing the risk that a witness will wrongfully implicate an innocent person. U.S.C.A. Const.Amend. 6.—Id.

Face-to-face confrontation with witnesses is not an indispensable element of the Sixth Amendment's guarantee of the right to confront one's accusers. U.S.C.A. Const.Amend. 6.—Id.

Ariz. 1989. Statute authorizing videotaped testimony of minor witnesses was applied in violation of defendant's right to confrontation when State was allowed to substitute videotaped testimony of defendant's children for face-to-face confrontational testimony in open court in the absence of particularized showing that defendant's children would be traumatized if they testified in open court, even though judge received letter from foster care review board expressing concern for emotional and psychological welfare of children; letter played no part in decision by judge that best interests of children required protection from face-to-face testimonial encounter, letter was not accepted by trial court as evidence, and letter offered no specific evidence concerning likely impact of courtroom testimony upon children. U.S.C.A. Const.Amend. 6; A.R.S. Const. Art. 2, § 24; A.R.S. § 13–4253.—State v. Vincent, 768 P.2d 150, 159 Ariz. 418.

Exception to right to confrontation exists under both State and Federal Constitution when State sustains its burden of proving by individualized showing to trial court that face-to-face testimony would so traumatize child witness as to prevent child from reasonably communicating; such a finding is tantamount to a finding of unavailability and would justify use of videotape procedure established by statute. U.S.C.A. Const.Amend. 6; A.R.S. Const. Art. 2, § 24; A.R.S. § 13–4253.—Id.

Ark. 1987. Confrontation clause provides the right to physically face those who testify against defendant and the right to conduct cross-examination. U.S.C.A. Const.Amend. 6.—Winfrey v. State, 738 S.W.2d 391, 293 Ark. 342.

Source: Dec. Digest, squibs "662.1," © West Publishing, used with permission.

contain a particular WEST key number or certain key words, or could limit the results to particular court(s) and to a particular time period.

The two key features on WESTLAW and LEXIS that allow for these additional search options, fields (WESTLAW) and segments (LEXIS), have been discussed above in searching the statutory databases. Although proximity requirements are

Exhibit 3–99

CRIMINAL LAW

662.1.—In general.

Cal.App. 2 Dist. 1995. "Confrontation" means more than being allowed to confront witness physically. U.S.C.A. Const.Amend. 6.—People v. Los Angeles County Superior Court (Piedrahita), 40 Cal.Rptr.2d 335, 34 C.A.4th 508, rehearing denied.

Sixth Amendment confrontation right is not absolute. U.S.C.A. Const.Amend. 6.—Id.

Minn.App. 1995. Defendant's right to confront accusatory witnesses may be satisfied absent physical, face-to-face confrontation at trial only where denial of such confrontation is necessary to further important public policy and only where reliability of testimony is otherwise assured. U.S.C.A. Const.Amend. 6.—State v. Peterson, 530 N.W.2d 843.

In certain circumstances, protecting child witnesses from trauma of face-to-face confrontation with defendant in child abuse case is sufficiently important to justify use of special procedure. U.S.C.A. Const.Amend. 6.—Id.

Source: Gen. Digest, squibs "662.1," © West Publishing, used with permission.

the same in the case law databases, the names of the fields or segments are different.

We start by using WESTLAW to look for cases bearing on our second hypothetical, regarding whether a person infected with either AIDS or HIV can be prosecuted for attempted murder or assault with a deadly weapon for biting or spitting on another person. As we construct our search strategy, we take advantage of the fact that West provides subject access to cases through its key number system. The West series of digests and their accompanying Descriptive Word Index volumes are not available on the WESTLAW system, but the key numbers and squibs that are found in the digests are compiled from the headnotes that appear at the beginning of each case appearing in the West case reporter series. Thus, if you know the key numbers that cover your search topic, a key number search strategy on WESTLAW can still be used. The trick is identifying the appropriate key numbers.

Here, the CALR researcher has two options. First, you can attempt to identify the appropriate key numbers through the printed digests using the procedures described above. (At this point you might ask, if you have to look up the key numbers in the printed sets, why should you not just continue the research process in the printed digests rather than turning to WESTLAW? The answer to that question will become apparent as we work our way through this example.) The second option is to use the **KEY** database on WESTLAW. The **KEY** database includes the topic outlines that correspond to the outlines found at the front of each topic in the printed versions of West's digests. We will use the **KEY** database as a starting place for developing our search strategy, although we caution the beginning legal researcher that this approach may be somewhat difficult because of the legal jargon used.

In this example we will be using the DOS version of WESTLAW, in contrast to the Windows version that we used in earlier examples. Once you have established an online connection with WESTLAW, **enter KEY** in order to access the **KEY** database. You will retrieve an alphabetical list of the 400-plus topics that comprise the West key number system. Page through the list until you come to a promising topic. In our example we are looking for topics that cover either assault with a deadly weapon or attempted murder. Remember that the topics deal with broad principles of law rather than specific facts, so you should not expect to find a topic for biting or spitting, or even for AIDS.

Exhibit 3–100 shows the screen from the alphabetical list that includes the reference to "**Assault and Battery**." Notice the number 37 that appears to the left of the topic. On WESTLAW each of the topics is assigned a number, which is used during the search process. The arrow to the left of the number is the hypertext feature.

Exhibit 3–100

Copyright (c) 1996 West Publishing Company
Key Number Service
Topic List

► 27 AMICUS CURIAE
► 28 ANIMALS
► 29 ANNUITIES
► 30 APPEAL AND ERROR
► 31 APPEARANCE
► 33 ARBITRATION
► 34 ARMED SERVICES
► 35 ARREST
► 36 ARSON
➤ ► 37 ASSAULT AND BATTERY
► 38 ASSIGNMENTS
► 40 ASSISTANCE, WRIT OF
► 41 ASSOCIATIONS
► 42 ASSUMPSIT, ACTION OF

If you wish to:
    View an item in more detail, select its jump marker
    View the next or previous page, type P or P– and press ENTER
    View the list of Key Number commands, type CMDS and press ENTER
    Go back to the previously accessed service, type GB and press ENTER

Source: KEY, table, © WESTLAW, used with permission.

If you move the cursor to the arrow to the left of "37" and press **enter**, you can proceed through the hierarchical outline for the topic "**Assault and Battery**," with the immediate result displayed in Exhibit 3–101. At the screen depicted in Exhibit 3–101, place the cursor on the arrow to the left of "**II. CRIMINAL RESPONSI-BILITY, k47-k100**," and press **enter**. We next come to a screen (Exhibit 3–102) that allows us to limit our search to "**OFFENSES, k47-k71**" by highlighting the arrow to the left and pressing **enter**. The next screen (Exhibit 3–103) lists the subheadings appearing under "**OFFENSES**." One of the subheadings is "**k56 Assault with dangerous or deadly weapon.**" Notice that there is no arrow to the left of this subheading, which indicates that there are no additional subheadings under that heading. Through this process we have found one of the same key numbers that we located through the use of the printed versions of the digests, Assault and Battery ⌐ 56. We can then use that key number to search any of the case reporter databases on WESTLAW to retrieve all the cases in the database in which that key number has been assigned to at least one headnote (squib) appearing at the beginning of the case.

We could repeat this process to locate the appropriate topic and number for attempted murder. We eventually would reach the screen depicted in Exhibit 3–104. Note the topic's name, "**Homicide**" and its corresponding number "203," and that under "**Homicide**" k256 is the number dealing with "**Attempt, threats,**

Exhibit 3–101

Copyright (c) 1996 West Publishing Company
Key Number Service

► Topic List   Mode: Outline
- -

►   37 ASSAULT AND BATTERY
                                        Analysis

►   I. CIVIL LIABILITY, k1–k46
►   II. CRIMINAL RESPONSIBILITY, k47–k100
END OF TOPIC

If you wish to:
　　View an item in more detail, select its jump marker
　　Run a search using a topic and key number, type DB followed by a database identifier and the number
　　　and press ENTER (e.g., DB ALLFEDS 410K196.1)

Source: KEY, "Assault," © WESTLAW, used with permission.

Exhibit 3–102

Copyright (c) 1996 West Publishing Company
Key Number Service

▶ Topic List   Mode: Outline
▶   37 ASSAULT AND BATTERY
▶   II. CRIMINAL RESPONSIBILITY, k47–k100
- -
▶   (A) OFFENSES, k47–k71
▶   (B) PROSECUTION, k72–k99
▶   (C) SENTENCE AND PUNISHMENT, k100–k100
END OF SECTION

If you wish to:
  View an item in more detail, select its jump marker
  Run a search using a topic and key number, type DB followed by a database identifier and the number
    and press ENTER (e.g., DB ALLFEDS 410k196.1)

Source: KEY, "Assault-offf," © WESTLAW, used with permission.

---

or solicitation to kill." Again, this is the same key number that we found using the printed digests.

Repeating essentially the same process we used with the printed digests, we can run a search using either or both of these key numbers to retrieve cases and then browse through our results to determine which, if any, deal with AIDS and biting or spitting. However, WESTLAW allows us to further refine our search strategy so that we not only retrieve the cases that have one of our key numbers, but that also address the particular factual circumstances we are investigating, AIDS or HIV that is transmitted by biting or spitting. Before we demonstrate this, we must describe the fields used in the case reporter databases on WESTLAW. Exhibit 3–105 identifies and provides a visual representation of those fields.

A brief description of the contents of each field follows:

Title—The complete names of all the parties involved in the case.

Date—The date of the decision.

Synopsis—A brief description of the case including a review of the facts, the court's holding, and the names of dissenting or concurring judges.

Topic—The West topic number (203), topic name (Homicide), key number (203k256), and text of the related key line (Attempt, threats, or solicitation to kill).

Exhibit 3–103

Copyright (c) 1996 West Publishing Company
Key Number Service

▶ Topic List   Mode: Outline
▶   37 ASSAULT AND BATTERY
▶   II. CRIMINAL RESPONSIBILITY, k47–k100
▶   (A) OFFENSES, k47–k71

- -

▶   k47 Nature and elements of criminal assault
            k54 Aggravated assault
            k55 Assault with intent to do great bodily harm
➡   k56 Assault with dangerous or deadly weapon
            k57 Shooting
            k58 Stabbing or other wounding
            k59 Indecent assault
            k60 Degrees
            k61 Attempts
▶   k62 Defenses
            k71 Persons liable
END OF SECTION

If you wish to:
   View an item in more detail, select its jump marker
   Run a search using a topic and key number, type DB followed by a database identifier and the number
      and press ENTER (e.g., DB ALLFEDS 410k196.1)

Source: KEY, "Assault-dangerous," © WESTLAW, used with permission.

Court—Identifies the court that decided the case.

Headnote—A part of the Digest field that contains the factual elements of a case related to the point of law associated with the key number assigned to the headnote (squib).

Digest—A composite of the topic, court, and headnote fields.

Attorney—The attorneys who represented the parties or who were involved in argument of the case.

Judge—The judge authoring the majority opinion in the case.

Opinion—The text of the opinion.

The three fields most likely to contain key words describing the factual circumstances of a case are **Synopsis**, **Headnote**, and **Opinion**. We have considerable flexibility in searching these fields. We can construct a strategy in which we

Exhibit 3–104

Copyright (c) 1996 West Publishing Company
Key Number Service

► Topic List   Mode: Outline
► 203 HOMICIDE
►   VII. EVIDENCE, k143–k257
►   (E) WEIGHT AND SUFFICIENCY, k228–k257

- -

          k247 Defense of property
          k248 Accident or misfortune
          k249 Principals and accessories
          k250 Degree of homicide in general
►   k251 Degree of murder
►   k255 Degree of manslaughter
     ➤   k256 Attempt, threats, or solicitation to kill
►   k257 Assault with intent to kill
END OF SECTION

If you wish to:
View an item in more detail, select its jump marker
Run a search using a topic and key number, type DB followed by a database identifier and the number and
     press ENTER (e.g., DB ALLFEDS 410k196.1)

Source: KEY, "Homicide," © WESTLAW, used with permission.

---

search more than one of the fields for the key words, or we can use a strategy that
looks for some key words in one field and other key words in another field. Deciding which fields to search and which key words to use requires some thought, and
perhaps some trial and error. In our example, we are looking for key words as they
relate to specific West key numbers. We will therefore limit our key word search
to the **Headnote** field that appears with each key number at the front of the case.
Remember, the **Headnote** contains the same information as the squibs we find in
the printed digests.

Earlier in this chapter we discussed in some detail the process of constructing a
search strategy in the **Terms and Connectors** search mode. We will not repeat
that instruction here. However, remember that the earlier discussion referred to the
Windows version of WESTLAW, and we are using the DOS version in our current
example. The basic commands are the same, but the location of information on the
screen that helps you construct your strategy is different. Note some of the
prompts that appear at the bottom of the screen depicted in Exhibit 3–106. The
Fields, Thesaurus, Help, and **Scope** prompts can all be used to help develop the
search strategy. With the help of the online **Thesaurus**, we constructed the following search strategy:

Exhibit 3–105

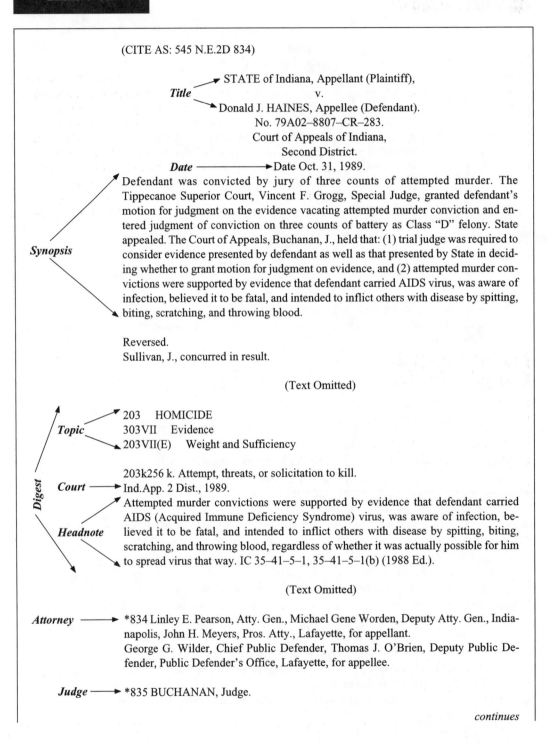

(CITE AS: 545 N.E.2D 834)

Title

STATE of Indiana, Appellant (Plaintiff),

v.

Donald J. HAINES, Appellee (Defendant).

No. 79A02–8807–CR–283.

Court of Appeals of Indiana,

Second District.

Date → Date Oct. 31, 1989.

Synopsis

Defendant was convicted by jury of three counts of attempted murder. The Tippecanoe Superior Court, Vincent F. Grogg, Special Judge, granted defendant's motion for judgment on the evidence vacating attempted murder conviction and entered judgment of conviction on three counts of battery as Class "D" felony. State appealed. The Court of Appeals, Buchanan, J., held that: (1) trial judge was required to consider evidence presented by defendant as well as that presented by State in deciding whether to grant motion for judgment on evidence, and (2) attempted murder convictions were supported by evidence that defendant carried AIDS virus, was aware of infection, believed it to be fatal, and intended to inflict others with disease by spitting, biting, scratching, and throwing blood.

Reversed.

Sullivan, J., concurred in result.

(Text Omitted)

Digest

Topic

203   HOMICIDE

303VII   Evidence

203VII(E)   Weight and Sufficiency

Court → 203k256 k. Attempt, threats, or solicitation to kill.

Ind.App. 2 Dist., 1989.

Headnote

Attempted murder convictions were supported by evidence that defendant carried AIDS (Acquired Immune Deficiency Syndrome) virus, was aware of infection, believed it to be fatal, and intended to inflict others with disease by spitting, biting, scratching, and throwing blood, regardless of whether it was actually possible for him to spread virus that way. IC 35–41–5–1, 35–41–5–1(b) (1988 Ed.).

(Text Omitted)

Attorney → *834 Linley E. Pearson, Atty. Gen., Michael Gene Worden, Deputy Atty. Gen., Indianapolis, John H. Meyers, Pros. Atty., Lafayette, for appellant.

George G. Wilder, Chief Public Defender, Thomas J. O'Brien, Deputy Public Defender, Public Defender's Office, Lafayette, for appellee.

Judge → *835 BUCHANAN, Judge.

continues

Exhibit 3–105   continued

(Text Omitted)

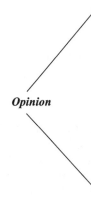

Opinion

FACTS

On August 6, 1987, Lafayette, Indiana, police officers John R. Dennis (Dennis) and Brad Hayworth drove to Haines' apartment in response to a radio call of a possible suicide. Haines was unconscious when they arrived and was lying face down in a pool of blood. Dennis attempted to revive Haines and noticed that Haines' wrists were slashed and bleeding. When Haines heard the paramedics arriving, he stood up, ran toward Dennis, and screamed that he should be left to die because he had AIDS. Dennis told Haines they were there to help him, but he continued yelling and stated he wanted to f—Dennis and "give it to him." Haines told Dennis that he would "use his wounds" and began jerking his arms at Dennis, causing blood to spray into Dennis' mouth and eyes. Throughout the incident, as the officers attempted to subdue him, Haines repeatedly yelled that he had AIDS, that he could not deal with it and that he was going to make Dennis deal with it.

Source: Fields, © WESTLAW, used with permission.

Exhibit 3–106

(37k56 203k256) & HE ((A.I.D.S. H.I.V. "Acquired immune" "Acquired immunodeficiency" "Human Immunodeficiency") /p (spit spitting bite bitten biting))

Please type your query as desired and press ENTER
Your database is ALLCASES (Federal & State Case Law)
Your search method uses Terms and Connectors

To simultaneously search federal and state decisions prior to 1945, use OLD as the database identifier, e.g., SDB OLD, QDB OLD, or DB OLD.

Use TAB to select a Jump marker ( ) or type a command and press ENTER

| ► Fields | ► Thesaurus | ► Help |
| Find by ► Citation or ► Title | ► Scope | ► Natural Language |

Source: ALLCASES, "AIDS" search, © WESTLAW, used with permission.

HE((A.I.D.S. H.I.V. "Acquired immune" "Acquired immunodefi-ciency" "Human immunodeficiency") /p (Spit spitting bite bitten biting)).

The **HE** (for headnote) indicates that the search is limited to looking for the terms, contained within the outer set of parentheses, appearing in the headnotes found at the front of each case in the database. If a document is to be retrieved, one of its headnotes must contain at least one of the terms included within the first set of inner parentheses—that is, **A.I.D.S.** or **H.I.V.** or "**Acquired immune**" or "**Acquired immunodeficiency**" or "**Human immunodeficiency**," and at least one of the terms appearing in the second set of inner parentheses—that is, "spit spitting bite bitten biting." The **/p** (for paragraph) symbol signifies that at least one of the terms on each side of the **/p** appear in the same paragraph. Remember that placing periods between letters limits the retrieval of a word to instances where it is used as an acronym or initialism (hence the periods in A.I.D.S. and H.I.V.), and that placing a phrase in quotes limits retrieval to documents where that exact phrase appears.

Our next step is to combine our key numbers and our headnote strategy into one search strategy. The result is depicted in Exhibit 3–106. The final strategy will retrieve documents that include either key number **37k56** or **203k256** *and* meet the requirements of the headnote portion of our search strategy. Our search strategy retrieves eight documents in the **ALLCASES** database. Cites to the first seven cases are listed in Exhibit 3–107. The researcher may want to use some of the

---

Exhibit 3–107

AUTHORIZED FOR EDUCATIONAL USE ONLY
CITATIONS LIST (Page 1)   Search Result Documents: 8
Database: ALLCASES

1. Weeks v. Scott, 55 F.3d 1059 (5th Cir. (Tex.), Jun 23, 1995) (No. 94–20838)

2. U.S v. Sturgis, 48 F.3d 784, 63 USLW 2607 (4th Cir. (Va.), Feb 21, 1995)(No. 94–5142)

3. Weeks v. Collins, 867 F.Supp. 544 (S.D.Tex., Oct 11, 1994) (No. Civ. A. 93–3708)

4. State v. Smith, 262 N.J.Super. 487, 621 A.2d 493, 61USLW 2642 (N.J.Super.A.D., Feb 17, 1993) (No. A–6363–89T1)

5. Weeks v. State, 834 S.W.2d 559 (Tex.App.–Eastland, Jul 09, 1992) (No. 11–90–045–CR)

6. State v. Haines, 545 N.E.2d 834, 58 USLW 2327 (Ind.App. 2 Dist., Oct 31, 1989) (No. 79A02–8807–CR–283)

7. Brock v. State, 555 So.2d 285 (Ala.Cr.App., Aug 25, 1989) (No. 8 DIV. 235)

Source: ALLCASES, "AIDS" list, © WESTLAW, used with permission.

browsing features we have discussed earlier to verify that these cases are relevant to our topic, but some familiar names clearly appear in the list.

Researchers who lack expertise in the use of field restrictions and proximity operators may prefer to try this search in **natural language** mode on WESTLAW. Indeed, even the expert researcher who has run a search in the **Terms and Connectors** mode may want to verify those results by running it in **natural language**. The latter technique is especially appropriate when key numbers have been used in the initial search strategy. Remember that the key numbers have been added to the case by the West editors. While the editors perform an excellent service, they are not perfect, and they may not always assign every appropriate key number to a case. Furthermore, recent cases may appear on WESTLAW before the editors have had a chance to add key numbers to them. The editors eventually add the key numbers, but in the interim a key number search will fail to retrieve those cases even if they are relevant to the search topic.

In the DOS version of WESTLAW a prompt appears at the bottom of the query entry screen that allows the researcher to switch to the **natural language** mode (see Exhibit 3–106). Alternatively, if the researcher is in **natural language** mode, the prompt at the bottom of the query entry screen will be **Terms and Connectors**. (See Exhibit 3–108.) Once we are in the **natural language** mode we simply type the key words that describe our search topic. (See Exhibit 3–108.) Remember that words are automatically truncated to pick up plural and other endings to the words, but also remember that it is important to look for synonymous

---

Exhibit 3–108

aids hiv spit bite assault deadly weapon attempted murder

Please type a description of your issue and press ENTER
Your database is ALLCASES (Federal & State Case Law)
Your search method uses Natural Language   Maximum Result: 20

To simultaneously search federal and state decisions prior to 1945, use OLD as the database identifier, e.g., SDB OLD, QDB OLD, or DB OLD.

Use TAB to select a Jump marker (▶) or type a command and press ENTER
▶ Restrict (e.g., Court, Date)   ▶ Thesaurus   ▶ Control Concepts
Find by ▶ Citation or ▶ Title   ▶ Scope   ▶ Terms and Connectors

Source: ALLCASES, "AIDS" nat., © WESTLAW, used with permission.

or "related" words by using the **Thesaurus** prompt. In our example, the thesaurus contains terms related to several that we used in our search strategy. (See Exhibit 3–109.) (We have previously demonstrated how to use the thesaurus and will not repeat that process here. Instead, we will skip ahead to the search entry screen that depicts our final search strategy. See Exhibit 3–110.) Remember that the parentheses in **Nat Lang** are interpreted as the Boolean operator **or**, so all the terms appearing in the parentheses immediately following the acronym "AIDS" will be treated as alternatives for retrieval purposes. Similarly, the word "homicide" is an alternative to the term "murder."

Recall that **natural language** is set to retrieve 20 documents and ranks the documents in order of relevancy. The cites to the first six cases are listed in Exhibit 3–111. A quick comparison indicates that some of the cases are the same as the ones retrieved using Terms and Connectors, but others are new. The difference in the results should reinforce the point that exhaustive legal research requires the pursuit of multiple avenues of inquiry rather than one-stop shopping.

Case Law Research with LEXIS

Because LEXIS lacks a comprehensive subject indexing for cases that is comparable to the West key number system, the researcher is limited to key word searching. Therefore, it is extremely important to give careful thought to the terms to be used in the search strategy. We have previously discussed the FREESTYLE

Exhibit 3–109

WESTLAW Thesaurus

Your description is:
AIDS HIV SPIT BITE ASSAULT DEADLY WEAPON ATTEMPTED MURDER

The following concepts in your description have related concepts. To view related concepts, type one or more concept numbers (e.g., 2, 3) and press ENTER.

1 AIDS
2 AID
3 HIV
4 BITE
5 ASSAULT
6 MURDER

     Use TAB to select a Jump marker (▶) or type a number and press ENTER
     ▶ Edit Description   ▶ Help

Source: ALLCASES, "AIDS" thes., © WESTLAW, used with permission.

Exhibit 3–110

AIDS ("ACQUIRED IMMUNE DEFICIENCY SYNDROME" "ACQUIRED IMMUNODEFICIENCY SYNDROME" A.I.D.S.) HIV ("HUMAN IMMUNODEFICIENCY VIRUS" H.I.V.) SPIT BITE AS-SAULT DEADLY WEAPON ATTEMPTED MURDER (HOMICIDE)

Please modify your description as desired and press ENTER
Your database is ALLCASES (Federal & State Case Law)
Your search method uses Natural Language   Maximum Result: 20

Note: Erase your existing description before typing a command from this screen.

Use TAB to select a Jump marker (▶) or type a command and press ENTER

▶ Restrict (e.g., Court, Date)   ▶ Thesaurus   ▶ Control Concepts
Find by ▶ Citation or ▶ Title   ▶ Scope   ▶ Terms and Connectors

Source: ALLCASES, "AIDS" edited, © WESTLAW, used with permission.

Exhibit 3–111

Copr. (C) West 1996 No claim to orig. U.S. govt. works
AUTHORIZED FOR EDUCATIONAL USE ONLY
CITATIONS LIST (Page 1)   Search Result Documents: 20
Database: ALLCASES

1. State v. Smith, 262 N.J.Super. 487, 621 A.2d 493, 61 USLW 2642 (N.J.Super.A.D., Feb 17, 1993) (NO. A–6363–89T1)

2. Scroggins v. State, 198 Ga.App. 29, 401 S.E.2d 13 (Ga.App., Nov 05, 1990) (NO. A90A1140, A90A1143, A90A1141, A90A1142, A90A1144)

3. Smallwood v. State, 106 Md.App. 1, 661 A.2d 747 (Md.App., Jul 13, 1995) (NO. 1678 SEPT. TERM 1994)

4. Smallwood v. State, 343 Md. 97, 680 A.2d 512, 65 USLW 2127 (Md., Aug 01, 1996) (NO. 122 SEPT. TERM 1995)

5. State v. Haines, 545 N.E.2d 834, 58 USLW 2327 (Ind.App. 2 Dist., Oct 31, 1989) (NO. 79A02–8807–CR–283)

6. U.S. v. Sturgis, 48 F.3d 784, 63 USLW 2607 (4th Cir. (Va.), Feb 21, 1995) (NO. 94–5142)

Source: ALLCASES, "AIDS" list, © WESTLAW, used with permission.

and **Boolean** search modes on LEXIS. The FREESTYLE mode is comparable to **natural language** searching on WESTLAW, and the **Boolean** mode is comparable to the **Terms and Connectors** mode on WESTLAW. While differences exist between the LEXIS and WESTLAW search systems, the techniques used in designing a search strategy and the capabilities of the systems are very similar.

As we apply these techniques and capabilities to develop a search strategy for our hypothetical Case 1, our first step is to determine which database we want to search. In Case 1 the defendant, John Winston, is accused of sexually abusing a child. The legal question raised is whether Winston's right to confront accusing witnesses is violated if the child is allowed to testify via closed-circuit television. The trial is taking place in Indiana, so we consider limiting our search to the LEXIS file that includes only cases decided in the state courts of Indiana. However, the issue also may have been argued in the federal courts, and the researcher may be interested in seeing how other state courts have ruled on this issue. While the rulings of other states' courts are not binding on Indiana's courts, the reasoning used to arrive at their decisions can be persuasive in constructing a legal argument. Therefore, we decide to search all state and federal court cases. Earlier in this chapter we went through the procedure used to identify **libraries** and **files** on the LEXIS system. There is a file on LEXIS that includes both federal and state court cases called **MEGA**.

Our next step is to determine whether we want to use the FREESTYLE or the **Boolean** search mode. Expert researchers consider many factors in choosing between these modes, but beginners will probably decide depending on their familiarity with the proximity requirements and field restrictions used in the **Boolean** mode. Most will probably opt for the FREESTYLE mode, which we emphasize here. The FREESTYLE mode also includes some special features that allow researchers to evaluate the adequacy of their search strategies. The one solid piece of advice we can offer is to be careful about including marginal terms in your search strategy.

We will briefly discuss the **Boolean** search process before proceeding to a more detailed explanation of the FREESTYLE search option. We previously discussed the **Boolean** search mode when we explained retrieving statutory documents via LEXIS. The same proximity operators are applicable on LEXIS regardless of the type of document being researched, but the segments are different in the case law files. The major segments are depicted in Exhibit 3–112.

The **opinion** segment is the one most likely to include our search terms. The **opinion** segment contains the text of the prevailing view of the court. The researcher might also consider searching the **concur** segment for the text of concurring opinions; the **dissent** segment for the text of dissenting opinions; or the **opinions** segment, which includes the **opinion**, **concur**, and **dissent** segments. Cases from some of the courts have supplemental segments. For example, U.S. Supreme Court decisions include a **syllabus** segment, which contains a summary of the history of a case and the key points of law addressed by the case. However, these segments are not present in all cases, so the researcher has to be careful about using them.

We want three major concepts or elements to be represented in the documents retrieved: child witness, the right to confront, and testimony given via television.

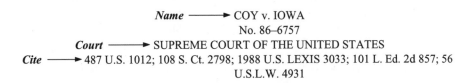

Exhibit 3–112

Name ⟶ COY v. IOWA
No. 86–6757
Court ⟶ SUPREME COURT OF THE UNITED STATES
Cite ⟶ 487 U.S. 1012; 108 S. Ct. 2798; 1988 U.S. LEXIS 3033; 101 L. Ed. 2d 857; 56
U.S.L.W. 4931

Date ⟶ { January 13, 1988, Argued
June 29, 1988, Decided

(Text Omitted)

Syllabus

SYLLABUS: Appellant was charged with sexually assaulting two 13-year-old girls. At appellant's jury trial, the court granted the State's motion, pursuant to a 1985 state statute intended to protect child victims of sexual abuse, to place a screen between appellant and the girls during their testimony, which blocked him from their sight but allowed him to see them dimly and to hear them. The court rejected appellant's argument that this procedure violated Confrontation Clause of the Sixth Amendment, which gives a defendant the right "to be confronted with the witnesses against him." Appellant was convicted of two counts of lascivious acts with a child, and the Iowa Supreme Court affirmed.

(Text Omitted)

Counsel

COUNSEL: Paul Papak, by appointment of the Court, 484 U.S. 810, argued the cause and filed briefs for appellant.

Gordon E. Allen, Deputy Attorney General of Iowa, argued the case for appellee. With him on the brief were Thomas J. Miller, Attorney General, and Roxann M. Ryan, Assistant Attorney General.*

*John L. Walker filed a brief for the National Association of Criminal Defense Lawyers as amicus curiae urging reversal.

(Text Omitted)

Judges

JUDGES: Scalia, J., delivered the opinion of the Court, in which Brennan, White, Marshall, Stevens, and O'Connor, JJ., joined. O'Connor, J., filed a concurring opinion, in which White, J., joined, post, p. 1022. Blackmun, J., filed a dissenting opinion, in which Rehnquist, C.J., joined, post, p. 1025. Kennedy, J., took no part in the consideration or decision of the case.

continues

Exhibit 3–112   continued

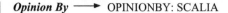

Opinion By ⟶ OPINIONBY: SCALIA

OPINION: JUSTICE SCALIA delivered the opinion of the Court.

Opinion ⟶   Appellant was convicted of two counts of lascivious acts with a child after a jury trial in which a screen placed between him and the two complaining witnesses blocked him from their sight. Appellant contends that this procedure, authorized by state statute, violated his Sixth Amendment right to confront the witnesses against him.

(Text Omitted)

Concur ⟶ CONCURBY: O'CONNOR
By

CONCUR: JUSTICE O'CONNOR, with whom JUSTICE WHITE joins, concurring.

OPINIONS

Concur ⟶   I agree with the Court that appellant's rights under the Confrontation Clause were violated in this case. I write separately only to note my view that those rights are not absolute but rather may give way in an appropriate case to other competing interests so as to permit the use of certain procedural devices designed to shield a child witness from the trauma of courtroom testimony.

(Text Omitted)

Dissent ⟶ DISSENT BY: BLACKMUN
By

DISSENT: JUSTICE BLACKMUN, with whom THE CHIEF JUSTICE joins, dissenting.

Dissent ⟶   Appellant was convicted by an Iowa jury on two counts of engaging in lascivious acts with a child. Because, in my view, the procedures employed at appellant's trial did not offend either the Confrontation Clause or the Due Process Clause, I would affirm his conviction. Accordingly, I respectfully dissent.

(Text Omitted)

Source: Segments. Reprinted with the permission of LEXIS-NEXIS, a division of Reed Elsevier Inc. LEXIS and NEXIS are registered trademarks of Reed Elsevier Properties Inc. FREESTYLE, KWIC, SuperKWIC, SuperKWIC and MEGA are trademarks of Reed Elsevier Properties Inc. SHEPARD'S and SHEPARDIZE are registered trademarks of Shepard's Company, a Partnership.

We design the search strategy depicted in Exhibit 3–113 below: "**opinion (child! w/5 witness! w/15 confront! w/15 televis!**)." Recall that the truncation symbol, "!" retrieves terms that include various end forms for a word. Therefore, "**child!**," retrieves documents containing the term "**children**" as well as "**child,**" and "**televis!**" retrieves "**televise,**" "**televising,**" and "**television.**" The "**w/#**" is a proximity requirement. The terms on either side of this command must appear within the designated number of words. Therefore, "**child! w/5 witness!**" retrieves docu-

Exhibit 3–113

Opinion (child! w/5 witness! w/15 confront! w/15 televis!)

Please type your search request then press the ENTER key.
What you enter will be Search Level 1.

Type .fr to enter a FREESTYLE (TM) search.

For further explanation, press the H key (for HELP) and then the ENTER key.

Source: MEGA, search, boolean. Reprinted with the permission of LEXIS-NEXIS, a division of Reed Elsevier Inc. LEXIS and NEXIS are registered trademarks of Reed Elsevier Properties Inc. FREESTYLE, KWIC, SuperKWIC, SuperKWIC and MEGA are trademarks of Reed Elsevier Properties Inc. SHEPARD'S and SHEPARDIZE are registered trademarks of Shepard's Company, a Partnership.

ments in which a word beginning with the root **"child"** appears within five words of a word beginning with the root **"witness."** The parentheses indicate that the included terms must appear in the segment indicated—in this instance, the **opinion** segment. The complete strategy would be interpreted as requiring that a term beginning with the root **"child"** appear within five words of a term beginning with the root **"witness,"** which in turn must appear within fifteen words of a term beginning with the root **"confront,"** which must appear within fifteen words of a term beginning with the root **"televis,"** and all the terms must appear within the **opinion** segment. When **entered**, the search strategy retrieves 45 cases, as indicated on the screen depicted in Exhibit 3–114.

Remember that in the **Boolean** mode a document must meet the specific requirement of the search strategy in order to be retrieved—for example, if a document meets all the other requirements but includes the term **"tv"** instead of **"television"** it will not be retrieved. Synonyms and related concepts can be included in the search strategy, but no thesaurus is available to help the researcher identify possible synonymous terms. Remember also that no attempt is made in the **Boolean** mode to rank the retrieved documents in order of importance. Consequently, a document that contains text meeting our search requirements just once could be listed before a document that satisfies our search strategy multiple times. It therefore is essential to browse all 45 documents to determine the adequacy of the search strategy.

Exhibit 3–114

OPINION (CHILD! W/15 WITNESS! W/15 CONFRONT! W/15 TELEVIS!)

Your search request has found 45 CASES through Level 1.
To DISPLAY these CASES press either the KWIC, FULL, CITE or SEGMTS key.
To MODIFY your search request, press the M key (for MODFY) and then the ENTER key.

For further explanation, press the H key (for HELP) and then the ENTER key.

Source: MEGA, result. Reprinted with the permission of LEXIS-NEXIS, a division of Reed Elsevier Inc. LEXIS and NEXIS are registered trademarks of Reed Elsevier Properties Inc. FREESTYLE, KWIC, SuperKWIC, SuperKWIC and MEGA are trademarks of Reed Elsevier Properties Inc. SHEPARD'S and SHEPARDIZE are registered trademarks of Shepard's Company, a Partnership.

The **KWIC** command allows the researcher to browse the documents by examining only those screens on which the search terms appear. The cases are presented sequentially, with the search terms highlighted on the displayed screens—for example, the terms **"child witnesses," "confrontation,"** and **"television"** are highlighted in the first document we retrieved. (See Exhibits 3–115 and 3–116.) Several citations are given for the case, but the official cite is 21 F.3d 885. Notice the ***892** preceding the text at the bottom of the screen. This figure indicates that, although the case begins at page 885 of volume 21 of the Federal Reporter 3d series, the text displayed is from page 892 of that volume.

Another way of evaluating the adequacy of the search strategy is to determine if our results include relevant cases we have already identified, such as *Maryland v. Craig*. The **cite** command displays a list of citations to the cases retrieved. We see that the twenty-first document is *Craig v. State*, the state court case that was reviewed by the U.S. Supreme Court and decided under the name *Maryland v. Craig*. (See Exhibit 3–117.)

To demonstrate the FREESTYLE feature, we modify our strategy to include a fourth element, **"testimony."** It would be reasonable to ask why a fifth element should not be added, such as **"sexual abuse,"** or even a sixth, such as **"trauma,"** since these terms also are important to the issue being researched. Indeed, we could try a search strategy with all six elements or with a different group of elements than we have chosen. However, it usually is safest to begin with a few

Exhibit 3–115

LEVEL 1 – 1 OF 45 CASES

UNITED STATES v. QUINTERO

No. 93–10217

UNITED STATES COURT OF APPEALS FOR THE NINTH CIRCUIT

21 F.3d 885; 1994 U.S. App. LEXIS 6568; 94 Cal. Daily Op. Service 2436; 94 Daily Journal DAR 4651

December 14, 1993, Argued, Submitted, San Francisco, California
April 7, 1994, Filed

OPINION:

. . . [*892] [**18] face-to-face confrontation at trial, a preference that must occasionally give way to considerations [**19] of public policy and the necessities of the case,'" <=45> id. at 849 (citations omitted). The Court held that "the state interest in protecting child witnesses from the trauma of testifying" is sufficiently important to justify procedures that depart from face-to-face confrontation with the defendant. <=46> Id. at 855. Allowing

Source: MEGA, display. Reprinted with the permission of LEXIS-NEXIS, a division of Reed Elsevier Inc. LEXIS and NEXIS are registered trademarks of Reed Elsevier Properties Inc. FREESTYLE, KWIC, SuperKWIC, SuperKWIC and MEGA are trademarks of Reed Elsevier Properties Inc. SHEPARD'S and SHEPARDIZE are registered trademarks of Shepard's Company, a Partnership.

Exhibit 3–116

21 F.3d 885, *892; 1994 U.S. App. LEXIS 6568, **19;
94 Cal. Daily Op. Service 2436; 94 Daily Journal DAR 4651

Remo to testify by closed-circuit television, after the trial court specifically found that testifying in court would be [*893] traumatic for him, was not plain error. n6 We therefore affirm.

– – – – – – – – – – – – – – – – – – – Footnotes –

n6 We note that the district court further protected the defendant's substantial rights . . .

Source: MEGA, display. Reprinted with the permission of LEXIS-NEXIS, a division of Reed Elsevier Inc. LEXIS and NEXIS are registered trademarks of Reed Elsevier Properties Inc. FREESTYLE, KWIC, SuperKWIC, SuperKWIC and MEGA are trademarks of Reed Elsevier Properties Inc. SHEPARD'S and SHEPARDIZE are registered trademarks of Shepard's Company, a Partnership.

Exhibit 3–117

LEVEL 1—45 CASES

17. STATE v. NIEHOFF, No. 60,666 NOT DESIGNATED FOR PUBLICATION, Supreme Court of Kansas, 771 P.2d 73; 1989 Kan. LEXIS 31, March 3, 1989, Filed

18. State v. Eaton, No. 60,991, Supreme Court of Kansas, 244 Kan. 370; 769 P.2d 1157; 1989 Kan. LEXIS 45, March 3, 1989, Opinion Filed

19. STATE v. ADAIR, No. 66,059, COURT OF APPEALS OF KANSAS, 1992 Kan. App. LEXIS 67, February 21, 1992, Filed, NOT DESIGNATED FOR PUBLICATION

20. State v. Albert, No. 62,413, Court of Appeals of Kansas, 13 Kan. App. 2d

671; 778 P.2d 386; 1989 Kan. App. LEXIS 589, August 25, 1989, Opinion Filed

21. CRAIG v. STATE, No. 110, September Term, 1988, Court of Appeals of Maryland, 316 Md. 551; 560 A.2d 1120; 1989 Md. LEXIS 109, July 24, 1989, As Amended July 25,1989.

22. People v. Staffney, Docket No. 102122, Court of Appeals of Michigan, 187 Mich. App. 660; 468 N.W.2d 238; 1990 Mich. App. LEXIS 528, August 7, 1990, Submitted, December 18, 1990, Decided

23. In re Vanidestine, Docket No. 120690, Court of Appeals of Michigan, 186 Mich. App. 205; 463 N.W.2d 225; 1990 Mich. App. LEXIS 439, May 16, 1990,

Source: MEGA, list. Reprinted with the permission of LEXIS-NEXIS, a division of Reed Elsevier Inc. LEXIS and NEXIS are registered trademarks of Reed Elsevier Properties Inc. FREESTYLE, KWIC, SuperKWIC, SuperKWIC and MEGA are trademarks of Reed Elsevier Properties Inc. SHEPARD'S and SHEPARDIZE are registered trademarks of Shepard's Company, a Partnership.

elements, keeping in mind that you can later modify your search to include additional elements. The search strategy we will be working with now includes four elements: the right to **"confront," "child witness," "testimony,"** and **"television."**

Before we run our search, we should quickly review how a search in the FREESTYLE feature works. First, we **enter** either a complete sentence or just the key words that represent the issues and facts we are looking for. The computer pulls terms or phrases from the description we have **entered**. It first looks for documents that contain terms or phrases representing all of the elements in the search strategy. Under the default option, the 25 most relevant documents are retrieved and are ranked in order of relevancy. Relevancy is determined by the number of times the search terms appear in a document, but with added importance given to search terms that appear relatively infrequently in the rest of the documents in the file. If there are not 25 documents that contain all elements of the search strategy, the computer will drop terms in order to retrieve the fixed number of documents.

In Exhibit 3–118 we depict the FREESTYLE query screen with our original search terms. We could have typed a complete sentence, such as "Is a defendant's right to confront his accusers violated when a child witness is allowed to provide testimony via closed circuit television?" and then left it to the computer to extract key terms or phrases from the sentence. A problem with this approach is that the computer will search for terms such as **"violated"** or **"circuit,"** which may lead us

Exhibit 3–118

child witness testimony confront television

Enter your FREESTYLE (TM) Search Description.
Enter phrases in quotation marks.
Example: What are the requirements for a "day care center" license?

Type .bool to exit FREESTYLE and run a Boolean search.

For further explanation, press the H key (for HELP) and then the ENTER key.

Source: MEGA, search, freestyle. Reprinted with the permission of LEXIS-NEXIS, a division of Reed Elsevier Inc. LEXIS and NEXIS are registered trademarks of Reed Elsevier Properties Inc. FREESTYLE, KWIC, SuperKWIC, SuperKWIC and MEGA are trademarks of Reed Elsevier Properties Inc. SHEPARD'S and SHEPARDIZE are registered trademarks of Shepard's Company, a Partnership.

astray from our true objective. It generally is preferable to enter only the terms or phrases you want searched, as we have done in Exhibit 3–118.

The FREESTYLE feature contains a thesaurus, which facilitates the process of identifying related terms. When we **enter** the search strategy we proceed to a screen (Exhibit 3–119) that provides an option for retrieving synonyms. Notice that the screen has placed the phrase "child witness" in quotes, which indicates that the phrase rather than the individual words "**child**" and "**witness**" will be used to retrieve documents. We **enter =4** to proceed to the next screen (Exhibit 3–120), which lists the terms from our original search strategy that have synonyms. The directions provided in the ensuing series of screens enable us to add terms. Exhibit 3–121 shows the screen providing synonyms for "**confront**." Note that we have decided to add the term "**face**" as a synonym for "**confront**." We repeat this process for all relevant terms until we have devised our final search strategy (see Exhibit 3–122). Note that the word "**face**" appears in parentheses immediately after "**confront**," and the term "**tv**" appears in parentheses immediately after "**television**." The parentheses in a FREESTYLE search indicate the Boolean operator **or** and refer to the term immediately preceding the term within parentheses. Thus, our search strategy retrieves documents that contain either "**confront**" or "**face**," "**television**" or "**tv**," as well as "**child witness**" and "**testimony**."

As mentioned above, if there are not 25 documents in the database that meet the requirements of the search strategy, the computer will drop terms from the search

Exhibit 3–119

=4_   FREESTYLE(TM) SEARCH OPTIONS

Press ENTER to start search.
To use a Search Option, enter an equal sign followed by the number.

Search Description:
    "CHILD WITNESS" TESTIMONY CONFRONT TELEVISION

Press ENTER to start search.
<=1>   Edit Search Description
<=2>   Enter/edit Mandatory Terms
<=3>   Enter/edit Restrictions (e.g., date)
<=4>   Synonyms and Related Concepts
<=5>   Change number of documents Current setting: 25

For further explanation, press the H key (for HELP) and then the ENTER key.

Source: MEGA, synonyms. Reprinted with the permission of LEXIS-NEXIS, a division of Reed Elsevier Inc. LEXIS and NEXIS are registered trademarks of Reed Elsevier Properties Inc. FREESTYLE, KWIC, SuperKWIC, SuperKWIC and MEGA are trademarks of Reed Elsevier Properties Inc. SHEPARD'S and SHEPARDIZE are registered trademarks of Shepard's Company, a Partnership.

Exhibit 3–120

2, 3, 4   TERM SELECTION

Enter numbers for related concepts or synonyms. Example: 1,2, 3, 6–10

<=1>   Return to Search Options

 1 Related concepts for your search description

Search Terms found in thesaurus
    2 TESTIMONY
    3 CONFRONT
    4 TELEVISION

For further explanation, press the H key (for HELP) and then the ENTER key.

Source: MEGA, synonym choices. Reprinted with the permission of LEXIS-NEXIS, a division of Reed Elsevier Inc. LEXIS and NEXIS are registered trademarks of Reed Elsevier Properties Inc. FREESTYLE, KWIC, SuperKWIC, SuperKWIC and MEGA are trademarks of Reed Elsevier Properties Inc. SHEPARD'S and SHEPARDIZE are registered trademarks of Shepard's Company, a Partnership.

Exhibit 3–121

---

3   SYNONYM SELECTION

Synonyms for: CONFRONT
Enter synonym numbers to include in search and press ENTER <e.g. 1, 2, 3–4>

<=1> Return to Search Options   <=2> Return to Term Selection
- -
  1 affront   2 encounter   3 face
  4 meet
- - - - - - - - - - - - - - - - - - Accost -
  5 face   6 front

For further explanation, press the H key (for HELP) and then the ENTER key.

Source: MEGA, synonyms "Confront." Reprinted with the permission of LEXIS-NEXIS, a division of Reed Elsevier Inc. LEXIS and NEXIS are registered trademarks of Reed Elsevier Properties Inc. FREESTYLE, KWIC, SuperKWIC, SuperKWIC and MEGA are trademarks of Reed Elsevier Properties Inc. SHEPARD'S and SHEPARDIZE are registered trademarks of Shepard's Company, a Partnership.

Exhibit 3–122

FREESTYLE(TM) SEARCH OPTIONS

Press ENTER to start search.
To use a Search Option, enter an equal sign followed by the number.

Search Description:
  "CHILD WITNESS" TESTIMONY CONFRONT (FACE) TELEVISION (TV)

Press ENTER to start search.
<=1> Edit Search Description
<=2> Enter/edit Mandatory Terms
<=3> Enter/edit Restrictions (e.g., date)
<=4> Synonyms and Related Concepts
<=5> Change number of documents Current setting: 25

For further explanation, press the H key (for HELP) and then the ENTER key.

Source: MEGA, edited "Witness." Reprinted with the permission of LEXIS-NEXIS, a division of Reed Elsevier Inc. LEXIS and NEXIS are registered trademarks of Reed Elsevier Properties Inc. FREESTYLE, KWIC, SuperKWIC, SuperKWIC and MEGA are trademarks of Reed Elsevier Properties Inc. SHEPARD'S and SHEPARDIZE are registered trademarks of Shepard's Company, a Partnership.

in order to meet its quota. Therefore, it is especially critical in FREESTYLE searching to validate your search results. There are several ways of doing this from the search screen that indicate that the search has been completed (Exhibit 3–123). One way is to browse the documents using the **key word in context** (**.KWIC**) or **SuperKWIC™**(**.SK**) options. Either of these browsing features allows the researcher to move quickly through retrieved documents, stopping only at the first page and the screens where at least one of the search terms appears. For example, we use the **.SK** command to display the citation information for our first document (see Exhibit 3–124), and then the first screen in the document where our search terms appeared (see Exhibit 3–125). The key words are highlighted as you browse, and helpful prompts appear at the bottom of the screen. As in the **Boolean** mode, the citations at the top of the screen identify where the cases are found in the printed reporter series, and also provide the exact page where the information displayed on the screen appears in the printed sets. For example, the reference to "***1034**" indicates that the immediately following text appears at page 1034 of volume 141 of the Miscellaneous Reporter 2d series. The first page of the case is on **1031**, as indicated by the citation, **141 Misc. 2d 1031**. Similarly, the number following the ** is the page where the text can be found in the second cited reporter series. Therefore, although *People v. Rivera* begins at page 909 of volume 535 of the New York Supplement 2d series (535 N.Y.S.2d 909), the text at the lower part of the screen (Exhibit 3–124) appears at page 911, as indicated by ****911**.

Exhibit 3–123

Your FREESTYLE search has retrieved the top 25 documents based on statistical ranking. Search terms are listed in order of importance.

"CHILD WITNESS" TV CONFRONT TELEVISION FACE TESTIMONY

Press ENTER to view documents in KWIC or use Full, Cite or Segmnt keys.

<=1> Browse documents in SuperKWIC (.SK)
<=2> Location of search terms in documents (.where)
<=3> Number of documents with search terms (.why)
<=4> Change document order (.sort)

For further explanation, press the H key (for HELP) and then the ENTER key.

Source: MEGA, "Witness" result. Reprinted with the permission of LEXIS-NEXIS, a division of Reed Elsevier Inc. LEXIS and NEXIS are registered trademarks of Reed Elsevier Properties Inc. FREESTYLE, KWIC, SuperKWIC, SuperKWIC and MEGA are trademarks of Reed Elsevier Properties Inc. SHEPARD'S and SHEPARDIZE are registered trademarks of Shepard's Company, a Partnership.

Exhibit 3–124

```
                    LEVEL 1—1 OF 25 CASES
                        People v. Rivera
                   [NO NUMBER IN ORIGINAL]
                Supreme Court of New York, Bronx County
         141 Misc. 2d 1031; 535 N.Y.S.2d 909; 1988 N.Y. Misc. LEXIS
                               740
                        December 2, 1988
```

OPINION:

. . . [*1034] [**911] [***6] not encourage the jury to draw an inference adverse to the interest of the defendant.

- -

.MORE	Next Page	.NP	Cite	.CI	Exit FREESTYLE	.BOOL	Print Doc	.PR
.WHERE	Prev Page	.PP	Kwic	.KW	New Search	.NS	Print All	.PA
.WHY	Next Doc	.ND	Full	.FU	Modify	.M	Cmds Off	.COF
.SORT	Prev Doc	.PD	SKWIC	.SK	Chg Library	.CL	Sign Off	.SO

Source: MEGA, "Witness" display. Reprinted with the permission of LEXIS-NEXIS, a division of Reed Elsevier Inc. LEXIS and NEXIS are registered trademarks of Reed Elsevier Properties Inc. FREESTYLE, KWIC, SuperKWIC, SuperKWIC and MEGA are trademarks of Reed Elsevier Properties Inc. SHEPARD'S and SHEPARDIZE are registered trademarks of Shepard's Company, a Partnership.

Exhibit 3–125

```
           141 Misc. 2d 1031, *1034; 535 N.Y.S.2d 909, **911;
                    1988 N.Y. Misc. LEXIS 740, ***6
```

"6. Upon request of the defendant, the court shall instruct the jury that they are to draw no inference from the use of live, two-way closed-circuit television in the examination of the vulnerable child witness.

"7. The vulnerable child witness shall testify under oath except as specified in subdivision two of section 60.20. The examination and cross-examination of the vulnerable child witness shall, in all other respects, be conducted in the same manner [***7] as if the vulnerable child witness had testified in the courtroom.

"8. When the testimony of the vulnerable child witness is transmitted from the testimonial room into the courtroom, the court stenographer shall record the testimony in the same manner as if the vulnerable child witness had testified in the courtroom.

Although this statute expressly authorizes the use of a "two-way" closed-circuit television setup, the constitutionality of the procedure was

- -

.MORE	Next Page	.NP	Cite	.CI	Exit FREESTYLE	.BOOL	Print Doc	.PR
.WHERE	Prev Page	.PP	Kwic	.KW	New Search	.NS	Print All	.PA
.WHY	Next Doc	.ND	Full	.FU	Modify	.M	Cmds Off	.COF
.SORT	Prev Doc	.PD	SKWIC	.SK	Chg Library	.CL	Sign Off	.SO

Source: MEGA, "Witness" browse. Reprinted with the permission of LEXIS-NEXIS, a division of Reed Elsevier Inc. LEXIS and NEXIS are registered trademarks of Reed Elsevier Properties Inc. FREESTYLE, KWIC, SuperKWIC, SuperKWIC and MEGA are trademarks of Reed Elsevier Properties Inc. SHEPARD'S and SHEPARDIZE are registered trademarks of Shepard's Company, a Partnership.

A second method for determining the validity of the search results involves using the **.why** command, which is a unique feature of FREESTYLE on LEXIS. The **.why** command appears both at the bottom of the screens being browsed (see Exhibits 3–124 and 3–125), and as an option on the original search result screen (Exhibit 3–123). The **.why** command reveals the presence of each search term in all of the retrieved documents (see Exhibit 3–126), and reports the total number of occurrences of each term throughout all the documents (see Exhibit 3–127). If we take a close look at Exhibit 3–126, we see that three of the four elements in our search strategy appear in all 25 documents retrieved, "**child witness**," "**television**," and "**testimony**." Both terms representing the fourth element, "confront" or "face," appear in all but three of the documents, and at least one of these terms appears in every document retrieved.

The terms are listed in descending order of importance, with "**child witness**" being the most important and "**testimony**" being the least. The relative weight assigned to each term is further explained on the screen depicted in Exhibit 3–127. Remember that terms that appear rarely throughout the text of all the documents in the file being searched are given extra weight in determining the ranking of documents. Thus, the phrase "**child witness**" is assigned a relatively high weight because it has only 2145 occurrences. Contrast the importance assigned to

Exhibit 3–126

LOCATION OF SEARCH TERMS IN DOCUMENTS (.where)

Document numbers are listed across the top of the chart.
Terms are listed down the side in order of importance.
Asterisks <*> indicate the existence of terms in documents.
To view a document, enter the document number.

	1	2	3	4	5	6	7	8	9	10	11	12	13	14	15	16	17	18	19	20	21	22	23	24	25
CHILD WITNESS	*	*	*	*	*	*	*	*	*	*	*	*	*	*	*	*	*	*	*	*	*	*	*	*	*
TV						*		*				*													
CONFRONT	*	*	*	*	*		*			*	*	*		*	*	*	*	*	*	*	*		*	*	*
TELEVISION	*	*	*	*	*	*	*	*	*	*	*	*	*	*	*	*	*	*	*	*	*	*	*	*	*
FACE	*	*			*	*	*	*	*	*	*	*	*	*	*	*	*	*	*	*	*	*	*		*
TESTIMONY	*	*	*	*	*	*	*	*	*	*	*	*	*	*	*	*	*	*	*	*	*	*	*	*	*

<=1>   Browse docs
<=2>   .why

For further explanation, press the H key (for HELP) and then the ENTER key.

Source: MEGA, "Witness" where. Reprinted with the permission of LEXIS-NEXIS, a division of Reed Elsevier Inc. LEXIS and NEXIS are registered trademarks of Reed Elsevier Properties Inc. FREESTYLE, KWIC, SuperKWIC, SuperKWIC and MEGA are trademarks of Reed Elsevier Properties Inc. SHEPARD'S and SHEPARDIZE are registered trademarks of Shepard's Company, a Partnership.

Exhibit 3–127

NUMBER OF DOCUMENTS WITH SEARCH TERMS (.why)

	Documents Retrieved	Documents Matched	Term Importance (0–100)
CHILD WITNESS	25	2145	33
TV	3	11360	24
CONFRONT	21	38113	17
TELEVISION	25	48180	16
FACE	23	413874	5
TESTIMONY	25	1124857	1

Total Retrieved: 25

<=1> Browse docs
<=2> .where

For further explanation, press the H key (for HELP) and then the ENTER key.

Source: MEGA, "Witness" why. Reprinted with the permission of LEXIS-NEXIS, a division of Reed Elsevier Inc. LEXIS and NEXIS are registered trademarks of Reed Elsevier Properties Inc. FREESTYLE, KWIC, SuperKWIC, SuperKWIC and MEGA are trademarks of Reed Elsevier Properties Inc. SHEPARD'S and SHEPARDIZE are registered trademarks of Shepard's Company, a Partnership.

"**child witness**" with that assigned to "**testimony**": "**testimony**" is given a low level of importance because it appears so frequently throughout the documents in the file. We might even consider removing "**testimony**" because it appears so frequently.

To find documents in addition to the initial 25, we use the **modify** prompt to return to our search entry screen. Option **=5** allows us to change the number of documents retrieved. (See Exhibit 3–128.) We can choose to retrieve anywhere from 1 to 1000 documents. We have opted for 45 documents (see Exhibit 3–129, upper lefthand corner), which is the same number we retrieved using the **Boolean** method. Another way of validating your search strategy is to determine whether you retrieve a significant number of the same cases using both modes.

A final way of evaluating our search strategy should be familiar from our **Boolean** example. We use the **cite** command to display a list of citations to the documents and then determine whether the list contains references to cases we have previously identified as being relevant. Using this procedure provides a reference to *Maryland v. Craig*, which gives us some confidence in our search strategy (Exhibit 3–130).

We now invite you to devise CALR search strategies to find cases dealing with our third hypothetical, which involves a prisoner's asserted right to be protected from the health risks associated with second-hand smoke. If you take time to analyze the issue, identify search terms, and persevere in learning the systems, we are

Exhibit 3–128

=5   FREESTYLE(TM) SEARCH OPTIONS

Press ENTER to start search.
Type .bool to exit FREESTYLE and run a Boolean search.

Search Description:
  "CHILD WITNESS" TESTIMONY CONFRONT (FACE) TELEVISION (TV)

Press ENTER to start search.
  <=1>   Edit Search Description
  <=2>   Enter/edit Mandatory Terms
  <=3>   Enter/edit Restrictions (e.g., date)
  <=4>   Synonyms and Related Concepts
  <=5>   Change number of documents Current setting: 25

For further explanation, press the H key (for HELP) and then the ENTER key.

Source: MEGA, "Witness" number. Reprinted with the permission of LEXIS-NEXIS, a division of Reed Elsevier Inc. LEXIS and NEXIS are registered trademarks of Reed Elsevier Properties Inc. FREESTYLE, KWIC, SuperKWIC, SuperKWIC and MEGA are trademarks of Reed Elsevier Properties Inc. SHEPARD'S and SHEPARDIZE are registered trademarks of Shepard's Company, a Partnership.

Exhibit 3–129

---

45   CHANGE NUMBER OF DOCUMENTS

Enter the maximum number of documents (1–1000) you wish to retrieve for this search. This number will apply to your searches until you sign off or change the setting.

Current setting: 25

For further explanation, press the H key (for HELP) and then the ENTER key.

Source: MEGA, "Witness" number. Reprinted with the permission of LEXIS-NEXIS, a division of Reed Elsevier Inc. LEXIS and NEXIS are registered trademarks of Reed Elsevier Properties Inc. FREESTYLE, KWIC, SuperKWIC, SuperKWIC and MEGA are trademarks of Reed Elsevier Properties Inc. SHEPARD'S and SHEPARDIZE are registered trademarks of Shepard's Company, a Partnership.

Exhibit 3–130

LEVEL 1–45 CASES

35. COMMONWEALTH v. LUDWIG, No. 34 E.D. Appeal Docket 1988, Supreme Court of Pennsylvania, 527 Pa. 472; 594 A.2d 281; 1991 Pa. LEXIS 111, May 3, 1989, Argued; January 8, 1991, Resubmitted, May 10, 1991, Decided

36. COMMONWEALTH v. TUFTS, No. N–4992, Supreme Judicial Court of Massachusetts, Norfolk, 405 Mass. 610; 542 N.E.2d 586; 1989 Mass. LEXIS 240, May 3, 1989, August 21, 1989

37. MARYLAND v. CRAIG, No. 89–478, SUPREME COURT OF THE UNITED STATES, 497 U.S. 836; 110 S. Ct. 3157; 1990 U.S. LEXIS 3457; 111 L. Ed. 2d 666; 58 U.S.L.W. 5044; 30 Fed. R. Evid. Serv. (Callaghan) 1, April 18, 1990, Argued, June 27, 1990, Decided

38. VIGIL v. TANSY, No. 89–2249, UNITED STATES COURT OF APPEALS FOR THE TENTH CIRCUIT, 917 F.2d 1277; 1990 U.S. App. LEXIS 18714; 31 Fed. R. Evid. Serv. (Callaghan) 689, October 26, 1990, Filed

- -

.MORE	Next Page	.NP	Cite	.CI	Exit FREESTYLE	.BOOL	Print Doc	.PR
.WHERE	Prev Page	.PP	Kwic	.KW	New Search	.NS	Print All	.PA
.WHY	Next Doc	.ND	Full	.FU	Modify	.M	Cmds Off	.COF
.SORT	Prev Doc	.PD	SKWIC	.SK	Chg Library	.CL	Sign Off	.SO

Source: MEGA, "Witness" result cites. Reprinted with the permission of LEXIS-NEXIS, a division of Reed Elsevier Inc. LEXIS and NEXIS are registered trademarks of Reed Elsevier Properties Inc. FREESTYLE, KWIC, SuperKWIC, SuperKWIC and MEGA are trademarks of Reed Elsevier Properties Inc. SHEPARD'S and SHEPARDIZE are registered trademarks of Shepard's Company, a Partnership.

---

confident that you will obtain some excellent search results. Our next step involves going to the case reporters to look at the cases and the links provided by the key number system.

The Case Reporters

We are finally ready to track down the judicial decisions to which we have been referred by so many different sources: secondary legal authorities, annotated statutes and constitutions, and finding tools such as Descriptive Word Indexes and case digests. The West key number system is central to the process of case law research. We will see how this system is superimposed on the judicial opinions published in West's case reporters. If it was not already apparent, you will gain a new appreciation of the usefulness of West's regional reporters (A.2d, N.E.2d, N.W.2d, P.2d, S.E.2d, S.W.2d, and So. 2d), Federal Supplement (F. Supp.), the Federal Reporter (including F.2d and F.3d), and the Supreme Court Reporter.

We start by examining *Maryland v. Craig*, 497 U.S. 836, 110 S. Ct. 3157, 111 L. Ed. 2d 666 (1990), the Supreme Court decision we have run across several times during our investigation of the confrontation issue in Case 1. In Chapter 1, we lamented that official case reports, including the U.S. Reports, do not facilitate legal research. Now we can demonstrate why, as we look up *Craig* in volume 497 of the U.S. Reports, beginning at page 836. (See Exhibit 3–131.) What we see initially is not part of the Supreme Court Justices' opinion; it is a **"syllabus"** pre-

---

Exhibit 3–131

Syllabus   497 U. S.

MARYLAND *v.* CRAIG

CERTIORARI TO THE COURT OF APPEALS OF MARYLAND

No. 89–478. Argued April 18, 1990—Decided June 27, 1990

Respondent Craig was tried in a Maryland court on several charges related to her alleged sexual abuse of a 6-year-old child. Before the trial began, the State sought to invoke a state statutory procedure permitting a judge to receive, by one-way closed circuit television, the testimony of an alleged child abuse victim upon determining that the child's courtroom testimony would result in the child suffering serious emotional distress, such that he or she could not reasonably communicate. If the procedure is invoked, the child, prosecutor, and defense counsel withdraw to another room, where the child is examined and cross-examined; the judge, jury, and defendant remain in the courtroom, where the testimony is displayed. Although the child cannot see the defendant, the defendant remains in electronic communication with counsel, and objections may be made and ruled on as if the witness were in the courtroom. The court rejected Craig's objection that the procedure's use violates the Confrontation Clause of the Sixth Amendment, ruling that Craig retained the essence of the right to confrontation. Based on expert testimony, the court also found that the alleged victim and other allegedly abused children who were witnesses would suffer serious emotional distress if they were required to testify in the courtroom, such that each would be unable to communicate. Finding that the children were competent to testify, the court permitted testimony under the procedure, and Craig was convicted. The State Court of Special Appeals affirmed, but the State Court of Appeals reversed. Although it rejected Craig's argument that the Clause requires in all cases a face-to-face courtroom encounter between the accused and accusers, it found that the State's showing was insufficient to reach the high threshold required by *Coy* v. *Iowa*, 487 U.S. 1012, before the procedure could be invoked. The court held that the procedure usually cannot be invoked unless the child initially is questioned in the defendant's presence and that, before using the one-way television procedure, the trial court must determine whether a child would suffer severe emotional distress if he or she were to testify by two-way television.

Held:

1. The Confrontation Clause does not guarantee criminal defendants an *absolute* right to a face-to-face meeting with the witnesses against them at trial. The Clause's central purpose, to ensure the reliability of the evidence against a defendant by subjecting it to rigorous testing in an adversary proceeding before the trier of fact, is served by the combined effects of the elements of confrontation: physical presence, oath, cross-examination, and observation of demeanor by the trier of fact. Although face-to-face confrontation forms the core of the Clause's values, it is not an indis-

continues

Exhibit 3–131 continued

pensable element of the confrontation right. If it were, the Clause would abrogate virtually every hearsay exception, a result long rejected as unintended and too extreme, *Ohio* v. *Roberts*, 448 U. S. 56, 63. Accordingly, the Clause must be interpreted in a manner sensitive to its purpose and to the necessities of trial and the adversary process. See, *e.g., Kirby* v. *United States*, 174 U. S. 47. Nonetheless, the right to confront accusatory witnesses may be satisfied absent a physical, face-to-face confrontation at trial only where denial of such confrontation is necessary to further an important public policy and only where the testimony's reliability is otherwise assured. *Coy, supra,* at 1021. Pp. 844–850.

2. Maryland's interest in protecting child witnesses from the trauma of testifying in a child abuse case is sufficiently important to justify the use of its special procedure, provided that the State makes an adequate showing of necessity in an individual case. Pp. 857–857.

(a) While Maryland's procedure prevents the child from seeing the . . .

Source: Reprinted from U.S. "Maryland v. Craig" official.

pared for the convenience of readers by a member of the Supreme Court's staff that summarizes the case issues and the Court's holding. Justice O'Connor's opinion for the Court does not begin until page 840. (See Exhibit 3–132.) From that point forward, the pages of the U.S. Reports are filled exclusively with Justice O'Connor's opinion, and then with Justice Scalia's dissenting opinion. There are no editorial devices to make the decision more accessible to legal researchers.

By way of contrast, consider how *Maryland v. Craig* is presented in West's Supreme Court Reporter. Without changing a word or a punctuation mark in what the Justices wrote, the West editors have added several important features to the case that make it possible to link the *Craig* decision to others presenting similar issues. At the outset, note that immediately above the case name we are provided with the parallel citations to *Maryland v. Craig* in the U.S. Reports and in Lawyers' Edition 2d. (See Exhibit 3–133.) Throughout the opinion, we are apprised of the corresponding pagination in the U.S. Reports (see, *e.g.*, the tiny "836" just before "Maryland"). This makes it possible to cite a specific page in the official report of the case without ever putting down the Supreme Court Reporter. This feature is available only in bound volumes of the Supreme Court Reporter, because West's editors must wait for the pagination to be finalized in the U.S. Reports before they can insert those page numbers into the opinions. The brief summary of the Court's holding that follows the case name, along with the citation to the lower court's decision, also are prepared by West. The numbered notes, which are called **headnotes** because they appear at the head of the case opinions, mark the first appearance of the key number system in the case reporter.

It is no coincidence, of course, that Criminal Law ☞ 662.1—the same topic and key number that proved useful in directing us to *Maryland v. Craig* when we

Exhibit 3–132

---

840   OCTOBER TERM, 1989

                     Opinion of the Court   497 U. S.

JUSTICE O'CONNOR delivered the opinion of the Court.

This case requires us to decide whether the Confrontation Clause of the Sixth Amendment categorically prohibits a child witness in a child abuse case from testifying against a defendant at trial, outside the defendant's physical presence, by one-way closed circuit television.

I

In October 1986, a Howard County grand jury charged respondent, Sandra Ann Craig, with child abuse, first and second degree sexual offenses, perverted sexual practice, assault, and battery. The named victim in each count was a 6-year-old girl who, from August 1984 to June 1986, had attended a kindergarten and prekindergarten center owned and operated by Craig.

Source: Reprinted from U.S. "Maryland v. Craig" opinion.

---

used the case digests—resurfaces in the headnotes in *Craig.* There is an exact correspondence between the case headnotes and the squibs that are reported in the digests. The case headnotes are created by West's editors; neither the judges who write the court opinions nor court staff have anything to do with their preparation. The headnotes do not come out of thin air, however. The content of each headnote is a summary of a statement of law made in the case's lead (majority or plurality) opinion. As West's editors read a case, they extract these summaries, assign them to the appropriate topic and key number, and then create the corresponding headnotes. As you can see, it is not unusual for a case to have several headnotes, depending on how long the case is and how many different issues it addresses.

You can locate the part of a court's opinion that corresponds to any particular headnote by using another feature of West's case reporters. To zero in on the part of the opinion in *Maryland v. Craig* that relates to headnote 6, we simply flip through the pages of the opinion until we come to a bold-faced bracketed "6." (See Exhibit 3–134.) The West editors slipped this bracketed number into Justice O'Connor's opinion, just as they did the other bracketed numbers associated with the other headnotes. Shortly after the inserted **[6]**, we come across statements in the opinion that generally correspond to the contents of headnote 6. We should reemphasize that West's editors in no way alter the wording or content of case opinions. The only changes they make in the body of an opinion are the nonsubstantive ones of adding the pagination that corresponds to the official case reporter and inserting the bracketed numbers that relate back to the headnotes.

West's key number system is remarkably versatile. Once you find one case that squarely addresses your research issue, you have the potential to find all related

Exhibit 3–133

3157

497 U.S. 836, 111 L.Ed.2d 666

836 MARYLAND, Petitioner

v.

Sandra Ann CRAIG.

No. 89–478.

Argued April 18, 1990.

Decided June 27, 1990.

Defendant was convicted in the Maryland Circuit Court, Howard County, Raymond J. Kane, Jr., J., of sexual offenses and assault and battery arising from her operation of preschool and abuse of preschool students, and defendant appealed. The Court of Special Appeals, affirmed, 76 Md.App. 250, 544 A.2d 784,. Defendant petitioned for writ of certiorari. The Court of Appeals, 316 Md. 551, 560 A.2d 1120, reversed and remanded. Certiorari was granted. The Supreme Court, Justice O'Connor, held that: (1) confrontation clause did not categorically prohibit child witness in child abuse case from testifying against defendant at trial, outside defendant's physical presence, by one-way closed circuit television; (2) finding of necessity for the use of one-way closed circuit television procedure had to be made on case specific basis; but (3) observation of child's behavior in defendant's presence and exploration of less restrictive alternatives to use of one-way closed circuit television procedure were not categorical prerequisites to use of one-way television procedure as a matter of federal constitutional law.

Vacated and remanded.

Justice Scalia filed a dissenting opinion, in which Justices Brennan, Marshall and Stevens joined.

Opinion on remand, 322 Md. 418, 588 A.2d 328.

1. Criminal Law ⌾⟶ **662.1**

The central concern of the confrontation clause is to ensure the reliability of the evidence against a criminal defendant by subjecting it to rigorous testing in the context of an adversary proceeding before the trier of fact. U.S.C.A. Const.Amend. 6.

2. Criminal Law ⌾⟶ **662.1**

A face-to-face confrontation enhances the accuracy of fact-finding by reducing the risk that a witness will wrongfully implicate an innocent person. U.S.C.A. Const.Amend. 6.

3. Criminal Law ⌾⟶ **662.8**

In narrow circumstances, the confrontation clause permits the admission of hearsay statements against a defendant despite the defendant's inability to confront the declarant at trial. U.S.C.A. Const.Amend. 6.

4. Criminal Law ⌾⟶ **662.1**

Face-to-face confrontation with witnesses is not an indispensable element of the Sixth Amendment's guarantee of the right to confront one's accusers. U.S.C.A. Const.Amend. 6.

5. Criminal Law ⌾⟶ **662.1, 662.65**
   Witnesses ⌾⟶ **228**

Child assault victim's testimony at trial of child abuse defendant through use of one-way closed circuit television procedure authorized by Maryland child witness protection statute did not impinge upon the truth seeking nor symbolic purposes of the confrontation clause; procedure required that child witness be competent to testify and testify under oath, defendant retained full opportunity for contemporaneous cross-examination, and judge, jury and defendant were able to view witness' demeanor and body by video monitor. Md.Code, Courts and Judicial Proceedings, § 9–102, U.S.C.A. Const.Amend. 6.

6. Criminal Law ⌾⟶ **662.1, 662.65**
   Witnesses ⌾⟶ **228**

If the State makes an adequate showing of necessity, the State's interest in protecting child witnesses from the trauma of testifying in a child abuse case is sufficiently important to justify the use of a special procedure permitting a child witness in abuse case to testify at trial in the absence of face-to-face confrontation with the defendant. U.S.C.A. Const.Amends. 6, 14.

Source: S.CT., "Maryland" headnotes, © West Publishing, used with permission.

Exhibit 3–134

Id., at 109, 110 S.Ct. at 1696 (quoting *Ferber, supra,* 458 U.S., at 756–757, 102 S.Ct., at 3354–55).

[6] We likewise conclude today that a State's interest in the physical and psychological well-being of child abuse victims may be sufficiently important to outweigh, at least in some cases, a defendant's right to face his or her accusers in court. That a significant majority of States have enacted statutes to protect child witnesses from the trauma of giving testimony in child abuse cases attests to the widespread belief in the importance of such a public policy. See *Coy,* 487 U.S., at 1022–1023, 108 S.Ct., at 2803–2804 (O'Connor, J., concurring) ("Many States have determined that a child victim may suffer trauma from exposure to the harsh atmosphere of the typical courtroom and have undertaken to shield the child through a vari-

ety of ameliorative measures"). Thirty-seven States, for example, permit the use of videotaped testimony of sexually abused children;[2] 24 States have authorized the use of . . .

---

victims from the emotional trauma of testifying. Accordingly, we hold that, if the State makes an adequate showing of necessity, the state interest in protecting child witnesses from the trauma of testifying in a child abuse case is sufficiently important to justify the use of a special procedure that permits a child witness in such cases to testify at trial against a defendant in the absence of face-to-face confrontation with the defendant.

[7] The requisite finding of necessity must of course be a case-specific one: The trial court must hear evidence and determine . . .

Source: S.CT., "Maryland" headnote in text, © West Publishing, used with permission.

cases. You have only to review the headnotes of the opinion to identify the topic and key number associated with the principle of law that is of interest to you. You can take that information to West's case digests and work backward and forward in time and in whatever jurisdictions you please to locate other relevant cases. This technique works whenever you have identified a good "starter" case, which you can do by using any of the references we have discussed up to this point.

All case reporters published by West conform to this same format. Recall how in Case 2 we used the Descriptive Word Index and the Tenth Decennial Digest, Part 1 to find a topic and key number—Homicide ☞ 256—that helped direct us to case law concerning whether assaults committed by HIV-infected offenders could be defined as attempted murder. We were referred to *State v. Haines,* 545 N.E.2d 834 (Ind. App. 1989), among other cases. When we look up this case in the North Eastern Reporter 2d, we are not surprised to find that one of the headnotes is classified under Homicide ☞ 256. (See Exhibit 3–135.) We locate the statement of law associated with headnote 5 in the *Haines* opinion by finding the [5] inserted by West's editors. (See Exhibit 3–136.)

A few wrinkles have been added to the headnotes at the beginning of the cases on WESTLAW that take advantage of the hypertext capabilities provided by WESTLAW and facilitate the search process. If we look at the WESTLAW version for the same headnote from *State v. Haines,* 545 N.E.2d 834, we recognize the hypertext features by the arrows appearing on the lefthand side of the screen. (See Exhibit 3–137.) The arrow at the upper left points to the bracketed [5], indi-

Exhibit 3–135

834 Ind.   545 NORTH EASTERN REPORTER, 2d SERIES

STATE of Indiana, Appellant
(Plaintiff),

v.

Donald J. HAINES, Appellee
(Defendant).

No. 79A02–8807–CR–283.

Court of Appeals of Indiana,

Second District

Oct. 31, 1989.

Defendant was convicted by jury of three counts of attempted murder. The Tippecanoe Superior Court, Vincent F. Grogg, Special Judge, granted defendant's motion for judgment on the evidence vacating attempted murder conviction and entered judgment of conviction on three counts of battery as Class "D" felony. State appealed. The Court of Appeals, Buchanan, J., held that: (1) trial judge was required to consider evidence presented by defendant as well as that presented by State in deciding whether to grant motion for judgment on evidence, and (2) attempted murder convictions were supported by evidence that defendant carried AIDS virus, was aware of infection, believed it to be fatal, and intended to inflict others with disease by spitting, biting, scratching, and throwing blood.

Reversed.

Sullivan, J., concurred in result.

1. Criminal Law ☞ 189

Double jeopardy did not bar reinstatement of jury's guilty verdicts on attempted murder counts after trial judge vacated those verdicts and substituted judgment of conviction on three counts of battery. U.S.C.A. Const.Amend. 5; IC 35-41-5-1, 35-42-2-1(2)(A) (1988 Ed.).

2. Criminal Law ☞ 977(1)

When trial court considers motion for judgment on the evidence subsequent to jury verdict, it must view all evidence in light most favorable to nonmoving party and may grant motion only if there is no substantial evidence or reasonable inference to be adduced therefrom to support essential element of claim; evidence must point unerringly to conclusion not reached by jury inasmuch as evidence is only susceptible of favoring judgment for moving party. Trial Procedure Rule 50(A).

3. Criminal Law ☞ 977(1)

Trial judge is prohibited from weighing evidence when considering whether to enter judgment contrary to verdict, and it is only when verdict against defendant is based on surmise, conjecture, or speculation as to one or more of necessary elements of claim that judgment on evidence for defendant should be upheld. Trial Procedure Rule 50(A).

4. Criminal Law ☞ 977(1)

Trial judge was required to consider all evidence presented at trial in deciding whether to grant defendant's motion for judgment on the evidence, regardless of whether that evidence was presented by State or defendant, and his failure to consider all evidence and weighing of evidence in deciding whether to grant judgment on evidence, as reflected by comment at sentencing hearing, was error. Trial Procedure Rule 50(A).

5. Homicide ☞ 256

Attempted murder convictions were supported by evidence that defendant carried AIDS (Acquired Immune Deficiency Syndrome) virus, was aware of infection, believed it to be fatal, and intended to inflict others with the disease by spitting, biting, scratching, and throwing blood, regardless of whether it was actually possible for him to spread virus that way. IC 35-41-5-1, 35-41-5-1(b) (1988 Ed.).

Source: NE2d, "Haines" headnote, © West Publishing, used with permission.

Exhibit 3–136

[5] Contrary to Haines' contention that the evidence did not support a reasonable inference that his conduct amounted to a substantial step toward murder, the record reflects otherwise. At trial, it was definitely established that Haines carried the AIDS virus, was aware of the infection, believed it to be fatal, and intended to inflict others with the disease by spitting, biting, scratching, and throwing blood. *Record* at 255, 266, 268–70, 304, 319, 331–37, 347–48, 355, 371, 383, 400, 441, 474, 478, 485, 494. His biological warfare with those attempting to help him is akin to a sinking ship firing on its rescuers.

Haines misconstrues the logic and effect of our attempt statute codified as Ind.Code 35-41-5-1. While he maintains that *the State* failed to meet its burden insofar as it did not present sufficient evidence regarding Haines' conduct which constituted a substantial step toward murder, *see Appellee's Brief* at 15–16, subsection (b) of IC 35-41-5-1 provides:

> "It is no defense that, because of a misapprehension of the circumstances, it would have been impossible for the accused person to commit the crime attempt."

In *Zickefoose v. State* (1979), 270 Ind. 618, 388 N.E.2d 507, our supreme court observed:

> "It is clear that section (b) of our statute rejects the defense of impossibility. *It is not necessary that there be a present ability to complete the crime, nor is it necessary that the crime be factually possible. When the defendant has done all that he believes necessary to cause the particular result, regardless of what is actually possible under existing circumstances, he has committed an attempt.* The liability of the defendant turns on his purpose as manifested

through his conduct. If the defendant's conduct in light of all relevant facts involved, constitutes a substantial step toward the commission of the crime and is done with the necessary specific intent, then the defendant has committed an attempt.

Previous Indiana cases have sometimes narrowly interpreted an attempt as conduct "'which will apparently result in the crime, unless interrupted by circumstances independent of the doer's will.'" *Jarman v. State* (1977), [267] Ind. [202], 368 N.E.2d 1348; *Williams v. State* (1973), 261 Ind. 385, 304 N.E.2d 311; *Herriman v. State* (1963), 243 Ind. 528, 188 N.E.2d 272. However, the new statute shows that this interpretation focusing on the result of the conduct is no longer applicable and that the law now focuses on the substantial step that the defendant has completed, not on what was left undone."

Id. at 623, 383 N.E.2d at 510 (emphasis supplied); *see also Kiper v. State* (1983), Ind., 445 N.E.2d 1353; *State v. Lewis* (1981), Ind., 429 N.E.2d 1110; *King v. State* (1984), Ind.App., 469 N.E.2d 1201, *trans. denied.*

In accordance with IC 35-41-5-1, the State was not required to prove that Haines' conduct could actually have killed. It was only necessary for the State to show that Haines did all that he believed necessary to bring about an intended result, *regardless* of what was *actually possible. See Zickefoose, supra.* Haines repeatedly announced that he had AIDS and desired to infect and kill others. At the hospital, Haines was expressly told by doctors that biting, spitting, and throwing blood was endangering others.

Source: NE2d, "Haines" headnote, © West Publishing, used with permission.

cating that we are looking at the fifth headnote assigned to the case corresponding to headnote 5 in the printed version. We proceed directly to the place in the decision where the text in headnote 5 appears by placing the cursor on the arrow in front of the number and pressing **enter**. (See Exhibit 3–138.) To return to the headnote screen **enter gb** (for "go back").

Exhibit 3–137

AUTHORIZED FOR EDUCATIONAL USE ONLY
545 N.E.2d 834   FOUND DOCUMENT   P 7 OF 28   NE   Page
(Cite as: 545 N.E.2d 834)
State v. Haines
▶ [5]
▶ 203   HOMICIDE
▶ 203VII   Evidence
▶ 203VII(E)   Weight and Sufficiency

▶ 203k256 k. Attempt, threats, or solicitation to kill.
Ind.App. 2 Dist., 1989.
Attempted murder convictions were supported by evidence that defendant carried AIDS (Acquired Immune Deficiency Syndrome) virus, was aware of infection, believed it to be fatal, and intended to inflict others with disease by spitting, biting, scratching, and throwing blood, regardless of whether it was actually possible for him to spread virus that way. IC 35-41-5-1, 35-41-5-1(b) (1988 Ed.).

Source: NE2d, "Haines" headnote, © West Publishing, used with permission.

Exhibit 3–138

AUTHORIZED FOR EDUCATIONAL USE ONLY
545 N.E.2d 834   FOUND DOCUMENT   P 19 OF 28   NE   Page
(Cite as: 545 N.E.2d 834, *838)
weighed the evidence in deciding whether to grant judgment on the evidence constituted error. See ▶ Huff, supra; ▶ Tancos, supra; T.R. 50(A); T.R. 50(A)(6).
▶ [5]Contrary to Haines' contention that the evidence did not support a reasonable inference that his conduct amounted to a substantial step toward murder, the record reflects otherwise. At trial, it was definitely established that Haines carried the AIDS virus, was aware of the infection, believed it to be fatal, and intended to inflict others with the disease by spitting, biting, scratching, and throwing blood. Record at 255, 266, 268–70, 304, 319, 331–37, 347–48, 355, 371, 383, 400, 441, 474, 478, 485, 494. His biological warfare with those attempting to help him is akin to a sinking ship firing on its rescuers.
Haines misconstrues the logic and effect of our attempt statute codified as Ind.Code 35-41-5-1. While he maintains that the State failed to meet its burden insofar as it did not present sufficient evidence regarding Haines' conduct which constituted a substantial step toward murder, see Appellee's Brief at 15-16, subsection (b) of IC 35-41-5-1 provides:
"It is no defense that, because of a misapprehension of the circumstances, it would have been impossible for the accused person to commit the crime attempt."
In ▶ Zickefoose v. State (1979), 270 Ind. 618, 388 N.E.2d 507, our supreme

Source: NE2d, "Haines" headnote in text, © West Publishing, used with permission.

The other arrows allow us to hypertext to places in the table of contents or outline for the digest topic associated with the headnote—in this case, homicide. Recall that on WESTLAW each digest topic is assigned a number. The number assigned to homicide is **203**; so **203k256** is the same as **Homicide** ⊙⟶ **256**. If we place the cursor on the arrow in front of **203k256** and press **enter**, we go to the portion of the topic outline for homicide that contains this particular key number. (See Exhibit 3–139.) The researcher may be able to identify other appropriate key numbers to add to her search strategy by using this process.

As you know from our consideration of the U.S. Reports and official state case reporters, you will not find the key number system outside of West's publications. Lawyers Cooperative Publishing Co. uses its own system for digesting and indexing Supreme Court decisions. The U.S. Supreme Court Digest, Lawyers' Edition helps direct researchers to U.S. Supreme Court decisions as well as to A.L.R. annotations in the Lawyers' Edition case reporters. We could use this digest to find many leads to Supreme Court decisions that may be of interest to the confron-

---

Exhibit 3–139

Key Number Service

► Topic List   Mode: All
► 203 HOMICIDE
►   VII. EVIDENCE, k143–K257
►   (E) WEIGHT AND SUFFICIENCY, k228–k257

          k256 Attempt, threats, or solicitation to kill
          k257 Assault with intent to kill
             (1) In general
             (2) Ability to execute intent
             (3) Showing as to grade of offense had death ensued
             (4) Sufficiency of circumstantial evidence in general
             (5) Conclusiveness of circumstances in general
             (6) False and improbable statements accompanied by other evidence
             (7) Threats accompanied by other evidence
             (8) Deadly character of assault

If you wish to:
   View an item in more detail, select its jump marker
   View the next or previous page, type P or P- and press ENTER
   Run a search using a topic and key number, type DB followed by a database identifier and the number and press ENTER (e.g., DB ALLFEDS 410k196.1)

Source: KEY, "Haines" key number to digest outline, © West Publishing, used with permission.

Exhibit 3–140

continues

Exhibit 3–140   continued

National Labor Relations Board order against successor employer, procedural safeguards as to, 38 L Ed 2d 388, 414 US 168, 94 S Ct 414

Object of right to guarantee opportunity to assess credibility of witnesses, 21 L Ed 2d 508, 393 US 314, 89 S Ct 540

One-way closed-circuit television used for receipt of testimony by child witness in child abuse action, 111 L Ed 2d 666, 110 S Ct 3157

Screen, at criminal trial, between child witnesses and defendant charged with lascivious acts with children, 101 L Ed 2d 857, 487 US 1012, 108 S Ct 2798

Source: U.S. Supreme Court Digest, Index "Child." Reprinted with the permission of LEXIS-NEXIS, a division of Reed Elsevier Inc. LEXIS and NEXIS are registered trademarks of Reed Elsevier Properties Inc. FREESTYLE, KWIC, SuperKWIC and MEGA are trademarks of Reed Elsevier Properties Inc. SHEPARD'S and SHEPARDIZE are registered trademarks of Shepard's Company, a Partnership.

tation issue presented in Case 1. (See Exhibit 3–140.) We repeatedly are referred to 111 L. Ed. 2d 666, 110 S. Ct. 3157, which of course is *Maryland v. Craig.*

When we look up *Maryland v. Craig* at 111 L. Ed. 2d 666, we find the parallel citations to the U.S. Reports and Supreme Court Reporter and the summary of the decision prepared by Lawyers Cooperative's editorial staff. (See Exhibit 3–141.) Instead of key numbers, the Lawyers' Edition uses topics and section numbers to categorize legal principles. This system ties into the U.S. Supreme Court Digest, Lawyers' Edition. Supreme Court opinions are introduced by a **Total Client-Service Library References** box, which provides handy citations to related Am. Jur. 2d articles, A.L.R. annotations, and other references published by Lawyers Cooperative. (See Exhibit 3–142.) The Lawyers Cooperative editors insert the pagination from the U.S. Reports into opinions when that information becomes available. They of course make no substantive changes in what the Justices have written. You may wish to look up Supreme Court opinions in Lawyers' Edition to take advantage of the Total Client-Service Library References feature, or to consult annotations at the rear of the volume, since neither the Supreme Court Reporter nor the U.S. Reports has these options.

Tables of cases are included at the beginning of all case reporters. Cases are arranged alphabetically by title, and they are separated by jurisdiction in West's regional reporters and in the Federal Reporter. Thus, if you are looking for a Supreme Court case that you know was decided in 1995 or an Arkansas Supreme Court case decided in 1996 and you have the name of one or both parties but you do not know the case citation, it is a simple enough task to thumb through the table of cases in the front of the appropriate case reporters until you find the case. Alternatively, you can consult the Table of Cases volume in the appropriate digest. Here you will find a cumulative listing of case names and citations for all decisions within the digest.

Although the online systems do not include tables of cases, it is easy to locate a case by using the names of the parties. On WESTLAW the names of the parties in a case are included in the **TITLE (TI)** field. (Refer back to Exhibit 3–105.) For example, if we were looking for *Maryland v. Craig*, we could use the search strategy **TI(Maryland & Craig)**. On LEXIS the same information is included in the

Exhibit 3–141

U.S. SUPREME COURT REPORTS 111 L Ed 2d

[497 US 836]
MARYLAND, Petitioner

v

SANDRA ANN CRAIG

497 US 836, 111 L Ed 2d 666, 110 S Ct 3157

[No. 89–478]

Argued April 18, 1990. Decided June 27, 1990.

Decision: Sixth Amendment's confrontation clause held not absolutely to prohibit Maryland from using one-way closed-circuit television for receipt of testimony by child witness in child abuse case.

SUMMARY

A Maryland statute permitted a trial judge to receive, by one-way closed-circuit television, the testimony of a child witness who was alleged to be a victim of child abuse, if the judge determined that testimony by the child in the courtroom would result in the child's suffering serious emotional distress such that the child could not reasonably communicate. Under the statute, (1) the witness, prosecutor, and defense counsel withdrew to a separate room, while the judge, jury, and defendant remained in the courtroom; (2) the witness was then examined and cross-examined in the separate room while a video monitor recorded and displayed the witness' testimony to those in the courtroom; and (3) during this time, the witness could not see the defendant, but the defendant remained in electronic communication with defense counsel, and objections could be made and ruled on as if the witness were testifying in the courtroom. Before the operator of a kindergarten and prekindergarten center was tried by a Maryland court on several charges related to the operator's alleged sexual abuse of a 6-year-old child who had attended the center, the state made a motion involving the statutory procedure, and the accused objected under the confrontation clause of the Federal Constitution's Sixth Amendment. The trial court, however, in denying the accused's objection, concluded that (1) under the procedure, an accused retained the essence of the right of confrontation; and (2) based upon expert testimony presented by the state, the testimony in a courtroom of the alleged victim and other children who were alleged to have been sexually abused by the accused would result in . . .

Source: LE2d, "Maryland v. Craig" summary. Reprinted with the permission of LEXIS-NEXIS, a division of Reed Elsevier Inc. LEXIS and NEXIS are registered trademarks of Reed Elsevier Properties Inc. FREESTYLE, KWIC, SuperKWIC and MEGA are trademarks of Reed Elsevier Properties Inc. SHEPARD'S and SHEPARDIZE are registered trademarks of Shepard's Company, a Partnership.

Exhibit 3–142

U.S. SUPREME COURT REPORTS   111 L Ed 2d

HEADNOTES

Classified to U.S. Supreme Court Digest, Lawyers' Edition

Criminal Law § 50—confrontation of witness— testimony by one-way closed-circuit tele- vison—right to face-to-face confrontation

1a-1g. The confrontation clause of the Federal Constitution's Sixth Amendment does not abso- lutely prohibit a child witness in a child abuse case from testifying against a defendant at a state crimi- nal trial, outside the defendant's physical pres- ence, by one-way closed-circuit television, be- cause (1) face-to-face confrontation with wit- nesses appearing at trial is not an indispensable element of the Sixth Amendment's confrontation guarantee, since (a) in certain narrow circum- stances, competing interests, if closely examined,

---

TOTAL CLIENT-SERVICE LIBRARY® REFERENCES

21A Am Jur 2d, Criminal Law §§ 720–723, 729, 731, 956–958, 960

9 Federal Procedure, L Ed, Criminal Procedure §§ 22:814, 22:814.3

24 Am Jur Proof of Facts 2d 515, Defense to Charges of Sex Offense

USCS, Constitution, Amendment 6

US L Ed Digest, Criminal Law § 50

Index to Annotations, Abuse of Persons; Closed-circuit Television; Confrontation of Witnesses

Auto-Cite®: Cases and annotations referred to herein can be further researched through the Auto- Cite® computer-assisted research service. Use Auto-Cite to check citations for form, parallel references, prior and later history, and annotation references.

ANNOTATION REFERENCES

Federal constitutional right to confront witnesses—Supreme Court cases. 98 L Ed 2d 1115.

Comment note.—What provisions of the Federal Constitution's Bill of Rights are applicable to the states. 18 L Ed 2d 1388, 23 L Ed 2d 985.

Closed-circuit television witness examination. 61 ALR4th 1155.

Necessity or permissibility of mental examination to determine competency or credibility of complainant in sexual offense prosecution. 45 ALR4th 310.

Instructions to jury as to credibility of child's testimony in criminal case. 32 ALR4th 1196.

Condition interfering with accused's view of witness as violation of right of confrontation. 19 ALR4th 1286.

Source: LE2d, "Maryland v. Craig" references. Reprinted with the permission of LEXIS-NEXIS, a division of Reed Elsevier Inc. LEXIS and NEXIS are registered trademarks of Reed Elsevier Properties Inc. FREESTYLE, KWIC, SuperKWIC and MEGA are trademarks of Reed Elsevier Properties Inc. SHEPARD'S and SHEPARDIZE are regis- tered trademarks of Shepard's Company, a Partnership.

NAME segment. (See Exhibit 3–112.) A search for the same case on LEXIS would look like this: **NAME(Maryland & Craig)**. If you know the date or ap- proximate date of the decision you can also add a date restriction to your WESTLAW or LEXIS search.

Shepard's Citators

In all of legal research, there is no feeling that quite compares to uncovering a case that is "on all fours" with the fact situation you are investigating. If you are advocating a position or are interested in a particular outcome, finding a case in your favor can be exhilarating, while finding a case contrary to your position can be extremely demoralizing. Even when you are indifferent about how a court rules and you just want to know what the decision is, it is quite satisfying to find judicial authority on point to your research issue. However, lest you become elated or dejected prematurely, heed well this caution: *Your work is still not finished.* Even when you find a case that appears definitively to resolve the issue you are researching, you **must**—and we emphasize that this is **not** optional—complete one additional step to make sure that the case remains good law (that is, it has not been over-

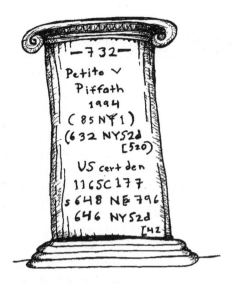

ruled, reversed, or significantly modified by subsequent case decisions), and is the most recent authority that is available. The step you must take is to Shepardize your case.

Shepard's Citators are available for cases, constitutions, statutes, agency decisions and regulations, law reviews, and a few other references. All Shepard's Citators provide similar information, which essentially consists of reporting where (and often why) the authority you are looking up has been cited in other sources. This is an invaluable service, as even a moment's reflection confirms. Shepard's allows you to go forward in time, starting with when the authority you are Shepardizing was published, to trace that authority's citation history in subsequent references.

The law is in constant evolution, and it is of utmost importance to respect this principle. You would not expect an authority from the 1800s to reflect the law accurately in the 1900s, and you should not necessarily expect last year's authority to be fully effective this year. Shepard's case citators serve a number of functions. While we focus here on Shepardizing judicial decisions, the techniques for Shepardizing other types of authority or references, and the reasons for doing so, are very similar. Below are the primary reasons for Shepardizing:

Shepard's *provides the parallel citations for judicial decisions.* Thus, if you know a citation to an official case reporter (for example, the U.S. Reports, or the North Carolina Reports) and you want to find the unofficial reporter (for example, the Supreme Court Reporter or Lawyers' Edition or the South Eastern Reporter), or if you only have the unofficial citation and you want the official cite, you can supply the missing information by Shepardizing the citation that you have.

Shepard's *gives the judicial history of a case.* This enables you to trace the path that a case has taken through the courts both before and after the decision you are Shepardizing. For example, by Shepardizing a state supreme court decision, you can produce the prior published opinions, if any, as the case worked its way up to the state high court. You also will be referred to any federal or state court opinions involving the case you are Shepardizing that were published after that decision was announced. For example, a court may have issued an opinion following an appeal, on petition for writ of certiorari, or after a remand from a higher court.

Shepard's *identifies all other court decisions in which the case you are Shepardizing has been cited.* When it is possible to make a classification, the editors of Shepard's also report why a case was cited in another decision. For example, Shepard's may indicate that the case was distinguished in another decision, criticized, called into question, cited in a dissenting opinion, affirmed, reversed, overruled, or received some other treatment. This feature also helps you determine whether the case you are Shepardizing remains valid or authoritative.

Shepard's *provides an efficient method for checking case citations.* Cases often involve multiple issues. You may be interested in the decisions that cite the case you are Shepardizing only for a limited purpose. For example, a court decision might first resolve a jurisdictional matter and then consider whether an item of evidence was properly admitted in a trial, whether the trial judge erred in instructing the jury on a point of law, or whether the statute under which the offender was sentenced is constitutional. If you are interested only in the last issue, you need not consult the cases that cite the decision for other reasons. Whenever possible, Shepard's will identify the headnote from the case you are Shepardizing that states the legal principle that was of interest to the court citing the decision, and you may elect not to look up cases when the headnote identified by the Shepard's corresponds to an issue that you do not find interesting.

Shepard's *refers you to a few nonjudicial authorities that have cited the case you are checking.* Shepard's includes A.L.R. annotations, select law reviews, and a few other sources that cite judicial decisions.

Different Shepard's Citators correspond to different case reporters. Each of the 50 states has its own Shepard's citator for state court decisions. A separate Shepard's also exists for each of the regional reporters in West's reporter system, which of course also contain state court cases. The major difference between a state citator and a regional citator is that the state Shepard's identifies cases from the particular state, as well as federal cases, that have cited the decision you are checking. The regional Shepard's captures cases from all jurisdictions that have cited the case. On the other hand, state citators provide a more comprehensive list of pertinent law review articles. In the federal court system, there are separate Shepard's for U.S. district court decisions published in the Federal Supplement, for U.S. court of appeals decisions in the Federal Reporter, and for U.S. Supreme Court decisions.

When the time comes to Shepardize a case, it is important first to find the set of Shepard's that is appropriate for the court system that decided the case and find all of the individual volumes of Shepard's within that set that include your decision. All sets of Shepard's are made up of multiple volumes, some of which are hardbound, and some paper. It is imperative that you begin with the *oldest* volume that contains the case you are Shepardizing, because successive volumes of Shepard's are not necessarily cumulative. You must check the outside front cover to find out the range of case reporters that is covered and then match the case you are Shepardizing accordingly. There may be one, more than one, or no hardbound volumes of Shepard's that apply, depending on how old the case is. After you collect all relevant hardbound volumes, begin gathering the more recently published paper volumes. You usually will find a fairly thick gold-colored paperback Shepard's, and often slimmer red or blue volumes, and a still-slimmer white volume. The progression of color-coded paper Shepard's represents the oldest to most recent supplements available. For newer decisions, you may be working exclusively with paperback Shepard's.

After you line up all of the necessary volumes of Shepard's, you can start the actual process of Shepardizing. This process is more easily illustrated than described. We begin with the Supreme Court decision that has surfaced so regularly during our research into the confrontation issue represented in Case 1, *Maryland v. Craig*, 497 U.S. 836, 110 S. Ct. 3157, 111 L. Ed. 2d 666 (1990).

Our first step is to identify the appropriate set of Shepard's for U.S. Supreme Court decisions, Shepard's United States Citations, Case Edition. We will use the Shepard's with Supreme Court Reporter citations, although equally appropriate would have been a Shepard's for the U.S. Reports or for Lawyers' Edition 2d.

Next, we assemble all of the individual Shepard's volumes that cover the case reporters in which *Maryland v. Craig* is found, starting with the oldest and working forward. At the time of this writing, we find two hardbound volumes (vol. 3.7 Shepard's United States Citations, Case Edition, 1994: Cases 99-114 S. Ct. and vol. 3.9 of that same series, 1996 Supplement, 99-116 S. Ct.), and two paper volumes (vol. 96 Shepard's United States Citations: Supreme Court Reporter, No. 15, August 1, 1997 (red), and vol. 96 Shepard's United States Citations: Supreme Court Reporter, Express Supplement No. 16, August 15, 1997 (blue)). The covers of the paper Shepard's list the Shepard's volumes that should be included in the library's collection. You can help ensure that you have covered the necessary territory by double-checking the books you have consulted against the lists on the front of the paper editions of Shepard's.

We finally are ready to Shepardize *Maryland v. Craig*. We start with the oldest applicable Shepard's, which is a hardbound volume. Using the Supreme Court Reporter cite to *Craig* (110 S. Ct. 3157), we eye the information provided at the top of each page as we flip through the Shepard's until we come to "Vol. 110," which is our cue that information will be provided for cases reported in volume 110 of the Supreme Court Reporter. Next, we look through the columns for the bold notation signifying the page number where the *Craig* opinion begins:—**3157**—. Once we have linked volume and page number, we can see the information the Shepard's provides about this case. (See Exhibit 3–143.)

The first two citations below the case name, which appear in parentheses—(497US836) and (111L£666)—are the parallel cites to the *Craig* decision. From our prior research, we already know that *Craig* is also reported at 497 U.S. 836 and at 111 L.Ed.2d 666. Nevertheless, one handy feature of Shepard's volumes is that you can always find the complete citation to a decision by looking up the case under either its official or unofficial cite.

We next learn about the judicial history of *Craig*. The seven citations following the parallel cites are all preceded by a lower case "s." The "s" means that the decision involves the "same case," but at a different stage of the proceedings: in this example, how the case worked its way up to the Supreme Court, and then case rulings following the Supreme Court's decision. If we were to track down the

Exhibit 3–143

Vol. 110		SUPREME COURT REPORTER			
136PaC464	423SE226	f 934F2d^{5}23	d 767FS1548	14CA4th1044	464NW248
580A2d787	425SE625	f 934F2d^{5}25	Cir. 10	15CA4th1264	489NW423
583A2d522	Wis	946F2d994	h 917F2d^{4}1279	2CaR2d168	488NW710
594A2d285	161Wis2d145	Cir. 3	h 917F2d^{5}1279	12CaR2d13	Kan
613A2d15	173Wis2d218	d 750FS^{8}728	e 937F2d^{7}522	12CaR2d289	250Kan154
613A2d1233	173Wis2d404	754FS^{1}65	h 962F2d1481	17CaR2d648	14KA2d670
615A2d434	467NW213	754FS^{3}65	962F2d^{6}1481	19CaR2d407	798P2d510
615A2d438	496NW176	754FS^{4}65	962F2d^{7}1481	820P2d270	825P2d149
S D	496NW633	Cir. 5	f 992F2d^{1}1124	Colo	Md
481NW251	Wyo	985 F2d^{4}781	f 992F2d^{7}1124	803P2d148	84MdA172
484NW886	847P2d1020	988F2d^{5}1368	9F3d868	803P2d160	85MdA447
492NW105	1992IlLR10	993F2d^{4}464	f 9F3d^{7}869	D C	578A2d307
501NW360	76MnL399	993F2d^{5}464	f 754FS^{6}834	581A2d776	584A2d128
540NW851	76MnL504	Cir. 6	754FS^{7}835	598A2d383	Mass
Tex	76MnL522	d 922F2d298	Cir. 11	Fla	409Mas446
802SW700	76MnL557	f 922F2d^{5}299	983F2d^{6}1548	564So2d568	29MaA939
815SW748	76MnL624	f 922F2d^{7}299	991F2d719	565So2d317	558NE20
828SW436	76MnL725	997F2d^{6}212	31MJ170	566So2d258	567NE893
831SW393		Cir. 7	31MJ173	568So2d123	Mich
842SW305	—3157—	922F2d^{3}393	31MJ174	581So2d923	438Mch691
Utah		3F3d^{1}1138	31MJ202	588So2d338	439Mch939
808P2d1053	Maryland	13F3d1004	31MJ210	592So2d275	442Mch110
848P2d683	v Craig	13F3d^{5}1005	31MJ880	597So2d409	442Mch124
853P2d870	1990	759FS^{3}435	32MJ181	602So2d1280	186McA209
Wash		824FS^{1}841	32MJ1050	608So2d36	187McA663
115Wsh2d725	(497US836)	Cir. 8	33MJ581	608So2d73	191McA356
119Wsh2d416	(111LE666)	923F2d1307	33MJ755	614So2d455	193McA659
59WAp681	s 110SC834	998F2d615	36MJ159	617So2d1094	198McA410
61WAp872	s 316Md551	f 998F2d^{7}616	36MJ879	617So2d1096	463NW227
65WAp437	s 322Md418	Cir. 9	37MJ290	617So2d1098	468NW239
800P2d841	s 76MdA250	915F2d^{8}1280	Ala	619So2d335	475NW758
800P2d842	s 544A2d784	d 921F2d^{6}931	582So2d584	Ga	478NW905
801P2d957	s 560A2d1120	d 921F2d^{9}932	Alk	199GaA434	480NW103
812P2d540	s 588A2d328	933F2d^{1}709	797P2d672	405SE285	485NW113
828P2d1126	110SC^{3}3146	j 962F2d1389	797P2d673	Ill	499NW757
832P2d85	Cir. 1	981F2d1061	807P2d1094	204IlA826	499NW763
844P2d427	982F2d20	982F2d^{1}385	839P2d409	562NE382	499NW789
W Va	805FS^{2}1060	j 982F2d1373	Calif	611NE82	Minn
183WV670	805FS^{6}1060	j 5F3d1237	54C3d1077	615NE14	459NW660
184WV414	Cir. 2	7F3d^{6}887	232CA3d866	Ind	Miss
398SE152	916F2d^{1}56	7F3d^{7}887	9CA4th641	575NE985	574So2d1374
400SE849	f 934F2d^{4}23	e 767FS^{5}1539	9CA4th1727	Iowa	574So2d1376

continues

Exhibit 3–143   continued

584So2d387	248NJS58	N D	412SE392	417SE893	66NYL1404
614So2d383	252NJS42	459NW820	414SE583	Wash	78VaL1535
Mo	258NJS54	Ohio	Tenn	59WAp681	1993WLR1547
806SW662	261NJS20	560S75	839SW392	61WAp334	61AL1155s ◀
806SW669	264NJS273	730A332	Tex	800P2d841	
829SW448	264NJS613	740A170	802SW700	810P2d72	—3177—
829SW463	577A2d485	564NE447	815SW748	W Va	
829SW466	589A2d1384	597NE173	818SW761	183WV675	**Lujan v**
Mont	599A2d189	598NE731	818SW766	398SE157	**National**
253Mt54	604A2d93	Okla	818SW770	432SE799	**Wildlife**
253Mt61	609A2d72	816P2d573	818SW777	Wis	**Federation**
253Mt62	617A2d674	816P2d575	822SW51	159Wis2d629	**1990**
257Mt461	617A2d1200	820P2d1348	822SW191	173Wis2d214	
830P2d1317	624A2d612	820P2d1350	828SW436	465NW211	(497US871)
850P2d290	625A2d506	Ore	831SW346	496NW174	(111LE695)
Nev	626A2d67	854P2d933	831SW350	500NW646	s 110SC834)
847P2d1367	N M	Pa	831SW393	Wyo	s 110SC3265
849P2d225	114NM529	527Pa474	837SW110	821P2d1290	s 835F2d305
N H	842P2d746	527Pa482	853SW90	837P2d1057	s 844F2d889
133NH793	N Y	527Pa486	856SW247	59ChL355	s 878F2d422
587A2d588	163NYAD552	527Pa497	856SW248	78Cor194	s 699FS327
N J	164NYAD947	594A2d281	860SW887	1992IlLR10	s 266ADC241
120NJ654	148NYM174	594A2d285	Vt	91McL603	s 268ADC15
127NJ256	558NYS2d614	594A2d288	156Vt637	76MnL399	s 269ADC271
130NJ562	560NYS2d55	594A2d293	588A2d621	76MnL588	s 278ADC320
132NJ495	560NYS2d184	So C	Va	76MnL630	

Source: Shepard's, Supreme Court Reporter, 110 sct 3157. Reproduced by permission of Shepard's. Further reproduction of any kind is strictly prohibited.

"same case" citations, we would find that the Supreme Court granted certiorari in *Craig* at 110 S. Ct. 834 (1990); the Maryland Court of Special Appeals decided the appeal of Craig's conviction at 76 Md. App. 250, 544 A.2d 784 (l988); the Maryland Court of Appeals then granted certiorari and issued the decision (3l6 Md. 551, 560 A.2d ll20 (1989)) that was reviewed by the U.S. Supreme Court; and finally, the Maryland Court of Appeals wrote an opinion on *Craig* at 322 Md. 418, 588 A.2d 328 (1991), after the case was remanded by the U.S. Supreme Court.

All of the other citations listed in the Shepard's are to other cases or secondary authorities that have cited *Craig*. These citations identify the specific *page* within an opinion where *Craig* is cited; they do not refer to the page where that decision begins. If you take a moment to study Exhibit 3–143, several of Shepard's organizational features will become apparent. Note that cases are listed by jurisdiction. The one Supreme Court case citing *Craig* is identified first: 110 S. Ct. 3l46. Next come federal cases in which *Craig* has been cited, which are arranged by federal circuit. U.S. court of appeals decisions ("F2d" and "F3d") precede cases decided by U.S. district courts sitting in the same circuit, which are reported in the Federal Supplement ("FS").

After all the federal cases and military cases ("MJ," or Military Justice Reporter) are reported, state court cases citing *Craig* are listed. At the very end, following the citations found in Wyoming cases in the Pacific 2d Reporter, Shepard's lists a few law review articles citing *Craig* as well as an annotation, "61 A.L.R.4th 1155s" (the "s" in this instance means that the case is cited in the annotation's paper supplement). If the abbreviations used in a Shepard's are unclear, simply consult the legend found near the beginning of the volume, which gives a complete explanation.

As you examine the Shepard's, you doubtlessly spot several mysterious notations. Some of the cites have a small raised numeral in between the abbreviation for the case reporter and the page number—for example, "110SC^{3}3146" and "805FS^{2}1060." Others are preceded by a lowercase letter other than "s" (same case), such as "f," "d," "j," "e," or "h." These are additional helpful features of a Shepard's — and they provide one way of responding to the most common lament of fledgling Shepardizers: "Do I have to read *all* of these cases?"

The raised numeral appearing immediately after the abbreviation for the case reporter corresponds to a headnote number from the cited case. As we discussed previously, it means that the case you are Shepardizing was cited in connection with the principle of law discussed in the identified headnote. This information is especially useful when you are Shepardizing a case involving multiple issues and you are interested in only one of them, or when the case has been cited in so many other decisions that you need a logical starting point for looking them up. For example, refer once again to Exhibit 3–133, which shows the headnotes for *Craig v. Maryland* as reported at 110 S.Ct. 3157. If we are interested in the legal principle discussed in headnote 6 ("[i]f the State makes an adequate showing of necessity . . . "), and how that principle has figured in other decisions, we can begin our search in Shepard's for cases with the superscript 6. As displayed in Exhibit 3–143, we should make note of "805FS^{6}1060" "997F2d^{6}212," "921F2d^{6}931," and the other decisions marked by the small, raised 6. When a citation in a Shepard's does not include a reference to a headnote, it means that the editors have not been able to match the principle for which the decision was cited with any of its headnotes.

The lowercase letters that appear before citations in a Shepard's signify something about the cited case's history—for example, it was affirmed ("a") or reversed ("r") or is the same case ("s"). The letters may also indicate the treatment it received in another decision—for example, it was criticized ("c"), distinguished ("d"), followed ("f"), or overruled ("o"). You do not have to guess about the meaning of these letters, because Shepard's volumes all have a table in the front explaining the abbreviation. (See Exhibit 3–144.) Of course, it is absolutely vital that you confirm that a case on which you hope to rely has not been reversed or overruled, but a Shepard's provides you with a wealth of additional information as well. If you refer again to Exhibit 3–143, where *Maryland v. Craig* is Shepardized, you see how quickly you are able to spot federal court of appeals decisions that have followed *Craig*, the decisions in which *Craig* has been explained or distinguished, and so forth.

Exhibit 3–144

HISTORY AND TREATMENT ABBREVIATIONS

Abbreviations have been assigned, where applicable, to each citing case to indicate the effect the citing case had on the case you are Shepardizing. The resulting "history" (affirmed, reversed, modified, etc.) or "treatment" (followed, criticized, explained, etc.) of the case you are Shepardizing is indicated by abbreviations preceding the citing case reference. For example, the reference "f434F2d872" means that there is language on page 872 of volume 434 of the *Federal Reporter*, Second Series, that indicates the court is "following" the case you are Shepardizing. Instances in which the citing reference occurs in a dissenting opinion are indicated in the same manner. The abbreviations used to reflect both history and treatment are as follows:

History of Case

a	(affirmed)	The decision in the case you are Shepardizing was affirmed or adhered to on appeal.
cc	(connected case)	Identifies a different case from the case you are Shepardizing, but one arising out of the same subject matter or in some manner intimately connected therewith.
m	(modified)	The decision in the case you are Shepardizing was changed in some way.
p	(parallel)	The citing case is substantially alike or on all fours, either in law or facts, with the case you are Shepardizing.
r	(reversed)	The decision in the case you are Shepardizing was reversed on appeal.
s	(same case)	The case you are Shepardizing involves the same litigation as the citing case, although at a different stage in the proceedings.
S	(superseded)	The citing case decision has been substituted for the decision in the case you are Shepardizing.
US reh den		Rehearing denied by the U.S. Supreme Court.
US reh dis		Rehearing dismissed by the U.S. Supreme Court.
v	(vacated)	The decision in the case you are Shepardizing has been vacated.

Treatment of Case

c	(criticized)	The citing case disagrees with the reasoning/decision of the case you are Shepardizing.
d	(distinguished)	The citing case is different either in law or fact, for reasons given, from the case you are Shepardizing.
e	(explained)	The case you are Shepardizing is interpreted in some significant way.
f	(followed)	The citing case refers to the case you are Shepardizing as controlling authority.
h	(harmonized)	An apparent inconsistency between the citing case and the case you are Shepardizing is explained and shown not to exist.
j	(dissenting opinion)	The case is cited in a dissenting opinion.
L	(limited)	The citing case refuses to extend the holding of the case you are Shepardizing beyond the precise issues involved.
o	(overruled)	The ruling in the case you are Shepardizing is expressly overruled.
q	(questioned)	The citing case questions the continuing validity or precedential value of the case you are Shepardizing.

Source: Shepard's, "History and Treatment." Reproduced by permission of Shepard's. Further reproduction of any kind is strictly prohibited.

We are not finished Shepardizing *Maryland v. Craig* until we have consulted all of the Shepard's Supreme Court volumes covering 497 U.S., or 110 S.Ct., or 111 L.Ed.2d, including the most recent paper issues. The same procedures are used to Shepardize a state case or a lower federal court decision. It is only necessary to consult the appropriate sets of Shepard's.

Before we leave this subject, we again point out that not only can judicial decisions be Shepardized, but so can statutes, constitutional provisions, administrative regulations, and law review articles, among other references. While Shepardizing the Eighth Amendment is not apt to be very helpful to find cases bearing on the "smoking cellmate" issue presented in Case 3 because of the sheer volume of citations that will be produced, you may have better luck locating useful decisions by Shepardizing specific state constitutional and statutory provisions. Shepard's also reports when statutes and constitutional provisions have been amended, repealed, or invalidated by court action.

Recall our earlier discussion of the New York statute, section 65.20 of the state Criminal Procedure Law, which set forth procedures for presenting children's testimony over closed-circuit television in criminal trials involving allegations of sexual abuse. We can Shepardize this statute, or subdivisions of it, to learn whether it has been altered by later legislative action, invalidated by judicial decision, or otherwise cited in court decisions. As with Shepardizing case law, we start with the oldest appropriate Shepard's, and work forward.

We use Shepard's New York Statute Citations, Consolidated Laws (Criminal Procedure Law–Labor Law), Part 2 (1988), and subsequent volumes. We immediately are provided with a number of court decisions that have cited the statute. There is no indication in any of the Shepard's that the statute has been changed or declared unconstitutional. (See Exhibit 3–145.) In fact, we learn from the Shepard's that the capital "C" in front of the citation means that the case citing the statute found it to be constitutional. (See Exhibit 3–146.) We repeat these steps in all applicable Shepard's volumes. Once you get the knack of Shepardizing a judicial decision or a statute, you can readily transfer those techniques to other search targets.

Exhibit 3–145

Criminal Procedure Law (Ch.11A)-(1970C996)					Ch.11A§ 65.20
474NYS2d	120NYAD	502NYS2d	80NYAD36	127NYAD	385NYS2d
[102	[949	[269	84NYAD42	[603	[763
474NYS2d	126NYAD	502NYS2d	85NYAD291	127NYAD	386NYS2d
[151	[668	[848	91NYAD	[968	[692
474NYS2d	129NYAD	511NYS2d96	[1184	128NYAD	388NYS2d
[928	[753	514NYS2d	99NYAD439	[569	[427
478NYS2d	68NYM 98	[516	101NYAD	128NYAD	390NYS2d
[1002	93NYM180	524NYS2d	[221	[716	[558
486NYS2d67	101NYM836	[987	103NYAD	128NYAD	390NYS2d
495NYS2d	104NYM11	425NE882	[941	[894	[859
[530	124NYM751	432NE791	103NYAD	77NYM108	400NYS2d
500NYS2d	326NYS2d	436NE487	[1033	86NYM348	[868
[449	[510	38BR350	104NYAD	99NYM1080	401NYS2d
502NYS2d	379NYS2d	§ 60.50	[898	97NYM994	[500
[848	[182	34NY919	104NYAD	99NYM1012	404NYS2d
503NYS2d	402NYS2d	36NY565	[951	112NYM887	[428
[580	[735	37NY629	105NYAD	114NYM539	412NYS2d
504NYS2d	407NYS2d	37NY906	[1031	120NYM885	[992
[824	[938	39NY811	105NYAD	123NYM339	413NYS2d
521NYS2d	421NYS2d	40NY327	[1112	123NYM366	[119
[937	[463	41NY50	106NYAD	125NYM395	413NYS2d
522NYS2d	422NYS2d	46NY186	[682	129NYM713	[748
[233	[298	46NY780	107NYAD	135NYM882	413NYS2d
523NYS2d36	427NYS2d	47NY917	[772	323NYS2d	[912
423NYS2d36	[568	49NY428	107NYAD	[291	417NYS2d
423NE37	430NYS2d	51NY984	[879	340NYS2d25	[672
423NE793	[319	54NY578	111NYAD	345NYS2d	419NYS2d
Cl.2	430NYS2d	57NY434	[822	[777	[486
54NY623	[606	57NY562	113NYAD	351NYS2d	422NYS2d
55NY147	433NYS2d	65NY273	[951	[828	[476
56NY66	[149	69NY186	113NYAD	359NYS2d	424NYS2d
51NYAD26	442NYS2d	37NYAD743	[971	[551	[569
64NYAD835	[494	40NYAD357	114NYAD	360NYS2d	426NYS2d
71NYAD80	447NYS2d	42NYAD144	[596	[101	[258
76NYAD600	[919	46NYAD698	120NYAD	362NYS2d	435NYS2d
77NYAD527	451NYS2d37	46NYAD984	[539	[281	[712
78NYAD831	467NYS2d	51NYAD861	121NYAD	369NYS2d	437NYS2d
97NYAD488	[840	54NYAD196	[396	[690	[988
104NYAD	478NYS2d	60NYAD134	121NYAD	376NYS2d	445NYS2d
[1002	[1000	60NYAD316	[562	[440	[661
116NYAD	480NYS2d	62NYAD	122NYAD	378NYS2d	446NYS2d
[212	[912	[1110	[178	[381	[917
117NYAD	499NYS2d	67NYAD	123NYAD	380NYS2d	447NYS2d
[969	[283	[1009	[453	[163	[505
120NYAD	500NYS2d	72NYAD205	124NYAD	380NYS2d	447NYS2d
[674	[447	73NYAD630	[861	[486	[658

continues

Exhibit 3–145   continued

452NYS2d	493NYS2d	[303	318NYS2d	§60.70	Subd. 5
[130	[848	523NYS2d	[712	Ad1973C750	C129NYM
456NYS2d	493NYS2d	[669	358NYS2d	Ad1984C954	[1018
[740	[958	524NYS2d16	[737	53NYAD787	C498NYS2d
457NYS2d	494NYS2d	316NE868	406NYS2d	384NYS2d	[979
[452	[441	330NE641	[802	[557	C503NYS2d
459NYS2d	502NYS2d45	339NE141	407NYS2d	§ 65.00	[240
[172	503NYS2d	340NE743	[744	C129NY et seq.	Subd. 6
466NYS2d	[107	351NE430	433NYS2d	[1017	C503NYS2d
[939	503NYS2d	353NE605	[752	C498NYS2d	[242
470NYS2d	[612	359NE369	444NYS2d	[978	Subd. 7
[617	504NYS2d	385NE1046	[855	C503NYS2d	C503NYS2d
473NYS2d	[716	386NE823	445NYS2d	[239	[241
[159	504NYS2d	393NE480	[1008	§§ 65.00 to	Subd. 9
473NYS2d	[790	402NE1160	452NYS2d	65.30	C503NYS2d
[920	506NYS2d	416NE1047	[344	Ad1985C505	[242
474NYS2d	[757	431NE278	454NYS2d	133NYM216	¶¶ a to I
[213	508NYS2d	442NE1251	[591	506NYS2d	C503NYS2d
475NYS2d	[299	443NE926	466NYS2d	[926	[242
[612	511NYS2d	461NE296	[750	§ 65.00	¶ a
478NYS2d	[402	480NE732	491NYS2d	Subd.1	C503NYS2d
[442	512NYS2d	505NE599	[563	C129NYM	[242
479NYS2d	[506	518NE911	267NE452	[1017	¶ b
[603	513NYS2d89	486F2d264	315NE788	C498NYS2d	C503NYS2d
479NYS2d	513NYS2d	530FS1390	413NE1167	[978	[242
[776	[209	38BR347	437NE1101	C503NYS2d	¶ c
480NYS2d	513NYS2d	3Hof567	38BR340	[241	C503NYS2d
[394	[310	23SR311	47BR1351	Subd. 4	[242
480NYS2d	513NYS2d	24SR7	23SR311	C503NYS2d	¶ d
[569	[819	34SR263	Subd. 2	[241	C503NYS2d
482NYS2d	516NYS2d	35SR353	111NYM695	Subd. 6	[242
[622	[117	45StJ390	114NYM409	C503NYS2d	¶ j
483NYS2d	516NYS2d	§ 60.55	444NYS2d	[243	C503NYS2d
[451	[972	A1980C548	[856	§ 65.10	[243
484NYS2d	517NYS2d	A1984C668	451NYS2d	et seq.	¶ l
[139	[371	27NY432	[643	C503NYS2d	C503NYS2d
484NYS2d	517NYS2d	35NY75	§60.60	[242	[242
[266	[574	51NY843	38BR340	§ 65.10	Subd. 10
484NYS2d	518NYS2d	56NY327	Subd. 1	Subd. 1	C129NYM
[608	[233	64NYAD526	35NYAD732	C503NYS2d	[1017
490NYS2d	519NYS2d65	64NYAD804	86NYM168	[241	C498NYS2d
[250	520NYS2d	96NYAD645	315NYS2d	§ 65.20	[978
491NYS2d	[302	111NYM694	[395	Subd. 1	Subd. 11
[143	521NYS2d	112NYM648	381NYS2d	C129NYM	C129NYM
493NYS2d	[288	115NYM733	[965	[1018	[1017
[642	523NYS2d	128NYM839	Subd. 2	C498NYS2d	
			23SR311	[979	

Source: Shepard's NY statutes, CPL 65.20. Reproduced by permission of Shepard's. Further reproduction of any kind is strictly prohibited.

Exhibit 3–146

ABBREVIATIONS—ANALYSIS

Form of Statute

Amend.	Amendment	No.	Number
App.	Appendix	p	Page
Art.	Article	¶	Paragraph
C or Ch.	Chapter	Proc.	Proclamation
Cl.	Clause	P.L.	Public Law
Ex. Ord.	Executive Order	§	Section
Ex	Extra Session	St.	Statutes at Large
Loc.	Local Laws	Subd.	Subdivision
(NYC)	New York City	Subsec.	Subsection

Operation of Statute

Legislative

A	(amended)	Statute amended.
Ad	(added)	New section added.
E	(extended)	Provisions of an existing statute extended in their application to a later statute, or allowance of additional time for performance of duties required by a statute within a limited time.
L	(limited)	Provisions of an existing statute declared not to be extended in their application to a later statute.
R	(repealed)	Abrogation of an existing statute.
Re-en	(re-enacted)	Statute re-enacted.
Rn	(renumbered)	Renumbering of existing sections.
Rp	(repealed in part)	Abrogation of part of an existing statute.
Rs	(repealed and superseded)	Abrogation of an existing statute and substitution of new legislation therefor.
Rv	(revised)	Statute revised.
S	(superseded)	Substitution of new legislation for an existing statute not expressly abrogated.
Sd	(suspended)	Statute suspended.
Sdp	(suspended in part)	Statute suspended in part.
Sg	(supplementing)	New matter added to an existing statute.
Sp	(superseded in part)	Substitution of new legislation for part of an existing statute not expressly abrogated.

Judicial

C	Constitutional.		V	Void or invalid.
U	Unconstitutional.		Va	Valid.
UP	Unconstitutional in part.		Vp	Void or invalid in part.

Treatment of Statute

i	Citing case interprets/construes statute
rt	Citing case discusses retroactivity of statute

Source: Shepard's statutes, "Analysis." Reproduced by permission of Shepard's. Further reproduction of any kind is strictly prohibited.

Using WESTLAW or LEXIS to Shepardize

There are several advantages to the CALR method of Shepardizing provided by WESTLAW or LEXIS: (1) the information is more current on the computer systems; (2) there is no need to check several bound volumes and supplements, because all the citations are presented in one listing; (3) it is easy to quickly retrieve cases that involve a particular treatment or that arise in a particular jurisdiction or that refer to a particular headnote in the cited case—or to find cases containing a mix of all three elements. Finally, the hypertext features allow the researcher to move quickly from the Shepard's reference to the text of the citing document. Shepardizing on LEXIS or WESTLAW can become an intricate and tricky process, and an exhaustive explanation of the process and the various capabilities of the two systems would require another book. Our intent is to familiarize you with some of the basic techniques, and alert you about a few areas where you should exercise caution.

For purposes of illustration, we will be Shepardizing the same case, *Maryland v. Craig*, on both systems. You can Shepardize a displayed document at any point on WESTLAW by **entering** the command **sh**. If you want to Shepardize a document not currently being displayed, **enter sh** followed by the citation to the document. Our case is cited as *Maryland v. Craig*, 497 U.S. 836, 110 S. Ct. 834, 111 L. Ed. 2d 666, which we can Shepardize on WESTLAW using any of the three citations. However, we must be careful if we want to make use of the headnote references provided by Shepard's. If we use the 111 L. Ed. 2d 666 cite, we get references to headnotes at the front of the case as reported in the Lawyers' Edition series. Remember that these headnotes are not the same as those used in West's Supreme Court Reporter and that they are not linked to the West key number system. To get the West headnotes we must Shepardize using the Supreme Court Reporter cite, "**sh 110 Sct 834**."

The arrangement, citations, and treatments assigned to citing documents are the same as those in printed Shepard's series, but a quick comparison of the screens depicted in Exhibits 3–147 and 3–143 shows several of the enhanced features provided by WESTLAW. Notice that treatment codes are spelled out as well as being indicated by a letter. Thus, the **J** preceding the twelfth document in the list is identified as a citation appearing in a dissenting opinion. The numbers to the left allow us to go to the text of a particular citation by **entering** the number. The headnote number being cited appears in a separate column on the right rather than being buried in the citation. The "**page 1 of 38**" message at the top righthand corner indicates that we are looking at the first of 38 screens of citations for our case. The prompts at the bottom of the screen allow us to refine our results.

We could page through all 38 screens looking for references to particular courts, headnotes, or citations with particular treatment codes. However, the **locate** command allows us to simplify this process. Assume we are looking for references to headnote number 5 in the Supreme Court Reporter version of *Maryland v. Craig*. (See Exhibit 3–148.) We further assume that we are interested in cases from the second federal circuit in which the treatment code given is **followed**. We can simply **enter** the command **locate** "**f,cir2,5**," or use the **Locate** prompt at the bottom of Exhibit 3–147 to retrieve and fill out the template displayed in Exhibit

Exhibit 3–147

FOR EDUCATIONAL USE ONLY   SHEPARD'S   Page 1 of 38
Citations to: 110 S.Ct. 3157
           Maryland v Craig 1990
Coverage: ▶ View coverage information for this result

Retrieval No.		Analysis	Citation	Headnote No.
1	▶ Shep	Same Text	(497 U.S. 836)	
2	▶ Shep	Same Text	(111 L.Ed.2d 666)	
3	SC	Same Case	110 S.Ct. 834	
4	SC	Same Case	316 Md. 551	
5	SC	Same Case	322 Md. 418	
6	SC	Same Case	76 Md.App. 250	
7	SC	Same Case	544 A.2d 784	
8	SC	Same Case	560 A.2d 1120	
9	SC	Same Case	588 A.2d 328	
10			110 S.Ct. 3139,3146	3
11			111 S.Ct. 1032, 1054	
12	J	Dissenting Opin	111 S.Ct. 1661, 1671	

▶ Display only **negative** history and treatment code references (LOC NEG)
▶ IC   ▶ SP   ▶ QC   ▶ Locate   ▶ Commands   ▶ SCOPE

Source: Shepard's, "110 S.Ct. 3157." Reproduced by permission of Shepard's. Further reproduction of any kind is strictly prohibited.

3–149. Either procedure will quickly limit our results to the case cited twice on the screen displayed in Exhibit 3–150.

Another word of caution is appropriate at this point. The original 38 screens contain references to federal and state courts as well as law review articles. In the federal series of Shepard's, the citing state court decisions do not include references to headnotes in the cited case. Therefore, when a headnote reference is included in our **locate** command, all references to state court decisions are eliminated.

Shepardizing™ on LEXIS involves a similar process. We can use any of the three parallel cites. We Shepardize® from a displayed document by entering the command **shep**. If a document is not currently displayed, simply enter **shep** followed by a citation to the document. For example, "**shep 497 U.S. 836**" retrieves citations listed under *Maryland v. Craig* appearing in the Shepard's series covering the official United States Reports. (See Exhibit 3–151.) Notice that the treatment codes are spelled out as on WESTLAW, and it is possible to hypertext to a particular citing document by **entering** its corresponding number. By using **.np**

Exhibit 3–148

Copr. © West 1997 No claim to orig. U.S. govt. works

AUTHORIZED FOR EDUCATIONAL USE ONLY

110 S.Ct. 3157   **FOUND DOCUMENT**   P 7 OF 78   SCT   **Page**

(Cite as: 497 U.S. 836, 110 S.Ct. 3157)

▶   [5]

▶   110 CRIMINAL LAW

▶   110XX Trial

▶   110XX(C) Reception of Evidence

▶   110k662 Right of Accused to Confront Witnesses

▶   110k662.1 k. In general.

U.S.Md.,1990.

Child assault victim's testimony at trial of child abuse defendant through use of one-way closed circuit television procedure authorized by Maryland child witness protection statute did not impinge upon the truth seeking nor symbolic purposes of the confrontation clause; procedure required that child witness be competent to testify and testify under oath, defendant retained full opportunity for contemporaneous cross-examination, and judge, jury and defendant were able to view witness' demeanor and body by video monitor, Md.Code, Courts and Judicial Proceedings, § 9–102, U.S.C.A. Const.Amend. 6.

Source: Shepard's, SCT, "Maryland" headnote. Reproduced by permission of Shepard's. Further reproduction of any kind is strictly prohibited.

Exhibit 3–149

Type your LOCATE request in the command line, or fill in one or more selections below and press **ENTER**. Press **TAB** to move the cursor.

You can restrict your Shepard's result using one or more of the following:

Headnote numbers:	5 _____ OR _____	OR _____	AND
Analysis codes:	f _____ OR _____	OR _____	AND
Jurisdiction abbrevs.:	cir2 _____ OR _____	OR _____	AND
Publication abbrevs.:	_____ OR _____	AND	
Added date:	AFTER _____ (date must be after 9/1/95)		

___ Auto: Save and apply this LOCATE request to other Shepard's results

Use **TAB** to select a Jump marker (▶) or type a **command** and press **ENTER**

▶ Cancel Locate (XL)   ▶ Analysis Codes   ▶ COURTS   ▶ PUBS Lists   ▶ Help

Source: Shepard's "Maryland" loc template. Reproduced by permission of Shepard's. Further reproduction of any kind is strictly prohibited.

Exhibit 3–150

FOR EDUCATIONAL USE ONLY   **SHEPARD'S**   Only Page

Citations to: **110 S.Ct. 3157**

               Maryland v Craig 1990

Coverage: ▶ View coverage information for this result

Located: F,CIR2,5

Retrieval No.	- - Analysis - -	- - - - Citation - - - -	Headnote No.
		Cir.2	
1	F	Followed	934 F.2d 19, 23	5
2	F	Followed	934 F.2d at 25	5

▶ Cancel LOCATE request (XLOC) and display complete result

▶ IC   ▶ SP   ▶ QC   ▶ Locate   ▶ Commands   ▶ SCOPE

Copyright (C) 1997 Shepard's; Copyright (C) 1997 West Publishing Co.

Source: Shepard's, "Maryland" loc result. Reproduced by permission of Shepard's. Further reproduction of any kind is strictly prohibited.

(for next page) command, we proceed to the next page of citations, where treatment codes are assigned to some of the cases. (See Exhibit 3–152.) There are no headnote references because the United States Reports volumes do not include headnotes. If we had used either of the parallel citations, we would have received references to headnotes as well as treatment codes.

Notice the message at the top of the screen: **"Document 1 (OF 3)."** This message marks an important distinction between LEXIS and WESTLAW. Three separate Shepard's series actually contain references to *Maryland v. Craig*: United States Citations, Shepard's Illinois Citations (see Exhibit 3–153), and Federal Law Citations in Selected Law Reviews (see Exhibit 3–154). On LEXIS, these three documents are treated separately, whereas on WESTLAW the citations from all the Shepard's series are merged into one list. Because the Shepard's United States Citations incorporates the references from the other two series, this is largely a moot point in our example, but you should be careful to check the coverage of the various Shepard's series on LEXIS, just as you would with the printed series.

To restrict results by treatment or headnote, **enter .se** (segments) from the Shepard's display screen and choose from the options provided. We have decided to limit the results in our example to citing cases involving the treatment code **followed** by **entering f**. (See Exhibit 3–155.) There is no option to restrict by jurisdiction so we necessarily retrieve more references than with our WESTLAW **locate** search. Note, however, that the first page of our results contains the cite to the same case we located on WESTLAW. (See Exhibit 3–156.)

Exhibit 3–151

(c) 1997 McGraw–Hill, Inc.—DOCUMENT 1 (OF 3)

CITATIONS TO: 497 U.S. 836
SERIES: Shepard's United States Citations
DIVISION: United States Reports
COVERAGE: All Shepard's Citations Through 01/97 Supplement.

NUMBER	ANALYSIS	CITING REFERENCE	SYLLABUS/HEADNOTE
		Maryland v. Craig (1990)	
1	parallel citation	(111 L.Ed.2d 666)	
2	parallel citation	(110 S.Ct. 3157)	
3	same case	316 Md. 551	
4	same case	322 Md. 418	
5	same case	76 Md.App. 250	
6	same case	544 A.2d 784	
7	same case	560 A.2d 1120	

To see the text of a citing case, press the citing reference NUMBER and then the ENTER key.
For further explanation, press the H key (for HELP) and then the ENTER key.
Press Alt-H for Help or .SO to End Session or Alt-Q to Quit Software.

Source: Shepard's, "497 U.S. 836." Reproduced by permission of Shepard's. Further reproduction of any kind is strictly prohibited.

Here, we cover in passing two special features on WESTLAW that allow the researcher to retrieve references to subsequent decisions that have had a negative impact on a cited case. The first of these is the **loc neg** option, which is positioned toward the bottom of the Shepard's display screens. (See Exhibit 3–147.) **Loc neg** will retrieve from Shepard's only those cases that have had a negative impact on the cited case. (See Exhibits 3–157 through 3–159.)

The other special feature is the **ic** or **insta-cite** command. The **insta-cite** command provides a **Direct History** of the case, with a description of the action taken at each step. It also provides a **Negative Indirect History**, which lists cases that have had a negative impact on the cited case and gives a brief description of the negative impact. The **insta-cite** command can be used at any point on WESTLAW by simply **entering ic** followed by the cite to a case. It can also be initiated from a displayed case by **entering** the command **ic**. It also is available as a prompt at the bottom of the Shepard's display screens. (See Exhibit 3–147.)

Exhibit 3–152

```
                    (c) 1997 McGraw-Hill, Inc.—DOCUMENT 1 (OF 3)
                                   497 U.S. 836

    NUMBER    ANALYSIS           CITING REFERENCE         SYLLABUS/HEADNOTE

       8      same case          588  A.2d 328
       9                         499  U.S. 39
      10                         502  U.S. 357
      11      distinguished      502  U.S. 358
      12                         502  U.S. 360
      13      dissenting opinion 123  L.Ed.2d 383
      14      dissenting opinion 113  S.Ct. 1730
      15                              Cir. 1
      16      dissenting opinion 982  F.2d 20
      17                          61  F.3d 63
      18                         805  F.Supp. 1060
      19                         857  F.Supp. 1005
      20      followed                Cir. 2
                                 916  F.2d 56
                                 934  F.2d 23

   To see the text of a citing case, press the citing reference NUMBER and then the ENTER key.
   For further explanation, press the H key (for HELP) and then the ENTER key.
   Press Alt-H for Help or .SO to End Session or Alt-Q to Quit Software.
```

Source: Shepard's, "497 U.S. 836 browse." Reproduced by permission of Shepard's. Further reproduction of any kind is strictly prohibited.

The command **ic 110 Sct 3157** will retrieve the **insta-cite** results for *Maryland v. Craig*. (See Exhibits 3–160 through 3–162.) On the first screen, notice that under **Direct History** there is a line indicating what the court did in the case cited below. For example, above the fifth cite is the notation, "**AND Judgment vacated by**." There is a similar brief description of the court's action under **Negative Indirect History**. For example, the language "**Not Followed on State Law Grounds**" precedes the references to the seventh and eighth citations. (See Exhibit 3–161.) You can go directly to the text of any of the listed cases by **entering** the corresponding number.

At the time this section was written, the capability to Shepardize statutes was not available for all jurisdictions on LEXIS. On WESTLAW, the process for Shepardizing statutes is similar to that for Shepardizing cases: You **enter sh**, followed by the cite to the statute, from any place on WESTLAW. From a screen displaying the statute, simply **enter sh**. In our example, we **entered sh ny cpl 65.20** and retrieved the same information we found in the printed sets. (Exhibit 3–163.)

Exhibit 3–153

```
                (c) 1997 McGraw-Hill, Inc.—DOCUMENT 2 (OF 3)

CITATIONS TO:  497 U.S. 836
SERIES:        Shepard's Illinois Citations
DIVISION:      United States Reports (Illinois Citing References)
COVERAGE:      All Shepard's Citations Through 12/96 Supplement.

NUMBER     ANALYSIS          CITING REFERENCE           SYLLABUS/HEADNOTE
- - - - - - - - - - - - - - - - - - - - - - - - - - - - - - - - - - - - - -
                               Maryland v.
                               Craig (1990)

  1        parallel citation  (111   L.Ed.2d 666)
  2        parallel citation  (110   S.Ct. 3157)
  3                            158    Ill.2d 365
  4                            158    Ill.2d 369
  5                            204    Ill.App.3d  826
  6                            243    Ill.App.3d  725
  7                            245    Ill.App.3d  589

To see the text of a citing case, press the citing reference NUMBER and then the ENTER key.
For further explanation, press the H key (for HELP) and then the ENTER key.
Press Alt-H for Help or .SO to End Session or Alt-Q to Quit Software.
```

Source: Shepard's, "497 U.S. 836." Reproduced by permission of Shepard's. Further reproduction of any kind is strictly prohibited.

Slip Opinions and Looseleaf Services

There is a considerable lag between the time judicial decisions are announced and when they are published in case reporters. Since decisions become effective immediately on announcement rather than after they are bound, shipped, and delivered to library shelves, you may want to read a court's opinion as soon as possible rather than having to wait weeks or months. Or you may want simply to satisfy your curiosity about a judicial decision you hear about on television or read about in the newspaper. Ideally, you may wish to develop a practice of routinely reviewing new court rulings, in order to keep abreast of emerging case law and stay on the cutting edge of legal developments. Our final business is to explain how to gain access to judicial decisions, and stay current with evolving case law without unnecessarily sacrificing time in the process.

A court decision is first made accessible to the public in printed form through a **slip opinion**. Slip opinions issue at the time decisions are announced. They report

Exhibit 3–154

(c) 1997 McGraw-Hill, Inc.—DOCUMENT 3 (OF 3)

CITATIONS TO: 497 U.S. 836
SERIES: Shepard's Federal Law Citations in Selected Law Reviews
DIVISION: United States Reports
COVERAGE: All Shepard's Citations Through 12/96 Supplement.

NUMBER	ANALYSIS	CITING REFERENCE	SYLLABUS/HEADNOTE
		Maryland v. Craig (1990)	
1		78 Cornell L.R. 194	
2		91 Mich. LR 603	
3		88 Nw.U.L.Rev. 692	
4		40 UCLA L. Rev. 99	
5		61 U.Chi.L.Rev. 1370	
6		1993 U.Ill.L.Rev. 695	
7		1993 Wis.L.Rev. 1547	

To see the text of a citing case, press the citing reference NUMBER and then the ENTER key.
For further explanation, press the H key (for HELP) and then the ENTER key.
Press Alt-H for Help or .SO to End Session or Alt-Q to Quit Software.

Source: Shepard's-law reviews, "497 U.S. 836." Reproduced by permission of Shepard's. Further reproduction of any kind is strictly prohibited.

---

single case decisions, without editorial adornments. You can acquire them directly from the court making the decision, or else wait until they are sent to your office or library. (Of course the slip opinion version will be accessible through WESTLAW and LEXIS almost immediately on its issuance.) We display the first page of Justice White's opinion for the Supreme Court in *Helling v. McKinney*, as it appeared in a slip opinion when this case was decided. (See Exhibit 3–164.) Notice that the opinion is subject to revision before it is reported in the official case reporter, the United States Reports. Slip opinions are issued by courts of all levels and jurisdictions, not just by the U.S. Supreme Court.

To keep up with changes in the law, you could make a habit of reading all slip opinions as they are published, but for most of us such constant vigilance would be impractical and not sufficiently focused on the subjects we find interesting. **Looseleaf services** perform an important function by collecting current judicial decisions, often editing them to case excerpts or summaries, and organizing them by subject matter. Opinions reported in looseleaf services are not quite as timely

Exhibit 3–155

f
You may restrict the display to those citations that Shepard's has designated (by editorial analysis) as dealing with the history or treatment of the cited case, or to those which refer to a specific paragraph of a syllabus or a headnote of the cited case.

EDITORIAL ANALYSIS ABBREVIATIONS

HISTORY OF CASE		TREATMENT OF CASE		TREATMENT OF CASE	
(	parallel citation	c	criticised	o	overruled
a	affirmed	d	distinguished	p	parallel
cc	connected case	e	explained	q	questioned
Dm	dismissed	Ex	examiner's decision	Va	valid
m	modified	f	followed	Vo	void
r	reversed	ha	harmonized	Vp	void in part
s	same case	j	dissenting opinion		
su	superseded	L	limited		
v	vacated	Lp	limited in part		

Please enter, separated by commas, the specific history or treatment ABBREVIATIONS (or the word ANY for citations with any editorial analysis abbreviation—history or treatment) and the NUMBERS of the paragraphs for the citations you want to display.

Source: Shepard's-analysis. Reproduced by permission of Shepard's. Further reproduction of any kind is strictly prohibited.

as slip opinions; there usually is a delay of at least one or two weeks between the decision of a case and when it becomes available in looseleaf form. However, this usually is fast enough for most people under most circumstances. These reports tend to be the best way to stay current with case law developments.

Looseleaf services get their names because the paper they use fits and is stored in the familiar two- or three-ring looseleaf binders that we all have been using since grade school. Different companies publish them, and they cover many different subjects. For criminal justice issues, you should be especially familiar with the **Criminal Law Reporter**, which is published by the Bureau of National Affairs (BNA). You also should become familiar with *United States Law Week* which is published by BNA, and the *United States Supreme Court Bulletin*, published by Commerce Clearing House (CCH).

The *Criminal Law Reporter* is published weekly. It reports in full all U.S. Supreme Court decisions that address criminal issues. It also publishes excerpts and summaries of select federal and state court criminal decisions. An index allows

Exhibit 3–156

(c) 1997 McGraw-Hill, Inc.—DOCUMENT 1 (OF 3)

CITATIONS TO: 497 U.S. 836
SERIES: Shepard's United States Citations
DIVISION: United States Reports
COVERAGE: All Shepard's Citations Through 01/97 Supplement.

RESTRICTIONS: F

NUMBER	ANALYSIS	CITING REFERENCE		SYLLABUS/HEADNOTE
1	parallel citation	(111	L.Ed.2d 666)	
2	parallel citation	(110	S.Ct. 3157)	
		Cir. 2		
20	followed	934	F. 2d 23	
		Cir. 4		
33	followed	49	F. 3d 1034	
		Cir. 5		
39	followed	160	F.R.D. 94	

To see the text of a citing case, press the citing reference NUMBER and then the ENTER key.
For further explanation, press the H key (for HELP) and then the ENTER key.

Source: Shepard's-analysis browse. Reproduced by permission of Shepard's. Further reproduction of any kind is strictly prohibited.

---

readers to find decisions by case name and by subject. As added features, the *Criminal Law Reporter* regularly reports on actions taken by the U.S. Supreme Court in criminal cases, including grants and denials of certiorari, and it occasionally summarizes oral arguments conducted before the Court.

Another useful feature of the *Criminal Law Reporter* is its annual review of Supreme Court decisions in criminal cases. This review covers the Court's decisions according to specific issues addressed. It appears each year over a period of several weeks shortly after the conclusion of a Term of Court. Each term of the Supreme Court extends from the first Monday in October until the justices wrap up their business, which normally is in late June or early July of the next year. We strongly recommend that you make a practice of reading the *Criminal Law Reporter* each week. The case excerpts are brief and usually quite interesting, and there is no better way to stay on top of important developments in case law. We reproduce the first page (Exhibit 3–165) and the Table of Contents and Topical

Exhibit 3–157

FOR EDUCATIONAL USE ONLY		**SHEPARD'S**		Page 1 of 3

Citations to: **110 S.Ct. 3157**
Maryland v Craig 1990

Coverage: ▶ View coverage information for this result
Located: NEGATIVE HISTORY/TREATMENT

Retrieval No.	- - Analysis - -	- - - - Citation - - - -	Headnote No.
		Cir. 3	
1	D Distinguished	750 F.Supp. 727, 728	8
		Cir. 4	
2	D Distinguished	49 F.3d 1024, 1036	7
		Cir. 6	
3	D Distinguished	922 F.2d 294, 298	
		Cir. 9	
4	D Distinguished	921 F.2d 928, 931	6

▶ Cancel LOCATE request (XLOC) and display complete result
▶ IC   ▶ SP   ▶ QC   ▶ Locate   ▶ Commands   ▶ SCOPE
Copyright (C) 1997 Shepard's; Copyright (C) 1997 West Publishing Co.

Source: Shepard's, "110 S.Ct. 3157," loc neg. Reproduced by permission of Shepard's. Further reproduction of any kind is strictly prohibited.

Summary (Exhibit 3–166) from the cases section of an issue of the *Criminal Law Reporter*, to give you a better idea of its contents.

United States Law Week (commonly referred to as *U.S. Law Week*) and the **United States Supreme Court Bulletin** (referred to as *Supreme Court Bulletin*) are similar in many respects to the *Criminal Law Reporter*, but their coverage is much broader. Each includes the full text of all U.S. Supreme Court decisions, not just those dealing with criminal matters, and each provides excerpts and summaries of lower court decisions that cover a broad spectrum of issues. While the lower court excerpts include criminal decisions, there are far fewer than you will find in the *Criminal Law Reporter*. Still, *U.S. Law Week* and the *Supreme Court Bulletin* are extremely useful if your interests extend beyond the criminal law. Like the *Criminal Law Reporter*, these other services are published weekly; they have their own indexing systems; and they are stored in looseleaf binders. We reproduce the first page of a Supreme Court decision as it appeared in *U.S. Law Week* almost immediately after the decision was announced. (See Exhibit 3–167.)

Exhibit 3–158

FOR EDUCATIONAL USE ONLY   SHEPARD'S   Page 2 of 3
Citations to: **110 S.Ct. 3157**
Located: NEGATIVE HISTORY/TREATMENT

Retrieval No.	- - Analysis - -		- - - - Citation - - - -	Headnote No.
			Cir. 9	
1	D	Distinguished	921 F.2d 928, 932	9
2	D	Distinguished	767 F.Supp. at 1548	
			Calif	
3	D	Distinguished	29 Cal.App.4th 454, 467	
4	D	Distinguished	34 Cal.App.4th 372, 385	
5	D	Distinguished	39 Cal.App.4th 284, 297	
6	D	Distinguished	34 Cal.Rptr.2d 761, 768	
			NC	
7	D	Distinguished	460 S.E.2d 173, 176	
			Wis	
8	D	Distinguished	202 Wis.2d 534, 554	

► IC   ► SP   ► QC   ► Locate   ► Commands   ► SCOPE
Copyright (C) 1997 Shepard's; Copyright (C) 1997 West Publishing Co.

Source: Shepard's, "110 S.Ct. 3157" loc neg. Reproduced by permission of Shepard's. Further reproduction of any kind is strictly prohibited.

Exhibit 3–159

FOR EDUCATIONAL USE ONLY   SHEPARD'S   Page 3 of 3
Citations to: **110 S.Ct. 3157**
Located: NEGATIVE HISTORY/TREATMENT

Retrieval No.	- - Analysis - -		- - - - Citation - - - -	Headnote No.
			Wis	
1	D	Distinguished	551 N.W.2d 830, 839	

► IC   ► SP   ► QC   ► Locate   ► Commands   ► SCOPE
Copyright (C) 1997 Shepard's; Copyright (C) 1997 West Publishing Co.

Source: Shepard's, "110 S.Ct. 3157" loc neg. Reproduced by permission of Shepard's. Further reproduction of any kind is strictly prohibited.

Exhibit 3–160

FOR EDUCATIONAL USE ONLY   **INSTA-CITE**   Page 1 of 3
CITATION: 110 S.Ct. 3157

Direct History

1   Craig v. State, 76 Md.App. 250, 544 A.2d 784, 57 USLW 2120
          (Md.App., Aug 03, 1988) (NO. 1547 SEPT. TERM 1987)
     Certiorari Granted by
2   Craig v. State, 314 Md. 458, 550 A.2d 1168 (Md., Dec 21, 1988) (TABLE,
          NO. 365 SEPT. TERM 1988)
     AND Judgment Reversed by
3   Craig v. State, 316 Md. 551, 560 A.2d 1120 (Md., Jul 24, 1989)
          (NO. 110 SEPT. TERM 1988)
     Certiorari Granted by
4   Maryland v. Craig, 493 U.S. 1041, 110 S.Ct. 834, 107 L.Ed.2d 830
          (U.S.Md., Jan 16, 1990) (NO. 89–478)
     AND Judgment Vacated by
=>5   **Maryland v. Craig,** 497 U.S. 836, 110 S.Ct. 3157, 111 L.Ed.2d 666,
          58 USLW 5044, 30 Fed. R. Evid. Serv. 1 (U.S.Md., Jun 27, 1990)
          (NO. 89–478)

▶ Shepard's   ▶ Shepard's PreView   ▶ QuickCite   ▶ Commands   ▶ SCOPE
(C) Copyright West Publishing Company 1997

Source: INSTA-CITE, "110," © West Publishing, used with permission.

Exhibit 3–161

FOR EDUCATIONAL USE ONLY   **INSTA-CITE**   Page 2 of 3
CITATION: 110 S.Ct. 3157

Direct History
        On Remand to
  6 Craig v. State, 322 Md. 418, 588 A.2d 328 (Md., Apr 08, 1991)
            (NO. 110 SEPT. TERM 1988, 69 SEPT. TERM 1990)

Negative Indirect History
Not Followed on State Law Grounds
  7 Com. v. Ludwig, 527 Pa. 472, 594 A.2d 281 (Pa., May 10, 1991)
            (NO. 34 E.D. 1988) (► Additional History)
  8 People v. Fitzpatrick, 158 Ill.2d 360, 633 N.E.2d 685.
            198 Ill.Dec. 844, 62 USLW 2551 (Ill., Feb 17, 1994) (NO. 74768)
Disagreement Recognized by
  9 State v. Deuter, 839 S.W.2d 391 (Tenn., Sep 14, 1992) (NO. 3) (► Additional History)

  ► Shepard's   ► Shepard's PreView   ► QuickCite   ► Commands   ► SCOPE
(C) Copyright West Publishing Company 1997

Source: INSTA-CITE, "110," © West Publishing, used with permission.

Exhibit 3–162

FOR EDUCATIONAL USE ONLY   **INSTA-CITE**   Page 3 of 3
CITATION: 110 S.Ct. 3157

Negative Indirect History
Declined to Extend by
  10 State v. Scott, 257 Mont. 454, 850 P.2d 286 (Mont., Apr 01, 1993)
            (NO. 92–052)
  11 People v. Dablon, 34 Cal.Rptr.2d 761 (Cal.App. 4 Dist., Oct 17, 1994)
            (NO. D018959) (► Additional History)

Secondary Sources
Corpus Juris Secundum (C.J.S.) References
            23 C.J.S. Criminal Law Sec.1116 Note 50.10 (Pocket Part)
            23A C.J.S. Criminal Law Sec.1210 Note 26 (Pocket part)
            98 C.J.S. Witnesses Sec.321 Note 26+ (Pocket Part)

  ► Shepard's   ► Shepard's PreView   ► QuickCite   ► Commands   ► SCOPE
(C) Copyright West Publishing Company 1997

Source: INSTA-CITE, "110" negative, © West Publishing, used with permission.

Exhibit 3–163

FOR EDUCATIONAL USE ONLY   **SHEPARD'S** (Rank 1 of 1)   Page 1 of 13

Citations to: **NY CRIM PRO s 65.20**
Citator: NEW YORK STATUTE CITATIONS
Division: McKinney's Criminal Procedure Law
Coverage: First Shepard's volume through Jan. 1997 Supplement

Retrieval No.	- - Analysis - -	- - - - Citation - - - - s 65.20
1		156 A.D.2d 92, 95
2		164 A.D.2d at 946
3		174 A.D.2d 696, 697
4		203 A.D.2d at 599
5	C Constitutional	552 N.Y.S.2d 68, 71
6		554 N.Y.S.2d 924, 925
7		560 N.Y.S.2d 53, 56
8		571 N.Y.S.2d 551, 552
9		611 N.Y.S.2d at 217
10	C Constitutional	551 N.E.2d 561, 564
		41 Syracuse L. Rev. at 252

Note: Use FIND to view the statute's historical notes. To view the result's table of contents, type **SHTOC** and press **ENTER**.

Copyright (C) 1997 Shepard's; Copyright (C) 1997 West Publishing Co.

Source: SHEPARD's NY 65.20. Reproduced by permission of Shepard's. Further reproduction of any kind is strictly prohibited.

Exhibit 3–164

NOTICE: This opinion is subject to formal revision before publication in the preliminary print of the United States Reports. Readers are requested to notify the
Reporter of Decisions, Supreme Court of the United States, Washington, D.C.
20543, of any typographical or other formal errors, in order that corrections may
be made before the preliminary print goes to press.

SUPREME COURT OF THE UNITED STATES

No. 91–1958

DONALD L. HELLING, ET AL., PETITIONERS *v.* WILLIAM McKINNEY

ON WRIT OF CERTIORARI TO THE UNITED STATES COURT OF APPEALS FOR THE NINTH COURT

[June 18, 1993]

JUSTICE WHITE delivered the opinion of the Court.

This case requires us to decide whether the health risk posed by involuntary exposure of a prison inmate to environmental tobacco smoke (ETS) can form the basis
of a claim for relief under the Eighth Amendment.

I

Respondent is serving a sentence of imprisonment in the Nevada prison system.
At the time that this case arose, respondent was an inmate in the Nevada State
Prison in Carson City, Nevada. Respondent filed a *pro se* civil rights complaint in
United States District Court under Rev. State. § 1979, 42 U. S. C. § 1983, naming as
defendants the director of the prison, the warden, the associate warden, a unit coun-
selor, and the manager of the prison store. The complaint, dated December 18,
1986, alleged that respondent was assigned to a cell with another inmate who
smoked five packs of cigarettes a day. App. 6. The complaint also stated that cigarettes were sold to inmates without properly informing of the health hazards a nonsmoking inmate would encounter by sharing a room with an inmate who smoked,
Id., at 7–8, and that certain cigarettes burned continuously, releasing some type of
chemical, *Id.*, at 9. Respondent complained of certain

Source: Slip opinion . . . Hellingly, U.S. Supreme Court.

Exhibit 3–165

The CRIMINAL
LAW REPORTER

Court Decisions
Legislative Action
Reports and Proposals

A Weekly Review of Developments in Criminal Law

| February 7, 1996 | THE BUREAU OF NATIONAL AFFAIRS, INC. | Volume 58, No. 18 |

IN THE COURTS

HIGHLIGHTS

Capital sentencing juries' alleged inaccuracy in predicting the future dangerousness of defendants is not relevant to the issue of the future dangerousness of any one particular defendant, the Texas Court of Criminal Appeals held. Accordingly, it excluded such evidence from death penalty proceedings after noting that the U.S. Supreme Court has already upheld the state's future dangerousness special circumstance against constitutional challenge. (Page 1395)

Electronic transmission of computer files containing obscene images is a violation of federal obscenity laws, the Sixth Circuit held. The court likened the defendant's activity to transmitting money by electronic means. It also held that, because the defendants had control over access to their bulletin board service and knew that they had a subscriber in the district where venue was laid, they could be held to the community standards of that district. (Page 1393)

Civil forfeitures were examined for their double jeopardy implications by two state supreme courts, with opposite results. The Illinois Supreme Court held that the state statute on forfeiture of a vehicle used to facilitate a drug crime imposes "punishment." The Louisiana Supreme Court said that forfeiture of a conveyance used to commit a crime is remedial and that, as applied to a 12-year-old car, was not disproportionate to the damages caused by the wrongdoer. (Pages 1397, 1398)

Warrantless activation of pagers' memories was addressed by a federal court in New York. In a comprehensive opinion, the court ruled on the law-

fulness, under both the Fourth Amendment and Electronic Communications Privacy Act, of warrantless police "searches" of three different pagers seized in separate circumstances. (Page 1403)

The sleeping compartment of a tractor trailer rig may have been the driver's "home" but it was also accessible from the cab and, therefore, could be searched incident to the driver's arrest, the Washington Supreme Court ruled. (Page 1400)

In this issue: Sections 1 (Courts and Legislatures), and 3 (Supreme Court Proceedings).

ALLOWING DOWNLOADING OF IMAGE FILES CAN BE PUNISHABLE UNDER FEDERAL OBSCENITY LAWS

CA 6 also rejects argument that relevant community ought to be broader than place where recipient of obscene images resided.

A bulletin board operator who distributed obscene pictures via a telephone line to his subscribers' personal computers was liable for violating federal obscenity laws, the U.S. Court of Appeals for the Sixth Circuit Court held January 29. The court rejected the defendant's argument that the obscenity statute, 18 USC 1465, did not apply to intangible objects like the computer image files. It also turned back a claim that the relevant community, for purposes of applying the obscenity test of *Miller v. California,* 413 U.S. 15 (1973), ought to be broader than the community in which the recipient of an obscene image resides. (U.S. v. Thomas, CA 6, Nos. 94-6648 etc., 1/29/96)

From their home in California, the defendants operated a computer bulletin board, which they

continues

Exhibit 3–165   continued

used to distribute digitized obscene pictures to their subscribers. They were arrested after a postal inspector, based in Tennessee, joined the bulletin board and thereafter downloaded a number of obscene pictures by placing a telephone call to the defendant's computer in California.

In rejecting the defendant's argument that Section 1465 has no application to intangible property—such as series of 1's and 0's that were transmitted from the defendant's computer to the postal inspector's computer—the court cited *U.S. v. Gilboe,* 684 F.2d 235 (CA 2 1982). *Gilboe* said that the fact that the defendant had transported money by wiring it, that is, by electronic impulses, did not preclude federal liability for unlawful transportation of money obtained by fraud. Similar

reasoning supports affirming the defendant's convictions under Section 1465, the court said; regardless of the manner of distribution, what the postal inspector received was an obscene picture that had been transported in interstate commerce.

Moreover, the distribution of the obscene picture was "knowing" even though it was the postal inspector who placed the call to the defendant's bulletin board system (BBS), thus causing the defendant's computer to distribute the obscene pictures to him. There was no need for the government to prove that the defendants had specific knowledge of each transmission of an obscene picture, the court said. Here, the defendants knew that they had a BBS subscriber in Tennessee who had paid them for

Section 1   **58 CrL 1393**

Source: Reprinted with permission from *Criminal Law Reporter,* Vol. 58, p. 1393 (Feb 7, 1996). Copyright 1996 by The Bureau of National Affairs, Inc. (800-372-1033) <http://www.bna.com>

Exhibit 3–166

Table of Cases

Topical Summary

continues

Exhibit 3–166 continued

Source: Reprinted with permission from *Criminal Law Reporter,* Vol. 58, p. 1416 (Feb 7, 1996). Copyright 1996 by The Bureau of National Affairs, Inc. (800-372-1033) <http://www.bna.com>

Exhibit 3–167

The United States
Law Week

Supreme Court
Opinions

January 23, 1996   THE BUREAU OF NATIONAL AFFAIRS, INC., WASHINGTON, D.C.   Volume 64, No. 27

OPINION ANNOUNCED JANUARY 22, 1996

The Supreme Court decided:

CRIMINAL LAW AND PROCEDURE—
Sentencing

U.S. Sentencing Commission's establishment of constructive or presumed weight of 0.4 milligram per dose of LSD, for use in determining weight of combinations of LSD and carrier mediums for purposes of setting sentences under federal Sentencing Guidelines, does not change rule established in *Chapman v. U.S.,* 500 U.S. 453, 59 LW 4530 (1991), that combined weight of LSD and its carrier medium determines whether defendant has crossed weight threshold established by 21 USC 841(b)(1) and is therefore subject to statutory mandatory minimum term of imprisonment. (*Neal v. U.S.,* No. 94-9088) . . . . Page 4077

Full Text of Opinion

No. 94-9088

MEIRL GILBERT NEAL, PETITIONER
v. UNITED STATES
ON WRIT OF CERTIORARI TO THE UNITED
STATES COURT OF APPEALS FOR THE
SEVENTH CIRCUIT

Syllabus

No. 94-9088. Argued December 4, 1995—Decided January 22, 1996
When the District Court first sentenced petitioner Neal on two plea-bargained convictions involving possession of LSD with intent to distribute,

the amount of LSD sold by a drug trafficker was determined, under both the federal statute directing minimum sentences and the United States Sentencing Commission's Guidelines Manual, by the whole weight of the blotter paper or other carrier medium containing the drug. Because the combined weight of the blotter paper and LSD actually sold by Neal was 109.51 grams, the court ruled, among other things, that he was subject to 21 U. S. C. § 841(b)(1)(A)(v), which imposes a 10-year mandatory minimum sentence on anyone convicted of trafficking in more than 10 grams of " a mixture of substance containing a detectable amount" of LSD. After the Commission revised the Guidelines' calculation method by instructing courts to give each dose of LSD on a carrier medium a constructive or presumed weight, Neal filed a motion to modify his sentence, contending that the weight of the LSD attributable to him under the amended Guidelines was only 4.58 grams, well short of § 841(b)(1)(A)(v)'s 10-gram requirement, and that the Guidelines' presumptive-weight method controlled the mandatory minimum calculation. The District Court followed *Chapman v. United States,* 500 U. S. 453, 468, in holding, *inter alia,* that the actual weight of the blotter paper, with its absorbed LSD, was determinative of whether Neal crossed the 10-gram threshold and that the 10-year mandatory minimum sentence still applied to him notwithstanding the Guidelines. In affirming, the en banc Seventh Circuit agreed with the District Court that a dual system now prevails in calculating LSD weights in cases like this.

continues

Exhibit 3–167   continued

Held: Section 841(b)(1) directs a sentencing court to take into account the actual weight of the blotter paper with its absorbed LSD, even though the Sentencing Guidelines require a different method of calculating the weight of an LSD mixture or substance. The Court rejects petitioner's contentions that the revised Guidelines are entitled to deference as a construction of § 841(b)(1) and that those Guidelines require reconsideration of the method used to determine statutory minimum sentences. While the Commission's expertise and the Guidelines' design may be of potential weight and relevance in other contexts, the Commission's choice of an alternative methodology for weighing LSD does not alter *Chapman's* interpretation of the statute. In any event, *stare decisis* requires that the Court adhere to *Chapman* in the absence of intervening statutory changes casting doubt on the case's interpretation. It is doubtful that the Commission intended the Guidelines to displace *Chapman's* actual-weight method for statutory minimum sentences, since the Commission's authoritative Guidelines commentary indicates that the new method is not an interpretation of the statute, but an independent calculation, and suggests that the statute controls if it conflicts with the Guidelines. Moreover, the Commission's dose-based method cannot be squared with *Chapman*. In these circumstances, this Court need not decide what, if any, deference is owed the Commission in order to reject its contrary interpretation. Once the Court has determined a statute's meaning, it adheres to its ruling under *stare decisis* and assesses an agency's later interpretation of the statute against that settled law. It is the responsibility of Congress, not this Court, to change statutes that are thought to be unwise or unfair.
46 F. 3d 1405, affirmed.

KENNEDY, J., delivered the opinion for a unanimous Court.

JUSTICE KENNEDY delivered the opinion of the Court.

The policy of sentencing drug offenders based on the amount of drugs involved, straightforward enough in its simplest formulation, gives rise to complexities, requiring us again to address the methods for calculating the weight of LSD sold by a drug trafficker. We reject petitioner's contention that the revised system for determining LSD amounts under the United States Sentencing Guidelines requires reconsideration of the method used to determine statutory minimum sentences, and we adhere to our former decision on the subject.

I

LSD (lysergic acid diethylamide) is such a powerful narcotic that the average dose contains only 0.05 milligrams of the pure drug. The

NOTE: Where it is deemed desirable, a syllabus (headnote) will be released * * * at the time the opinion is issued. The syllabus constitutes no part of the opinion of the Court but has been prepared by the Reporter of Decisions for the convenience of the reader. See *United States v. Detroit Lumber Co.,* 200 U.S. 321, 337.

NOTICE: These opinions are subject to formal revision before publication in the preliminary print of the United States Reports. Readers are requested to notify the Reporter of Decisions, Supreme Court of the United States, Washington, D.C. 20543, of any typographical or other formal errors, in order that corrections may be made before the preliminary print goes to press.

Source: Reprinted with permission from *The United States Law Week,* Vol. 64, p. 4077 (Jan 23, 1996). Copyright 1996 by The Bureau of National Affairs, Inc. (800-372-1033) <http://www.bna.com>

CONCLUSION

The focus of this chapter has been primary legal authorities: what they are, how to find those that are useful for your research purposes, and how to ensure that the ones you find are still valid and up to date. We dwelled at greatest length on research techniques for case law because judicial decisions are of such importance in giving meaning to other sources of law, including administrative regulations, statutes, and federal and state constitutional provisions.

Our discussion of Shepardizing may have added a new word to your vocabulary and a new and obligatory technique to your research repertoire. It is absolutely imperative that you Shepardize judicial decisions to make sure the cases you want to rely on are still good law. Although Shepardizing an administrative regulation, a statute, a constitutional provision, or a case can help confirm the law's continuing validity, this is just one reason to use a Shepard's: Another is to find judicial decisions that have cited the authority you are Shepardizing and that may have significantly altered the interpretation of the original citation.

Administrative regulations are of indisputable importance to some criminal justice matters, but they are the form of primary legal authority that you are least likely to encounter in the course of your research. Legislation and constitutional provisions figure so prominently in the criminal justice process that you should expect to use statutory and constitutional reference materials regularly. Annotated statutes and constitutions are a tremendous help in directing your research to relevant judicial decisions. Once you find a single case that is on point to your topic,

which you can do through a variety of techniques—including relying on other primary authorities and secondary authorities and on descriptive word indexes and case digests—you are off and running. You have this capacity because of the remarkable structure of West Publishing Company's key number system.

The key number system is a marvelous creation. It allows you to use case digests to navigate backward and forward in time and through any jurisdictions you choose in order to collect cases that are relevant to the issue you are researching. The system imposes order on case law research conducted with printed sources. Before the arrival of computerized legal research, using West's key number system was the only practical, organized method for conducting case law research. However, computerized databases such as WESTLAW and LEXIS now offer many invaluable alternatives for pursuing legal research.

4 Putting It All Together: Sample Problems and Research Strategies

INTRODUCTION

The law is sometimes described as a seamless web. The rules and principles that affect different issues of law are frequently related, if not interdependent. The boundary lines between subfields of law—for example, criminal law, mental health law, corrections law, juvenile justice, and civil forfeiture, to name a few relevant to criminal justice—often are more illusory than real.

In their own way, legal research strategies mimic the fluid qualities of the substantive law. It would be a mistake to march resolutely through a legal research problem as if there were one fixed recipe for success or a single, established sequence of search procedures. Successful legal research requires flexibility, ingenuity, adaptability, and persistence. No one research strategy applies invariably to all research problems. Although the procedures we have discussed thus far should be useful guides, we cannot in good conscience represent that a "follow the steps outlined in the preceding chapters" advisory will guarantee answers to your questions of law. From time to time, you doubtless will have to be willing to depart from and redesign original blueprints for completing legal research.

This chapter offers us a chance to try to integrate and apply the information we have assembled thus far. We do so by creating a few more hypothetical research problems and exploring different strategies for solving them. Keep in mind that our way of tackling these research questions is just one possible way. They are not the only methods for trying to solve them, and they are not necessarily the means that will work best for you.

We have constructed a variety of research problems for this exercise. Some are questions that lend themselves to traditional legal analysis—that is, they are the types of questions that lawyers regularly are called on to answer by producing legal authorities that identify governing rules, doctrines, or principles and applying those authorities to the facts of a particular case.

Other issues are of a slightly different nature. They are not the types of questions that classically have demanded the attention of lawyers, but rather are the kinds of issues that interest those who study, administer, or conduct research that relates to the law. These issues tend to be questions about the law and its application, rather than questions answered by reference to "black letter" legal rules.

We do not wish to attribute undue significance to the distinction made between traditional legal research issues that focus on legal rules and doctrine, and those that concern law and its application. The differences are a matter of degree; they do not represent qualitatively different dimensions of legal inquiry. For our purposes, these different types of questions are of interest because they help determine which legal research strategies can most effectively be used to explore them.

We would consider the following examples to involve traditional questions of law: Can an offender who committed murder in Kentucky at age 16 be sentenced to death on conviction for that crime? Does a defense lawyer who appears at trial intoxicated thereby provide constitutionally ineffective assistance of counsel to his or her client? Does a federal prisoner have the right to marry while incarcerated? Is expert testimony about the battered-woman syndrome admissible in the New Jersey courts to help support a murder defendant's claim that she killed in self-defense? Is the good faith exception to the exclusionary rule recognized in Connecticut?

Examples of questions about the law that may be of special interest to academics, criminal justice administrators, law enforcement officials, policymakers, and other researchers, include the following: How did former Justice Blackmun's views about the death penalty, as evidenced by the opinions he wrote in capital cases, change during the 23 years he served on the Supreme Court? How often have claims of constitutionally ineffective assistance of counsel been considered in published federal court opinions over the last 10 years? What has been the volume of prisoners' habeas corpus filings in the federal courts over the past 20 years? How often has the research or testimony of Dr. Lenore Walker been cited in published decisions involving the battered-woman syndrome? How frequently have "exceptions to the good faith exception" to the exclusionary rule been recognized in reported decisions, resulting in the suppression of evidence seized under the authority of invalid search warrants?

The type of question with which you are confronted is likely to help dictate your legal research strategy. In particular, you may conclude that computerized techniques are preferable to printed sources for investigating some types of issues, and vice versa. We consider a variety of research strategies as we work through the examples that follow.

PRACTICE RESEARCH PROBLEMS AND ILLUSTRATIVE STRATEGIES FOR THEIR SOLUTION

Problem 1

Mac Travis, age 16, robbed a convenience store in Louisville and killed the store clerk during the course of the robbery. If he is tried as an adult and convicted of first-degree (capital) murder, can he be sentenced to death?

To answer this question, we first have to determine whether Kentucky statutes authorize capital punishment under the described facts. If they do, we must investigate whether there are any constitutional barriers—either federal or state—to executing an offender who was just 16 at the time of his crime.

We begin with the index to the **Kentucky Revised Statutes**. Terms such as **"murder,"** **"death penalty,"** **"juveniles,"** and others might help direct us to statutory provisions that apply to this problem. Under the heading "**Capital Punishment**," we spot a subtopic that looks promising: *"***Youthful Offenders. Prohibited, §640.040.***"* (See Exhibit 4–1.) We are referred to this same section of the statute under the subtopic "**Sentences. . . . Youthful offenders.**"

We check Kentucky Revised Statute § 640.040 in the main volume, and we learn that "[a] youthful offender may be sentenced to capital punishment if he was sixteen (16) years of age or older at the time of the commission of the offense." (See Exhibit 4–2.) The lone case annotation involves *Stanford v. Kentucky,* 492

Exhibit 4–1

GENERAL INDEX (A—I)

CAPITAL PUNISHMENT—Cont'd
Radio and television media.
  Audio-visual equipment.
    Prohibited during execution, §431.250.
  Persons who may attend executions, §431.250.
Secretary of corrections cabinet.
  Persons who may attend executions, §431.250.
Sentences.
  Death sentence.
    Commitment when death sentence
      imposed, §532.100.
  Mental retardation.
    Seriously mentally retarded defendants,
      §532.140.
  Racial bias in capital sentencing.
    Study on, §17.1531.
  Supreme court.
    Review of death sentence by supreme court,
      §532.075.

Youthful offenders.
  Sentence to capital punishment prohibited,
    §640.040.
Sheriffs.
  Persons who may attend execution, §431.250.
Supreme court.
  Death sentence.
  Review by supreme court, §532.075.
Time, §431.240.
  Day of execution, §431.218.
  Governor.
    Duty to fix time in case judgment not
      executed on day appointed, §431.240.
    Duty to fix time in case of insanity,
      pregnancy or escape, §431.240.
Warden of penitentiary.
  Persons who may attend execution, §431.250.
  Return on judgment, §431.260.
Youthful offenders.
  Prohibited, §640.040.

Source: The statutes reprinted or quoted verbatim in the following pages are taken from the Kentucky Revised Statutes Annotated, Copyright by Michie, a division of Reed Elsevier Inc. and Reed Elsevier Properties Inc., and are reprinted with the permission of Michie. All rights reserved.

Exhibit 4–2

YOUTHFUL OFFENDERS

640.040. Capital punishment and other prohibited dispositions.—(1) No youthful offender who has been convicted of a capital offense who was under the age of sixteen (16) years at the time of the commission of the offense shall be sentenced to capital punishment. A youthful offender may be sentenced to capital punishment if he was sixteen (16) years of age or older at the time of the commission of the offense. A youthful offender convicted of a capital offense regardless of age may be sentenced to a term of imprisonment appropriate for one who has committed a Class A felony and may be sentenced to life imprisonment without benefit of parole for twenty-five (25) years.

(2) No youthful offender shall be subject to persistent felony offender sentencing under the provisions of KRS 532.080 for offenses committed before the age of eighteen (18) years.

(3) No youthful offender shall be subject to limitations on probation, parole or conditional discharge as provided for in KRS 533.060.

(4) Any youthful offender convicted of a misdemeanor or any felony offense which would exempt him from KRS 635.020(2), (3), (4), (5) or (6) shall be disposed of by the circuit court in accordance with the provisions of KRS 635.060. (Enact. Acts 1986, ch. 423, § 137, effective July 1, 1987.)

NOTES TO DECISIONS

1. Capital Punishment.

There is neither a historical nor a modern societal consensus forbidding the imposition of capital punishment on any person who murders at 16 or 17 years of age, and such punishment does not offend the constitutional prohibition against cruel and unusual punishment. Stanford v. Kentucky, 492 U.S.—, 109 S. Ct. 2969, 106 L. Ed. 2d 306 (1989).

Research References. Petrilli, Kentucky Family Law, Juvenile Court, § 32.24.

Collateral References. 43 C.J.S., Infants, §§ 206, 210.

Source: The statutes reprinted or quoted verbatim in the following pages are taken from the Kentucky Revised Statutes Annotated, Copyright by Michie, a division of Reed Elsevier Inc. and Reed Elsevier Properties Inc., and are reprinted with the permission of Michie. All rights reserved.

U.S. [361], 109 S. Ct. 2969, 106 L. Ed. 2d 306 (1989), a U.S. Supreme Court decision that is described as holding that the federal Constitution's cruel and unusual punishments clause is not offended by the execution of 16- or 17-year-old offenders. We check the pocket part to see if there are possible changes in the statute and to secure any later case decisions that may be relevant. (In fact, the statute was changed in part, but the changes did not affect our problem; nor did we find other case squibs related to our issue.) We also have the option of Shepardizing Ky. Rev. Stat. § 640.040(1) to determine whether the legislation has been altered in any way and to seek out cases citing this provision.

Since we have been given a lead to *Stanford v. Kentucky,* we can look it up in one of the U.S. Supreme Court case reporters. We choose to consult the **Supreme Court Reporter**, volume 109, page 2969, because we know it uses West's key number system. As we read the case, we confirm the accuracy of the squib presented with the annotated statute. The Supreme Court ruled in a plurality decision (only four of the nine Justices agreed with the reasoning of the lead opinion and Justice O'Connor concurred only in the result) that there is no federal constitutional prohibition against the capital punishment of 16-year-old murderers. The headnotes that West's editors have created for *Stanford* allow us to identify topics and key numbers that will connect us to other cases considering similar issues. (See Exhibit 4–3.) When we Shepardize *Stanford* to confirm its continuing valid-

Exhibit 4–3

492 U.S. 361, 106 L.Ed.2d 306

361 **Kevin N. STANFORD, Petitioner**

v.

KENTUCKY.

Heath A. WILKINS, Petitioner

v.

MISSOURI.

Nos. 87–5765, 87–6026.

Argued March 27, 1989.

Decided June 26, 1989.

Rehearing Denied Aug. 30, 1989.

See 492 U.S. 937, 110 S.Ct. 23.

A defendant who was approximately 17 years and 4 months old at time he committed a murder in Kentucky was convicted of murder, sodomy, robbery and receiving stolen property and was sentenced to death by the Jefferson Circuit Court, Charles M. Leibson, J. Defendant appealed. The Supreme Court of Kentucky, 734 S.W.2d 781, affirmed. In another case, a defendant who was approximately 16 years and 6 months old when he committed a murder in Missouri was certified for trial as an adult. He was convicted in the Circuit Court, Clay County, Glennon E. McFarland, J., of first-degree murder and sentenced to death, and he appealed. The Supreme Court of Missouri, 736 S.W.2d 409, Billings, C.J., affirmed. On certiorari, the Supreme Court, Justice Scalia, held that imposition of capital punishment on an individual for a crime committed at 16 or 17 years of age did not violate evolving standards of decency and thus did not constitute cruel and unusual punishment under the Eighth Amendment.

Affirmed.

Justice O'Connor filed an opinion concurring in part and concurring in judgment.

Justice Brennan filed a dissenting opinion in which Justices Marshall, Blackmun and Stevens joined.

1. Criminal Law ⚷ **1213.8(8)**
Capital punishment for individuals who committed capital crime at 16 or 17 years of age was not one of those modes or acts of punishment considered cruel or unusual at time the Bill of Rights was adopted in light of common law's rebuttable presumption of incapacity to commit felony at age 14. U.S.C.A. Const.Amend. 8.

2. Criminal Law ⚷ **1213.8(8)**
Question of whether imposition of capital punishment for crime committed at 16 or 17 years of age was cruel and unusual had to be determined by examining whether punishment was contrary to evolving standards of decency that mark the progress of a maturing society. U.S.C.A. Const.Amend. 8.

3. Criminal Law ⚷ **1213.8(8)**
The fact that, of the 37 states whose laws permitted capital punishment, 15 declined to impose it on 16 year olds and 12 . . .

Source: Supreme Court Reporter, 109 S.Ct. 2969 "Stanford," © West Publishing, used with permission.

ity, we expect to find additional cases that have cited this decision on similar issues.

Even though executing a 16-year-old offender does not violate the federal
Constitution's cruel and unusual punishments clause, we still must ascertain
whether this practice violates the Kentucky Constitution. By consulting the index
within the Constitutions volume of the **Kentucky Revised Statutes**, we learn that
section 17 of the Kentucky Constitution prohibits "cruel and unusual punishment." (See Exhibit 4–4.) On turning to this provision we discover that the state
constitution actually bans cruel punishment, and thus is textually different from
the federal Constitution's prohibition against cruel and unusual punishments. (See
Exhibit 4–5.) However, the Notes to Decisions in the main volume (see Exhibit
4–6) and the pocket part (see Exhibit 4–7) do not indicate that section 17 of the
Kentucky Constitution provides any different protections from the Eighth Amendment in this context. We would have to read and Shepardize the cited cases (*Ice v.*

Exhibit 4–4

CONSTITUTION OF KENTUCKY INDEX

CRIMINAL PROCEDURE.
  Bail.
    Right to bail, Const. Ky., §16.
  Change of venue, Const. Ky., §11.
  Double jeopardy, Const. Ky., §13.
  Indictments.
    Indictable offense not to be prosecuted by
    information, Const. Ky., §12.

Rights of accused, Const. Ky., §11.
CRUEL AND UNUSUAL PUNISHMENT.
  Prohibited, Const. Ky., §17.

D

DAMAGES.
  Wrongful death.
    Recovery, Const. Ky., §241.

Source: The statutes reprinted or quoted verbatim in the following pages are taken from the Kentucky Revised Statutes Annotated, Copyright by Michie, a division of Reed Elsevier Inc. and Reed Elsevier Properties Inc., and are reprinted with the permission of Michie. All rights reserved.

Exhibit 4–5

CONSTITUTION OF KENTUCKY

§ 17. Excessive bail or fine, or cruel punishment, prohibited.—Excessive bail shall not be required,
nor excessive fines imposed, nor cruel punishment inflicted.

Source: The statutes reprinted or quoted verbatim in the following pages are taken from the Kentucky Revised Statutes Annotated, Copyright by Michie, a division of Reed Elsevier Inc. and Reed Elsevier Properties Inc., and are reprinted with the permission of Michie. All rights reserved.

Exhibit 4–6

NOTES TO DECISIONS

ANALYSIS

1. In general.
2. Application.
3. Bail.
4. —Excessive.
5. —Habeas corpus.
6. —Peace bonds.
7. Common law.
8. Fines.
9. —Excessive.
10. —Imprisonment.
11. Punishment.
12. —Appeal.
13. —Death penalty.
14. —Juvenile status.
15. —Deterrence.
16. —Disproportionate.
17. —Habitual criminals.
18. —Imprisonment.
19. —Judgment.
20. —Jurisdiction.
21. —Life imprisonment.
22. ——Without parole.
23. —Prisoners.
24. —Verdicts.
25. —License revocations.

---

14. ——Juvenile Status.
   While defendant's young age at the time of murder was an important factor that should have been given serious consideration at both the transfer hearing in juvenile court and as a mitigating circumstance at the sentencing phase in circuit court, it was not a constitutional distinction since the United States Supreme Court has not yet decided that juvenile status puts the death penalty in conflict with the Eighth Amendment. Ice v. Commonwealth, 667 S.W.2d 671 (Ky.), cert. denied, 469 U.S. 860, 105 S. Ct. 192, 83 L. Ed. 2d 125 (1984).

Source: The statutes reprinted or quoted verbatim in the following pages are taken from the Kentucky Revised Statutes Annotated, Copyright by Michie, a division of Reed Elsevier Inc. and Reed Elsevier Properties Inc., and are reprinted with the permission of Michie. All rights reserved.

Commonwealth and *Stanford v. Kentucky*) to ascertain whether the decisions specifically involved an interpretation of the state constitution, whether the offenders were as young as age 16, and whether the holdings may have been undermined by later developments.

Finally, we turn to the **Descriptive Word Index** of the **Kentucky Digest 2d** to determine whether we can unearth any other relevant state court decisions. The heading "**Capital Punishment**" refers us to "**Criminal Law**" (see Exhibit 4–8), and we find several references to capital punishment under that heading. (See Exhibit 4–9.) As directed, we then consult **Criminal Law** ⟶ **1208.1(4,5)** and **Homicide** ⟶ **354**. Criminal Law and its key number prove to be most helpful. The main volume of the Kentucky Digest 2d, under Criminal Law ⟶ 1208.1(4), indicates that the Kentucky Supreme Court held, in *Salisbury v. Commonwealth*, 417 S.W.2d 244 (1967), that a death sentence "is not in violation of state or federal constitutions." (See Exhibit 4–10.) We will want to read this case to learn more about the facts and issues it involved. The only relevant case noted in the pocket part is *Stanford v. Kentucky*. (See Exhibit 4–11.)

Exhibit 4–7

---

§ 17. Excessive bail or fine, or cruel punishment, prohibited.

NOTES TO DECISIONS

ANALYSIS

11. Punishment.
13. —Death penalty.
14. ——Juvenile status.
16.1. —Not disproportionate.
27. Jury instruction.
11. **Punishment.**
13. **—Death Penalty.**

14. ——Juvenile Status.
There is neither a historical nor a modern societal consensus forbidding the imposition of capital punishment on any person who murders at 16 or 17 years of age, and such punishment does not offend the constitutional prohibition against cruel and unusual punishment. Stanford v. Kentucky, 492 U.S. 361, 109 S. Ct. 2969, 106 L. Ed. 2d 306 (1989).

Source: The statutes reprinted or quoted verbatim in the following pages are taken from the Kentucky Revised Statutes Annotated, Copyright by Michie, a division of Reed Elsevier Inc. and Reed Elsevier Properties Inc., and are reprinted with the permission of Michie. All rights reserved.

Exhibit 4–8

CAPIAS   **24 Ky D 2d—286**

References are to Digest Topics and Key Numbers

CAPITAL—Cont'd
TAXATION—Cont'd
  Capital stock representing corporate property. Tax 120
  Foreign public service corporations, going concern value under governmental license. Tax 398
TRUSTS and trustees, see this index Trusts and Trustees
TURNPIKE and toll road companies. Turnpikes 7
WATER or waterworks company. Waters 186
WILLS—
  Description in will. Wills 573
  Legacies payable from capital. Wills 732(5)
  Limitation over of principal after life estate in income. Wills 634(20)

  Testamentary trusts. Wills 684.1–684.10
WITHDRAWAL of capital from limited partnership. Partners 364
CAPITAL CASE
SEPARATION of jury, waiver of rights to have jurors kept together. Crim Law 868
CAPITAL OFFENSE
BAIL, right to release on. Bail 43
CAPITAL PLANNING AND ZONING COMMISSION
APPEALS, city ordinance empowering. Zoning 29
CAPITAL PUNISHMENT
CRIMINAL law, see this index Criminal Law
CAPITAL STOCK TAX

Source: Kentucky Digest 2d, DWI, "Capital Punishment," © West Publishing, used with permission.

Exhibit 4–9

CRIMINAL LAW—Cont'd
ATTACHMENT, criminal act in incurring liability as ground for. Attach 33
ATTEMPTS, see this index Attempt
ATTENDANCE and conduct of officers at trial. Crim Law 644
ATTORNEY and client, counsel for accused, see Counsel for accused, post
AUTHORITY—

CAPITAL punishment—
  Generally—
    Crim Law 1208.1(4, 5)
    Homic 354
  Juror's disqualifying scruples—
    Exclusion in selection of jury. Jury 33(2.1)
  Sentence and punishment, post
CASE certified or reserved, see this index Certified, Reserved, or Reported Cases or Questions
CASE or statement of facts for purpose of review, see this index Case on Appeal

SENTENCE and punishment. Crim Law 977–1003, 1205–1219
  Administrative agency fixing punishment. Crim Law 1216(3.6)
  Advice and warnings. Crim Law 986(1)
  Affidavits—
    Postconviction relief. Crim Law 998(16)
  Affirmance of conviction—
    Postconviction relief. Crim Law 998(13)
  Aggravating or mitigating circumstances—
    Crim Law 1208.1(5), 1208.6
    Homic 354
  Aggravation or mitigation, matters in. Crim Law 986.2
  Amendment. Crim Law 996
  Armed conduct, enhancement of punishment. Crim Law 1208.6(2)
  Armed offense increasing punishment. Crim Law 1208.6(2)
  Attempt, see this index Attempt
  Banishment. Crim Law 1208.4(1)
  Banishment condition of probation violated. Crim Law 982.9(1)
  Bifurcated trial, see this index Separate Trials
  Capital punishment—
    Generally—
      Crim Law 1208.1(4, 5)
      Homic 354
  Bail, accused entitled to. Bail 43

Source: Kentucky Digest 2d, DWI, "Criminal Law," © West Publishing, used with permission.

Finding authorities that apply to a question of law is just part of the battle. The other part involves careful reading and analysis of what you have found. The statute we produced leaves no doubt about the state legislature's views regarding death-penalty eligibility for 16-year-olds, and *Stanford v. Kentucky* unambiguously holds that the capital punishment of such offenders does not violate the Eighth Amendment to the U.S. Constitution. Nevertheless, unless and until you produce a Kentucky Supreme Court case specifically ruling that the state constitution's ban on cruel punishments is not offended by the execution of 16-year-old offenders, it would be hazardous to assume that this necessarily is the

Exhibit 4–10

⌖ **1208.1(4). Death sentence.**
See also Homicide ⌖ 354.

Ky. 1984. A review of a sentence of death in a capital case requires the Supreme Court to consider whether the sentence is excessive or disproportionate to penalty imposed in similar cases, as required by statute, as well as the circumstances of the crime and all the evidence surrounding the defendant and his background. KRS 532.075(6) (a-c).

McQueen v. Com., 669 S.W.2d 519, certiorari denied 105 S.Ct. 269, 83 L.Ed.2d 205.

Ky. 1980. Death penalty shall not be imposed unless at least one of statutory aggravating circumstances is found to exist. KRS 532.025.
Smith v. Com., 599 S.W.2d 900.

Ky. 1967. Death sentence is not in violation of state or federal constitutions.
Salisbury v. Com., 417 S.W.2d 244.

Source: Kentucky Digest 2d, "Crim. Law 1208.1," © West Publishing, used with permission.

Exhibit 4–11

⌖ **1208.1(4). Death sentence.**
See also Homicide ⌖ 354.
See ⌖ 1208.1(4.1).

⌖ **1208.1(4.1)—In general.**
U.S.Ky. 1989. In determining whether there was a settled consensus that imposition of capital punishment for crime committed at 16 or 17 years of age was inhumane, ages at which the states permitted their particularized capital punishment system to be applied were more relevant than statutes setting 18 or more as legal age for engaging in activities including driving, drinking alcoholic beverages, and voting. (Per Justice Scalia with Chief Justice Rehnquist, Justices White and Kennedy concurring, and Justice O'Connor concurring in result.) U.S.C.A. Const.Amend. 8.—Stanford v. Kentucky, 109 S.Ct. 2969, 492 U.S. 361, 106 L.Ed.2d 306, rehearing denied 110 S.Ct. 23, 492 U.S. 937, 106 L.Ed.2d 635, rehearing denied Wilkins v. Missouri, 110 S.Ct. 23, 492 U.S. 937, 106 L.Ed.2d 635.
Evidence including public opinion polls, the views of interest groups and positions adopted by professional association, rather than laws and the

applications of laws, could not be used to establish a national consensus concerning the inhumanity of imposing capital punishment for crime committed at 16 or 17 years of age. (Per Justice Scalia with Chief Justice Rehnquist, Justices White and Kennedy concurring, and Justice O'Connor concurring in result.) U.S.C.A. Const.Amend. 8.—Id.
Ky. 1992. Accomplice to capital offense is eligible for the same punishment as principal.—Humphrey v. Com., 836 S.W.2d 865.
Ky. 1990. Death sentence for defendants convicted of robbery and murder was not excessive. KRS 532.075.—Epperson v. Com., 809 S.W.2d 835, certiorari denied Hodge v. Kentucky, 112 S.Ct. 885, 502 U.S. 1037, 116 L.Ed.2d 789, certiorari denied 112 S.Ct. 955, 502 U.S. 1065, L.Ed.2d 122.
Ky. 1985. Death sentence was not imposed under influence of passion, prejudice, or any other arbitrary factor. KRS 532.025(2)(a).—Harper v. Com., 694 S.W.2d 665, certiorari denied 106 S.Ct. 2906, 476 U.S. 1178, 90 L.Ed.2d 992.

⌖ **1208.1(5).—Aggravating or mitigating circumstances.**

Source: Kentucky Digest 2d, "Crim. Law 1208.1 (4.1)" pocket, © West Publishing, used with permission.

law. It would be interesting to explore other state courts' consideration of this issue by consulting **American Law Reports (A.L.R.)** annotations, law reviews, the **Decennial Digest** series, and other authorities and finding tools. However, only a Kentucky Supreme Court decision interpreting the Kentucky Constitution will definitively resolve this issue.

Problem 2

Harry A. Blackmun served as an Associate Justice on the U.S. Supreme Court from 1971 through 1994. During his tenure on the Court, Justice Blackmun participated in the decision of several death penalty cases. Assume that we are interested in tracing Justice Blackmun's views about the death penalty, as expressed in all of his written opinions on that subject while he was on the Supreme Court. How do we go about collecting these opinions?

The most direct approach would be to obtain a list of U.S. Supreme Court decisions involving the death penalty from 1971 through 1994 and then to cull from those cases Justice Blackmun's written decisions. In Chapter 3 we learned about subject indexes to cases, which are called **digests.** For this particular problem we could consult either Lawyers Cooperative's **United States Supreme Court Digest** or West's **Federal Practice Digest** series to find the relevant opinions, but that would be a very time consuming and labor-intensive process.

WESTLAW and **LEXIS** offer a more efficient way of answering this question. Each has a database that includes only U.S. Supreme Court opinions, and each allows the researcher to look for cases by relying on both subject matter and the name of the judge authoring the opinion. Therefore, we should be able to construct a computer search strategy that retrieves the full text of Justice Blackmun's Supreme Court opinions regarding the death penalty.

An alternative approach would be to look for recent law review articles that analyze Justice Blackmun's views on the death penalty and use the references in those articles to find relevant opinions. Either the **Index to Legal Periodicals (ILP)** or **LegalTrac** could help us find references to appropriate articles, but if Blackmun's name does not appear in the title or as a subject descriptor assigned to the document, we might miss important articles. Thus, a search of the full-text law review databases on WESTLAW or LEXIS is more likely to retrieve references to relevant articles.

We also should note that WESTLAW has separate files containing the full text of individual U.S. Supreme Court Justices' opinions and other writings. These files include only selected writings of the Justices; they are not comprehensive. Nevertheless, they can prove useful, especially in accessing the full text of law review articles written by Justices. The file for Justice Blackmun is, obviously enough, **Blackmun**.

Let us begin our research by looking for law review articles on Justice Blackmun in the WESTLAW database for legal periodical articles, **JLR**. We will use the **Natural Language** mode because its search retrieval method is particularly well suited to this problem. **Natural language** ranks the retrieved documents in order of relevancy based on the number of occurrences of terms or phrases appearing in the documents and by giving added weight to terms or phrases that occur less frequently throughout the database. Thus, the **natural language** search method should be particularly effective in retrieving articles that repeatedly refer to Justice Blackmun, as opposed to mentioning him just a few times.

We begin our search by typing in the terms "**Blackmun death penalty capital punishment**". (See Exhibit 4–12.) We have used both "**capital punishment**" and "**death penalty**" because they are synonymous phrases (or related concepts), and we want articles that refer to either phrase as well as to "**Blackmun**." Terms entered in the **natural language** search mode may be treated either as individual

Exhibit 4–12

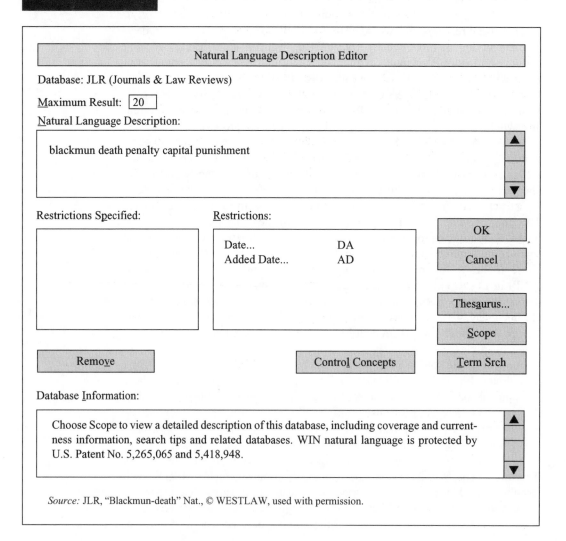

Natural Language Description Editor

Database: JLR (Journals & Law Reviews)

Maximum Result: [20]

Natural Language Description:

blackmun death penalty capital punishment

Restrictions Specified:

Restrictions:

| Date... | DA |
| Added Date... | AD |

OK

Cancel

Thesaurus...

Scope

Remove

Control Concepts

Term Srch

Database Information:

Choose Scope to view a detailed description of this database, including coverage and currentness information, search tips and related databases. WIN natural language is protected by U.S. Patent No. 5,265,065 and 5,418,948.

Source: JLR, "Blackmun-death" Nat., © WESTLAW, used with permission.

terms or as phrases. The computer compares the chosen terms against terms and phrases contained in a thesaurus. If two or more consecutive terms match a phrase included in the thesaurus, WESTLAW will treat these terms as a phrase and search for them accordingly. If the computer defines the search terms as a phrase, it will enclose the phrase in quotation marks as the search is being run. In this example it treats **"death penalty"** and **"capital punishment"** as phrases. (See Exhibit 4–13.)

Recall that **natural language** will retrieve 20 documents ranked in order of relevancy, with the most relevant document ranked **R 1**. The first page of the first document retrieved will be displayed. (See Exhibit 4–14.) Documents retrieved via **natural language** are browsed in the **term** mode. In the **term** mode, only the first screen of each document and the screens where our search terms appear are displayed when we hit the **enter** key. We could browse through the documents to determine how relevant they are to our question. However, in the **JLR** database it may be more efficient to use the **list** command (**l**), which displays bibliographic information about the articles. By employing the **list** command, we quickly come to a screen revealing two articles that include **"Blackmun"** in their titles. (See Exhibit 4–15.) Both of these articles were published in 1995 and therefore are likely to cover Blackmun's complete tenure on the Court. We can view the text of the articles by **entering** the numbers indicating their rankings in the list. The articles should provide us with cites to relevant opinions authored by Justice Blackmun.

We now turn to the more direct way of finding Justice Blackmun's death penalty opinions. Using the Supreme Court database (**SCT**) in WESTLAW, we enter the following **Terms and Connectors** search strategy: **"sy(blackmun +5 (held**

Exhibit 4–13

Your description is:
BLACKMUN DEATH PENALTY CAPITAL PUNISHMENT

Concepts in your description:
BLACKMUN "DEATH PENALTY" "CAPITAL PUNISHMENT"

Your database is **JLR**

Your search is proceeding.

| . . . . |

To cancel your search, type **X** and press **Enter**

Source: JLR, "Blackmun-death" searching, © WESTLAW, used with permission.

Exhibit 4–14

Copr. © West 1996 No claim to orig. U.S. govt. works

AUTHORIZED FOR EDUCATIONAL USE ONLY

Citation	Rank (R)	Page (P)	Database	Mode
24 STMLJ 1	R 1 OF **20**	P 1 OF 257	JLR	**Term**

(Cite as: 24 St. Mary's L.J. 1)

Saint Mary's Law Journal

1992

*1 CAPITAL PUNISHMENT: A CRITIQUE OF THE POLITICAL AND PHILOSOPHICAL
THOUGHT SUPPORTING THE JUSTICES' POSITIONS

Samuel J.M Donnelly [FNa1]

Copyright (c) 1992 by the St. Mary's University of San Antonio; Samuel J.M. Donnelly

I. Introduction

II. Brennan, Marshall, and Rawls: A Rawlsian Theory of Capital Punishment

A. A Rawlsian Theory of Capital Punishment

B. Brennan Compared

C. Marshall, Hart, and Utilitarianism

III. The Plurality: Burger and Pre-McCleskey Rehnquist Courts

A. Democratic Theory

Source: JLR, "Blackmun-death" result, © WESTLAW and St. Mary's Law Journal, vol. 24, p. 1, used with permission.

Exhibit 4–15

Copr. © West 1996 No claim to orig. U.S. govt. works

AUTHORIZED FOR EDUCATIONAL USE ONLY

CITATIONS LIST (Page 2)   Search Result Documents: 20

Database: JLR

5. 45 Am. U. L. Rev. 239 December, 1995 Conference THE DEATH PENALTY IN THE TWENTY-FIRST CENTURY Stephen B. Bright   Edward Chikofsky   Laurie Estrand   Harriet C. Ganson Paul D. Kamenar   Robert E. Morin   William G. Otis   Jamin Raskin   Ira P. Robbins   Douglas G. Robinson   Diann Rust-Tierney

6. 43 U. Kan. L. Rev. 367 January 1995 MARKING THE PROGRESS OF A HUMAN JUSTICE: [FNA] HARRY BLACKMUN'S DEATH PENALTY EPIPHANY Randall Coyne [FNaa]

7. 28 Akron L. Rev. 125 Fall/Winter, 1995 JUSTICE BLACKMUN AND CRIMINAL JUSTICE: A MODEST OVERVIEW [FNa1] Stephen L. Wasby [FNaa1]

8. 16 Pepp. L. Rev. 737 July, 1989 Note THOMPSON v. OKLAHOMA: DEBATING THE CONSTITU-TIONALITY OF JUVENILE EXECUTIONS Susan M. Simmons

Source: JLR, "Blackmun-death" list citations, © WESTLAW, used with permission.

concur! dissent!)) & sy,di(death capital +1 penalty punishment)" (see Exhibit 4–16). This search strategy will retrieve documents in which **"Blackmun"** appears in the **synopsis** field and precedes by five or fewer words either **"held,"** or a word beginning with the root **"concur!"**, or a word beginning with the root **"dissent!"**. In addition, in either the **synopsis** or **digest** field the retrieved documents must contain either **"death"** or **"capital,"** which must precede by one word either **"penalty"** or **"punishment."** This is another way of searching for the phrases **"capital punishment"** or **"death penalty"**. This search strategy requires consid-

Exhibit 4–16

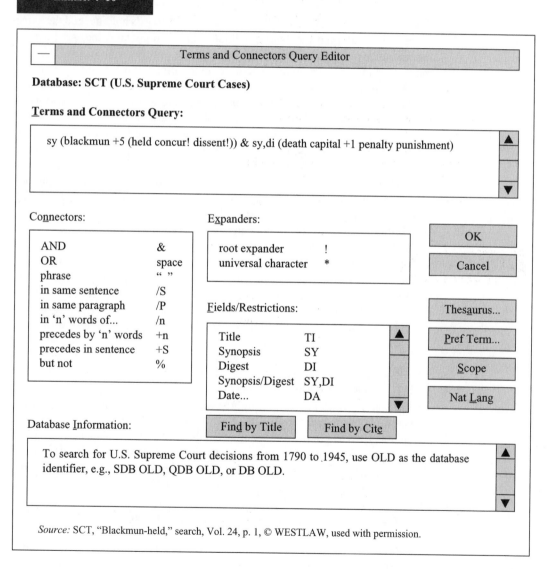

—	Terms and Connectors Query Editor

Database: SCT (U.S. Supreme Court Cases)

Terms and Connectors Query:

> sy (blackmun +5 (held concur! dissent!)) & sy,di (death capital +1 penalty punishment)

Connectors:

AND	&
OR	space
phrase	" "
in same sentence	/S
in same paragraph	/P
in 'n' words of...	/n
precedes by 'n' words	+n
precedes in sentence	+S
but not	%

Expanders:

root expander	!
universal character	*

Fields/Restrictions:

Title	TI
Synopsis	SY
Digest	DI
Synopsis/Digest	SY,DI
Date...	DA

OK

Cancel

Thesaurus...

Pref Term...

Scope

Nat Lang

Database Information:

Find by Title   Find by Cite

> To search for U.S. Supreme Court decisions from 1790 to 1945, use OLD as the database identifier, e.g., SDB OLD, QDB OLD, or DB OLD.

Source: SCT, "Blackmun-held," search, Vol. 24, p. 1, © WESTLAW, used with permission.

erable familiarity with WESTLAW and with the terminology used in court decisions. Researchers should not rely on the **Terms and Connectors** search mode until they have practiced it enough to develop the necessary knowledge base.

We could have tried a **natural language** search using the same strategy we used in the law review database. However, that would not have been a very satisfactory strategy in WESTLAW's U.S. Supreme Court database because "**Blackmun**" is certain to appear in a high percentage of documents, and our other terms, "**capital punishment**" and "**death penalty**," will appear in many cases not germane to our research question. Since WESTLAW and LEXIS allow us to search for decisions by the name of the authoring judge, we would be foolish not to take advantage of this option.

In the WESTLAW system, the **synopsis** field (**sy***)* contains a brief history of the case, the major issue(s) presented, a summary of the court's holding, and the name of the judge authoring the court's opinion. This field contains the same text as appears in the printed Supreme Court Reporter series between the title of the case and the headnotes and opinion. As before, in the first part of our search strategy, "**sy(Blackmun +5 (held concur! dissent!))**", we are looking for Justice Blackmun's name to appear prior to and within five words of either "**held, concur!**" or "**dissent!**". Requiring "**Blackmun**" to appear within five words of and prior to "**held**" will retrieve cases in which Blackmun authored the Court's prevailing opinion. Typically, the wording will be similar to "The Supreme Court, Justice Blackmun, held that. . . . ."

We also want to access Justice Blackmun's dissenting and concurring opinions regarding the death penalty. References to Justices' concurring and dissenting opinions also appear in the **synopsis** field. Here, searching becomes more difficult. Justices frequently author concurring and dissenting opinions, but they often just add their names to join an opinion written by another Justice. We must design a strategy that retrieves the dissenting and concurring opinions filed by Justice Blackmun, but not those in which he merely joined a dissenting or concurring opinion written by another Justice. We can accomplish this if we are familiar with how the editors at West refer to dissenting and concurring opinions in the **synopsis** field and are confident that the phrasing has remained fairly consistent over time.

If you are not certain about this phrasing, you can consult a few **synopsis** fields in the printed Supreme Court Reporter. The wording typically will be similar to the following: "Justice Blackmun filed a dissenting opinion in which Justices Stevens, Souter, and Ginsburg joined." We used the truncation symbol (!) to pick up decisions in which the editors might have used the word "**dissented**" instead of "**dissenting**," or "**concurred**" instead of "**concurring.**" By using the proximity requirement, **+5**, we retrieve cases in which the above phrasing appears but avoid retrieving cases in which Blackmun merely joined another Justice's opinion. If we use the proximity requirement **/5**, which would require merely that "**Blackmun**" appear within five words of "**dissent!**" or "**concur!**" but not necessarily that it precede either of those terms, we also would have retrieved cases with wording similar to the following: "Justice Stevens filed a dissenting opinion in which Justice Blackmun joined."

The second part of our search strategy, "**sy,di(death capital +1 penalty punishment)**", is designed to limit the cases we retrieve to death penalty decisions.

We are limiting our search to the terms appearing in either the **synopsis** or **digest** fields. Recall that the **digest** field includes the headnotes to the case, which summarize the significant points of law discussed in the Court's decision. We do not want to look for our terms in the **opinion** field, which contains the full text of the Justices' opinions, because the Justices may make reference to the death penalty even in cases that do not deal directly with capital punishment. Confining our search to the **digest** and **synopsis** fields will produce much more precise results.

The most recently decided cases are listed first in the **Terms and Connectors** mode. We are in **term** mode, so the first screen displayed will be the first page of the first document, *Tuilaepa v. California,* 114 S. Ct. 2630. (See Exhibit 4–17.) Note the "**R 1 OF 40**" near the top of the screen, which indicates that this is the first of 40 documents retrieved.

The first screen contains the initial part of the **synopsis** field, which includes information indicating that this is a death penalty case. However, we are provided no information identifying the Justices who filed opinions in the case, until we **enter p,** and proceed to the second screen. (See Exhibit 4–18.) Here, we learn that **"Justice Blackmun filed a dissenting opinion."** Note that our search terms are highlighted. As we continue browsing the document, we will come to the place in

---

Exhibit 4–17

Copr. © West 1996 No claim to orig. U.S. govt. works

AUTHORIZED FOR EDUCATIONAL USE ONLY

Citation	Rank (R)	Page (P)	Database	Mode
114 S.Ct. 2630	R 1 OF **40**	P 1 OF 82	SCT	**Term**
129 L.Ed.2d 750, 62 USLW 4720				

(Cite as: 114 S.Ct. 2630)

Paul Palalaua TUILAEPA, Petitioner,

v.

CALIFORNIA,

William Arnold PROCTOR, Petitioner,

v.

CALIFORNIA,

Nos. 93–5131, 93–5161.

Argued March 22, 1994.

Decided June 30, 1994.

Defendant was convicted of first-degree murder, forcible rape and first-degree burglary, in the Superior Court, Shasta County, Joseph H. Redmon, J., and, following change of venue, was sentenced to death, in the Superior Court, Sacramento County, Sheldon H. Grossfeld, J. On automatic appeal, the California Supreme Court, ▶ 4 Cal.4th 499, 15 Cal.Rptr.2d 340, 842 P.2d 1100, affirmed. In separate case, another defendant was convicted of first-degree murder and attempted robbery, in the Supreme Court, Los Angeles County, Elsworth M. Beam, J., and was sentenced to death. On automatic appeal, the California Supreme Court, ▶ 4 Cal.4th 569, 15 Cal.Rptr.2d 382, 842 P.2d 1142, affirmed.

Source: SCT, "Blackmun-held" result, © WESTLAW, used with permission.

Exhibit 4–18

Copr. © West 1996 No claim to orig. U.S. govt. works

AUTHORIZED FOR EDUCATIONAL USE ONLY

114 S. Ct. 2630   R 1 OF **40**   P 2 OF 82   SCT   **Page**

(Cite as: 114 S.Ct. 2630)

Petitions for writs of certiorari were granted in both cases. The United States Supreme Court, Justice Kennedy, held that: (1) California **death penalty** special circumstance, requiring sentencer to consider circumstances of crime, was not unconstitutionally vague; (2) special circumstance, requiring sentencer to consider defendant's prior criminal activity, was not unconstitutionally vague; and (3) special circumstance, requiring sentencer to consider defendant's age at time of crime, was not unconstitutionally vague.
  Affirmed.
  Justice Scalia and Justice Souter filed concurring opinions.
  Justice Stevens filed an opinion concurring in the judgment, in which Justice Ginsburg joined.
  Justice **Blackmun** filed a **dissenting** opinion.

  Source: SCT, "Blackmun-dissenting" displayed, © WESTLAW, used with permission.

the **digest** field where the other part of our search strategy appears. (See Exhibit 4–19.)

If we proceed to the second retrieved case, *Simmons v. South Carolina,* 114 S. Ct. 2187, we see that Justice Blackmun authored the prevailing decision: "**Justice Blackmun held . . . .**" (See Exhibit 4–20.) The information in the **synopsis** indicates that this is a death penalty case. If we continue to browse this document we soon arrive at the place in the **digest** field where the second part of our search strategy is satisfied. We could continue browsing the 40 documents to determine that all meet our search requirements. The thorough researcher will rely not only on this search result, but also will check the retrieved cases against other information, such as the previously garnered cites from relevant law review articles.

Finally, we should check the **Blackmun** database on WESTLAW, which contains a selection of Justice Blackmun's opinions and other writings. We could run the following strategy in this database in order to find relevant law review articles authored by Justice Blackmun: "**so(law) & (death capital +1 penalty punishment).**" (See Exhibit 4–21.) The "**so**" refers to the **source** field. By entering "**law**" in this field we are limiting our results to law review articles. We are using the **Terms and Connectors** mode again because that is the only way to restrict our initial results to law review articles. The rest of the strategy, "**death capital +1 penalty punishment,**" is similar to the one we used in the case law search. However, in the present search we are not using field restrictions; here we are interested in retrieving articles in which Justice Blackmun made reference to the death penalty or capital punishment in either the title or text of the articles. This strategy retrieves two documents, one of which was an article that Justice Blackmun wrote for the Yale Law Journal in 1994. (See Exhibit 4–22.) The article deals with international law and at first glance does not appear to be related to capital punishment.

Exhibit 4–19

Copr. © West 1997 No claim to orig. U.S. govt. works
AUTHORIZED FOR EDUCATIONAL USE ONLY

114 S.Ct. 2630	R 1 OF **40**	P 3 OF 83	SCT	**Page**

(Cite as: 114 S.Ct. 2630)
Tuilaepa v. California
▶ [1]
▶ 110 CRIMINAL LAW
▶ 110XXVI Punishment of Crime
▶ 110k1208 Extent of Punishment in General
▶ 110k1208.1 In General
▶ 110k1208.1(4) Death Sentence

▶ 110k1208.1(4.1) k. In general.
U.S. Cal., 1994.
To be eligible for **death penalty**, defendant must be convicted of crime for which **death penalty** is proportionate punishment.

Source: SCT, "Blackmun-dissenting" browse-headnote, © WESTLAW, used with permission.

Exhibit 4–20

Copr. © West 1996 No claim to orig. U.S. govt. works
AUTHORIZED FOR EDUCATIONAL USE ONLY

Citation	Rank (R)	Page (P)	Database	Mode
114 S.Ct. 2187	R 2 OF **40**	P 1 OF 77	SCT	**Term**
129 L.Ed.2d 133, 62 USLW 4509				

(Cite as: 114 S.Ct. 2187)

Jonathan Dale SIMMONS, Petitioner,

v.

SOUTH CAROLINA.

No. 92–9059.

Argued Jan. 18, 1994.

Decided June 17, 1994.

Defendant was convicted by the South Carolina Circuit Court, Richland County, Ralph King Anderson, Jr., J., of murder, first-degree burglary and petty larceny, and jury sentenced defendant to death. Defendant appealed. The South Carolina Supreme Court, Moore, J., ▶ 427 S.E.2d 175, affirmed. Defendant's petition for certiorari was granted. The Supreme Court, Justice **Blackmun, held** that: (1) state, which had raised specter of defendant's future dangerousness, violated defendant's due process rights by refusing to instruct jury that, as alternative to death sentence, sentence of life imprisonment carried with it no possibility of parole, and (2) trial court's jury instruction that life imprisonment was to be given its ordinary meaning and that jury was not to consider parole did not satisfy in substance defendant's request for jury charge on parole ineligibility.

Source: SCT, "Blackmun-held," result-browse, © WESTLAW, used with permission.

Exhibit 4–21

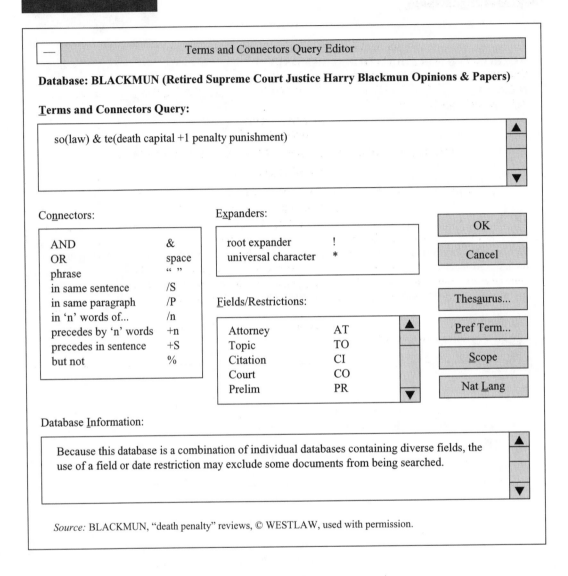

Terms and Connectors Query Editor

Database: BLACKMUN (Retired Supreme Court Justice Harry Blackmun Opinions & Papers)

Terms and Connectors Query:

so(law) & te(death capital +1 penalty punishment)

Connectors:

AND	&
OR	space
phrase	" "
in same sentence	/S
in same paragraph	/P
in 'n' words of...	/n
precedes by 'n' words	+n
precedes in sentence	+S
but not	%

Expanders:

root expander	!
universal character	*

Fields/Restrictions:

Attorney	AT
Topic	TO
Citation	CI
Court	CO
Prelim	PR

OK

Cancel

Thesaurus...

Pref Term...

Scope

Nat Lang

Database Information:

Because this database is a combination of individual databases containing diverse fields, the use of a field or date restriction may exclude some documents from being searched.

Source: BLACKMUN, "death penalty" reviews, © WESTLAW, used with permission.

However, if we browse the document in **term** mode we will run across a section that discusses the death penalty in the context of international law, which provides additional insight into the Justice's thinking on the topic. (See Exhibit 4–23.) Thus, our search strategies have enabled us to gather a significant list of cases and law review articles that are helpful in tracing Justice Blackmun's written opinions dealing with the death penalty.

Exhibit 4–22

3   Copr. © West 1996 No claim to orig. U.S. govt. works
AUTHORIZED FOR EDUCATIONAL USE ONLY

Citation	Rank(R)	Page(P)	Database	Mode
104 YLJ 39	R 1 OF 2	P 1 OF 29	BLACKMUN	**Page**

(Cite as: 104 Yale L.J. 39)

Yale **Law** Journal
October, 1994

*39 THE SUPREME COURT AND THE LAW OF NATIONS [FNa1]

Harry A. Blackmun [FNd1]

Copyright (c) 1994 by The Yale **Law** Journal Company, Inc.; Harry A. Blackmun

I. FIRST PRINCIPLES

The Declaration of Independence opens with the following memorable passage: When in the Course of human Events, it becomes necessary for one People to dissolve the Political Bands which have connected them with another, and to assume among the Powers of the Earth, the separate and equal Station to which the Laws of Nature and of Nature's God entitle them, a decent Respect to the Opinions of Mankind requires that they should declare the causes which impel them to the Separation. [FN1]
As Professor Louis Henkin has noted, the early architects of our Nation . . .

Source: Reprinted by permission of The Yale Law Journal Company and Fred B. Rothman & Company from *The Yale Law Journal*, Vol. *104*, pages *39–49*.

Exhibit 4–23

Copr. © West 1996 No claim to orig. U.S. govt. works
AUTHORIZED FOR EDUCATIONAL USE ONLY
104 YLJ 39   R 1 OF **2**   P 13 OF 29   BLACKMUN   **Term**
(Cite as: 104 Yale L.J. 39, *45)
Amendment's Cruel and Unusual Punishments Clause "must draw its meaning from evolving standards of decency that mark the progress of a maturing society." [FN43] The drafters of the Amendment were concerned, at root, ***46** with "the dignity of man," [FN44] and understood that "evolving standards of decency" should be measured, in part, against international norms. Thus, in cases striking down the **death penalty** as a punishment for rape [FN45] or for unintentional killings, [FN46] the Court has looked to both domestic custom and the "climate of international opinion" to determine what punishments are cruel and unusual. [FN47]

Taking international law seriously where the **death penalty** is concerned, of course, draws into question the United States' entire **capital punishment** enterprise. According to Amnesty International, more than fifty countries (including almost all of Western Europe) have abolished the **death penalty** entirely, and thirty-seven others either have ceased imposing it or have limited its imposition to extraordinary crimes. [FN48] Even those countries that continue to impose the **death penalty** almost universally condemn the execution of juvenile offenders. [FN49] They do so in recognition of the fact that juveniles are too young, and too capable of growth and development, to act with the culpability necessary to justify society's ultimate punishment. The United States, however, persistently has defended its "right" to sentence . . .

Source: Reprinted by permission of The Yale Law Journal Company and Fred B. Rothman & Company from *The Yale Law Journal*, Vol. *104*, pages *39–49*.

Problem 3

Judy Reid was tried, convicted, and sentenced to a term of imprisonment for automobile theft in a North Carolina state court. She maintained that she had not stolen the car but rather that she had received the owner's permission to drive it. The owner denied she had given Judy permission to drive the car and so testified at the trial. Judy's court-appointed counsel was perceptibly intoxicated during the trial proceedings. Assuming that these facts can be established, has Judy made the case that she has been provided with constitutionally ineffective assistance of counsel?

Let us begin by consulting secondary authorities in order to identify the general legal standards under which claims of ineffective assistance of counsel are assessed and to get an idea about how these standards are applied to different factual situations. **American Law Reports (A.L.R.)** annotations and law review articles may prove to be especially useful for these purposes. They also can be expected to refer us to relevant case law.

A.L.R. annotations, which collect judicial decisions and describe their holdings, supply an excellent overview of legal issues. We search the A.L.R. Index to A.L.R. 2d, 3d, 4th, 5th, Federal, and L. Ed. 2d for annotations that may speak to the problem of intoxicated criminal defense lawyers. Among the different terms that help describe this issue, we have the most luck by looking under "**Attorney**" ("Attorney or Assistance of Attorney") in the index. We find two likely annotations. The first, under the subheading "**Constitutional right to counsel** . . .— deficient representation, when is attorney's representation of criminal defendant so deficient as to constitute denial of federal constitutional right to effective assistance of counsel," is found at 83 L. Ed. 2d 1112. The second, under the subheading "**Due process** . . . —assistance of accused, generally," appears at 2 A.L.R.4th 27. (See Exhibit 4–24.)

The first annotation, at 83 L. Ed. 2d, discusses Supreme Court case law regarding constitutionally ineffective assistance of counsel and, while it provides a wealth of information about the leading Supreme Court decisions, the federal standards by which ineffective assistance of counsel claims are judged, and related annotations and encyclopedia articles, it does not directly address the issue of inebriated lawyers. The annotation provided in 2 A.L.R.4th 27 does discuss this particular matter. Either from the topical outline ("§ 13. Factors peculiar to counsel : . . . [d] Health; use of alcohol or drugs"), or the index preceding the annotation ("Alcohol use by counsel, generally, § 13[d]"), we are referred to the appropriate section, Section 13[d]. (See Exhibit 4–25.)

Exhibit 4–24

ALR INDEX

ATTORNEY OR ASSISTANCE OF ATTORNEY—Cont'd

Conflict of Laws (this index)

Conservatorship, negligence, inattention, or professional incompetence in handling client's affairs in estate or probate matters as ground for disciplinary action—modern cases, 66 ALR4th 342

Constitutional right to counsel

generally, 2 L Ed 2d 1644; 9 L Ed 2d 1260; 18 L Ed 2d 1420

–appeal, Supreme Court's views as to accused's federal constitutional right to counsel on appeal, 102 L Ed 2d 1049

–deficient representation, when is attorney's representation of criminal defendant so deficient as to constitute denial of federal constitutional right to effective assistance of counsel, 83 L Ed 2d 1112

–due process, see group Due process in this topic

Drafting of Instruments (this index)

Due process

–advising accused of right to counsel, lack of, 3 ALR2d 1003

–assistance of accused, generally, 2 ALR4th 27

–judge, disqualification for relationship to attorney, 50 ALR2d 151

–students, propriety and effect of law students acting as counsel in court suit, 3 ALR4th 358

–unlicensed person, representation by, 68 ALR2d 1141

Dues, use of compulsory bar association dues or fees for activities from which particular members dissent, 40 ALR4th 672

Source: ALR INDEX, "Constitutional Right to Counsel. Permission has been granted by the current copyright holder, West Group. Further reproduction of any kind is strictly prohibited. For additional information, please contact West Group Customer Services representative at 1-800-328-4880.

Exhibit 4–25

ANNOTATION
MODERN STATUS OF RULES AND STANDARDS IN STATE COURTS AS TO ADEQUACY
OF DEFENSE COUNSEL'S REPRESENTATION OF CRIMINAL CLIENT

by

Gregory G. Sarno, J.D.

IV. ILLUSTRATIVE FACTORS

§ 11. Evidence of guilt or innocence:
[a] Effect of substantial incriminating evidence—as undermining ineffectiveness argument
[b] —As not negating ineffectiveness argument
[c] —As supporting ineffectiveness argument
[d] Effect of weak incriminating evidence
[e] Effect of purportedly exculpatory evidence

§ 12. Success or failure:
[a] In general
[b] Effect of conviction on fewer than all charges, or of lesser-induced offense
[c] Effect of imprisonment on shorter than maximum sentence
[d] Effect of other factors

§ 13. Factors peculiar to counsel:
[a] Status as appointed or retained
[b] Extent of legal experience
[c] Membership in local bar; possession of attorney's license, generally
[d] Health; use of alcohol or drugs
[e] Contempt citation
[f] Prior ineffectiveness determination; subsequent disbarment; criminal conviction
[g] Assistance by associate counsel
[h] Ability to foretell legal developments
[i] Other factors

§ 14. Factors peculiar to defendant

§ 15. Factors peculiar to attorney-client relationship:
[a] Compatibility
[b] Amount of consultation time
[c] Attorneys' fees; royalty contracts
[d] Other factors

§ 16. Other matters

INDEX

ABA standards, § 4
Abduction, §§ 7[a], 8[a, b, g], 9[a], 11[a], 12,
    13[b, i], 14
Ability of counsel to foretell legal developments,
    § 13[h]
Abortion, charge of performing, § 7[a]

Abrasive conduct of counsel, § 13[e]
Absence of witnesses, § 6[a]
Absentia, defendant being tried in, § 10
Accidental killing, §§ 10, 16
Accident to counsel, effect of, § 13[d]
Accomplice, §§ 6[a], 9[b], 11[d], 12[a], 13[b]

continues

Exhibit 4–25	continued

Acquaintance of victim on jury, § 16
Acquittal, failure to move for, § 8[a]
Addiction to drugs, matters pertaining to, §§ 5[a],
  9[b], 11[a], 12[a]
Address of witnesses, failure to supply, § 8[f]
Admonition of accused, judge's failure to comply

  with rule regarding, § 9[a]
Affidavit, failure to produce, § 5[a]
Affray, §§ 7[b], 10
Aggravated assault, § 3[a]
Alcohol use by counsel, generally, §13[d]

Source: ALR annotation, 2 ALR4th 27, references & content. Permission has been granted by the current copyright holder, West Group. Further reproduction of any kind is strictly prohibited. For additional information, please contact West Group Customer Services representative at 1-800-328-4880.

After checking the main volume, we refer to the pocket supplement of the annotation under Section 13[d]. There we find descriptions of several cases involving claims of ineffective assistance of counsel and attorneys' use of alcohol or other drugs. (See Exhibit 4–26.) One of these cases, *State v. Moorman,* 320 N.C. 387, 358 S.E.2d 502 (1987), is a North Carolina Supreme Court decision. It appears that the court considered a defense lawyer's use of pain-killing drugs and his drowsiness and inattentiveness during trial to help support its conclusion that a criminal defendant was denied the effective assistance of counsel.

This annotation offers additional guidance. It discusses the test for ineffective assistance of counsel announced by the Supreme Court in *Strickland v. Washington,* 466 U.S. 668, 104 S. Ct. 2052, 80 L. Ed. 2d 674 (1984), and it provides citations to North Carolina cases adhering to the *Strickland* standard. (See Exhibit 4–27.)

Exhibit 4–26

[d] Health; use of alcohol or drugs

Ineffective assistance was not established by allegation that lead trial counsel was taking "great quantity" of narcotics pills and was behaving in "generally slovenly manner," exhibited "some slurred speech" and "lack of coordination," and had saliva "running out the sides of his mouth." Allen v State (1979, Ala Crim) 380 So 2d 313, cert den (1980 Ala) 380 So 2d 341.

Though counsel was alcoholic at time of trial and later died of disease, there was no per se rule of deficiency with respect to alcoholic attorneys, and, in view of evidence that counsel had provided competent representation, defendant's claim of ineffective assistance of counsel failed. People v Garrison (1989) 47 Cal 3d 746, 254 Cal Rptr 257, 765 P2d 419, reh den.

Though counsel was subject to disbarment proceedings, mental illness and narcotic use during his representation of defendant, prejudice was not presumed where circumstances of case, which did not involve possibility of death penalty or complicated or voluminous evidentiary matters, did not mandate inference that counsel was unable to discharge his duties; and prejudice to defendant was not affirmatively shown on the record where none of counsel's alleged failures pertained to matters that would have raised reasonable doubt of defendant's guilt. People v Bernardo (1988, 1st Dist) 171 Ill App 3d 652, 121 Ill Dec 550, 525 NE2d 857.

Ill health of 70-year-old defense counsel who suffered heart attack before original trial date (causing its postponement), who was taking five medications during trial, and who was required to lie down in the judge's chamber to rest during trial, was considered to have undermined the quality of his representation, which was found to be incompetent. White v State (1981, Ind App) 414 NE2d 973.

Facts that appointed trial counsel was released from hospital on January 19 and was compelled to use wheelchair and crutches at times during trial which began on February 27, and that numerous recesses were required, did not deprive defendant of effective assistance on ground of counsel's impaired physical condition. State v Rice (1980, Kan) 607 P2d 489.

See People v Storch (1989) 176 Mich App 414, 440 NW2d 14, § 16.

There was no merit to an ineffective assistance contention posed by a defendant convicted of the bombing murder of his wife, and predicated partly on his trial attorney's allegedly being distraught over the death of his own wife just prior to trial. Parsons v State (1980, Mo App) 607 SW2d 844.

Absent any specific error for conduct identified in trial that affected trial's outcome, counsel's cocaine abuse was irrelevant to issue of ineffective assistance. State v Coates (1990, Mont) 786 P2d 1182.

Defendant in rape prosecution was afforded ineffective assistance of counsel where, inter alia, case was close on the facts, where counsel promised in opening statement to produce critical piece of evidence which he never presented and apparently had no reason to think he could present, where he stated in closing that his client's testimony was not worthy of belief, and where he regularly used combination of pain-killing drugs, had frequent migraine headaches, and was drowsy and inattentive during portions of trial. State v Moorman (1987) 320 NC 387, 358 SE2d 502.

See Carter v State (1989, Okla Crim) 773 P2d 373, § 13[f].

Neither since-deceased trial counsel's comment that he had consumed intoxicants every morning for the last 40 years, nor trial judge's . . .

Source: ALR, 2 ALR4th 27–244, "Moorman." Permission has been granted by the current copyright holder, West Group. Further reproduction of any kind is strictly prohibited. For additional information, please contact West Group Customer Services representative at 1-800-328-4880.

Exhibit 4–27

Supplement   2 ALR4th 27–244

. . . not result of informed decision but, rather, of counsel's inadequate preparation. Briones v State (1993, Hawaii) 848 P2d 966.

§ 6.5[New] Strickland test

Editor's note—In Strickland, the United States Supreme court announced a new test for determining the validity of a claim of ineffective assistance of counsel, requiring defendant to show (1) that counsel's performance was deficient, (requiring showing that counsel was not functioning as counsel guaranteed defendant by Sixth Amendment); and (2) that deficient performance prejudiced defense, (requiring showing that counsel's errors were so serious as to deprive defendant of fair trial). Strickland v Washington (1984) 466 US 668, 80 L Ed 2d 674, 104 S Ct 2052, on remand (CA11 Fla) 737 F2d 894, habeas corpus proceeding (SD Fla) 587 F Supp 525, affd (CA11 Fla) 737 F2d 922 and later proceeding (Fla) 453 So 2d 389.

Also adopting Strickland test:

US—Hill v Lockhart (1985, US) 88 L Ed 2d 203, 106 S Ct 366; Guam v Santos (1994, DC Guam) 856 F Supp 572; Lowry v Lewis (1994, CA9 Ariz) 21 F3d 344, 94 CDOS 2435, 94 Daily Journal DAR 4649, petition for certiorari filed . . .

. . . NJ Super 15, 643 A2d 18, certif den (NJ) 645 A2d 141; State v Johnson (1994, App Div) 274 NJ Super 137, 643 A2d 631; State v Moore (1994, NJ Super Ct App Div) 273 NJ Super 118, 641 A2d 268, certif den (NJ) 645 A2d 139 (by implication).

NM—State v Richardson (1992, App) 114 NM 725, 845 P2d 819, cert den 114 NM 550, 844 P2d 130 (adding that standard for judging claim is not strictly objective).

NY—People v Ploss (1984, 3d Dept) 105 App Div 2d 1031, 483 NYS2d 449; People v Kroemer (1994, App Div, 4th Dept) 613 NYS2d 304, app den 84 NY2d 828; People v Vigilante (1992, Sup) 581 NYS2d 261.

NC—State v Harbison (1985) 315 NC 175, 337 SE2d 504, cert den (US) 90 L Ed 2d 672, 106 S Ct 1992; State v Seagroves (1985) 78 NC App 49, 336 SE2d 684, review den 316 NC 384, 342 SE2d 905.

ND—State v Ronngren (1985, ND) 361 NW2d 224; Woehlhoff v State (1992, ND) 487 NW2d 16; Hoffarth v State (1994, ND) 515 NW2d 146; State v Lefthand (1994, ND) 523 NW2d 63.

Source: ALR, 2 ALR4th 27–244 supp. "Strickland." Permission has been granted by the current copyright holder, West Group. Further reproduction of any kind is strictly prohibited. For additional information, please contact West Group Customer Services representative at 1-800-328-4880.

Law review articles may provide us with a more critical analysis of ineffective-assistance-of-counsel issues. We consult a volume (September 1994–August 1995) of the ILP. If we are uncertain about what topic to consult in the ILP, we can glance at the List of Subject Headings at the front of the volume. On checking for "**Effective assistance of counsel**," we are referred to "**Right to counsel**." (See Exhibit 4–28.) Citations refer us to several articles under that heading, including one that looks especially promising: "A decade of Strickland's . . . tin horn: doctrinal and practical undermining of the right to counsel," by W.S. Geimer. This

Exhibit 4–28

LIST OF SUBJECT HEADINGS

Duty of confidentiality *See* Confidentiality
Duty to rescue *See* Rescue doctrine
DWI *See* Driving while intoxicated
Easements
    See also Adjoining landowners; Conservation
    easements; Dedication to public use; Equi-
    table servitudes; Prescription; Rights of
    way; Usufruct
Eavesdropping *See* Electronic surveillance
Ecclesiastical law
    See also Religious organizations; Subsidiarity
Ecology
Economic conditions
Economic development
    See also Sustainable development
Economic jurisprudence
    See also Coase theorem; Public choice theory
Economic loss
Economic policy
    See also Indexation
Economic sanctions
    See also Blockade; Boycotts; Embargo
Economics
    See also Capitalism; Cost benefit analysis;
    Economic conditions; Economic develop-
ment; Economic policy; Finance; Financial
services; Indians/Economic development;
Individualism; Inflation; International eco-
nomic relations; Land use; Money; Political
science; Socialism; Valuation
EDI *See* Electronic data interchange
Education
    See also Colleges and universities; Home
    schooling; Legal education; Right to educa-
    tion; Schools and school districts
Education/Finance
    See also Colleges and universities/Finance;
    Schools and school districts/Finance
Educational malpractice
Effective assistance of counsel *See* Right to
counsel
Eighth amendment protections
    See also Cruel and unusual punishment; Exces-
    sive fines clause
Elderly *See* Aged
Election finance *See* Campaign funds
Elections
    See also Campaign funds; Domicile and resi-
    dence; . . .

Source: Index to Legal Periodicals, 1994/95 List of Subject Headings, © H.W. Wilson Company, used with permission.

---

article was published in 4 William & Mary Bill of Rights Journal 91 (1995). (See Exhibit 4–29.) We also consult the table of cases at the back of the ILP, where we are referred to another law review article (see Exhibit 4–30) that discusses *Strickland v. Washington*. (Somewhat curiously, Professor Geimer's article is not included; see Exhibit 4–30.)

Now we are in a better position to move to case law. Looking up *Strickland v. Washington* in the Supreme Court Reporter helps us use West's key number system, and referring to Lawyers' Edition 2d allows us to use the Total Client-Service Library References feature and to get cites to Am. Jur. 2d articles, A.L.R. annota-

Exhibit 4–29

SUBJECT AND AUTHOR INDEX

Right to confront witnesses *See* Confrontation clause

Right to counsel

See also

Custodial interrogation

Pro bono representation

Pro se litigants

Constitutional law—People v. Griggs [604 N.E.2d 257 (Ill. 1992)]: Illinois ignores Moran v. Burbine [106 S. Ct. 1135 (1986)] to expand a suspect's Miranda [Miranda v. Arizona, 86 S. Ct. 1602 (1966)] rights. C. I. Roumeliotis, student author. 16 *W. New Eng. L. Rev.* 329-64 '94

Constitutional law—sixth amendment—right to counsel—ambiguous requests—the United States Supreme Court held that after a knowing and voluntary waiver of the Miranda [Miranda v. Arizona, 86 S. Ct. 1602 (1966)] rights, officers may continue questioning until the suspect makes an unambiguous request for counsel. J. T. Reed, student author. 33 *Duq. L. Rev.* 1109-28 Summ '95

Criminal trials—subpoenas—Connecticut adopts a "compelling need" test to limit the ability of prosecutors to subpoena defense attorneys.—

Ullmann v. State, 647 A.2d 324 (Conn. 1994). 108 *Harv. L. Rev.* 775-80 Ja '95

Davis v. United States [114 S. Ct. 2350 (1994)]: clarification regarding ambiguous counsel requests, and an invitation to revisit Miranda [Miranda v. Arizona, 86 S. Ct. 1602 (1966)] R. Kohlmann. 1995 *Army Law.* 26-32 Mr '95

Davis v. United States [114 S. Ct. 2350 (1994)]: the Supreme Court rejects a third layer of prophylaxis. N. M. Kennelly, student author. 26 *Loy. U. Chi. L.J.* 589-630 Spr '95

The death of fairness? Counsel competency and due process in death penalty cases [panel discussion] 31 *Hous. L. Rev.* 1105-204 Wint '94

A decade of Strickland's [Strickland v. Washington, 104 S. Ct. 2052 (1984)] tin horn: doctrinal and practical undermining of the right to counsel. W. S. Geimer. 4 *Wm. & Mary Bill Rts. J.* 91-178 Summ '95 ◄━━━

The eighth amendment and ineffective assistance of counsel in capital trials. 107 *Harv. L. Rev.* 1923-40 Je '94

Getting what you paid for: the judicial cap on attorneys' fees for the representation of the condemned in the Supreme Court. B. R. Braun. 69 *N.Y.U. L. Rev.* 1014-45 O/N '94

Source: Index to Legal Periodicals, 1994/95 p. 753 "Right to Counsel," © H.W. Wilson Company, used with permission.

---

tions, and other references. We also want to Shepardize *Strickland*, paying particular attention to North Carolina cases that have cited it.

The A.L.R. annotation refers us to the North Carolina decision, *State v. Moorman*, which appears to be somewhat relevant to our fact situation. When we read this decision in the **South Eastern Reporter 2d**, we learn that the attorney's alleged impairment from pain-killing medications plays a very minor part in the court's conclusion that the defendant was denied effective assistance of counsel. Nevertheless, the attorney's condition is discussed in *Moorman,* so we Shepardize that decision. We start with the first—that is, the oldest **Shepard's**

Exhibit 4–30

Story v. Advance Bank Austl. Ltd., [1993] 10
  A.C.S.R. 699
  4 *Austl. J. Corp. L.* 264-74 Je '94
Strauss; Qutb v., 11 F.3d 488 (1993)
  69 *Tul. L. Rev.* 308-18 N '94
Street, Estate of 974 F.2d 723 (1992)
  26 *Tax Adviser* 228-35 Ap '95
  72 *Taxes* 412-19 Jl '94
Stretton v. Disciplinary Bd., 944 F.2d 137 (1991)
  51 *Wash. & Lee L. Rev.* 1085-123 Summ '94
Strickland v. Washington, 104 S. Ct. 2052
  (1984)
  97 *W. Va. L. Rev.* 1-51 Fall '94
Stringer v. Ashley, No. 321906/88U Jan. 27,
  1994, Ont. Ct. (Gen. Div.)
  16 *Advoc. Q.* 506-10 N '94

Sub Sea Int'l, Inc.; Brickham v., 617 So. 2d 483
  (La. 1993)
  55 *La. L. Rev.* 447-68 N '94
Subafilms, Ltd. v. MGM-Pathe Communications
  Co., 24 F.3d 1088 (1994)
  20 *N.C. J. Int'l L. & Com. Reg.* 435-56 Wint
  '95
Subranni (In re Atlantic Business & Commu-
  nity Dev. Corp.); IRS v., 994 F.2d 1069
  (1993)
  67 *Temp. L. Rev.* 435-49 Spr '94
Sullivan v. Boston Gas Co., 605 N.E.2d 805
  (Mass. 1993)
  27 *Suffolk U. L. Rev.* 997-1008 Fall '93
Sullivan; Rust v. 111 S. Ct. 1759 (1991)
  74 *B.U. L. Rev.* 201-66 Mr '94

Source: Index to Legal Periodicals, 1994/95, p. 994 "Table of Cases," © H.W. Wilson Company, used with permission.

South Eastern Reporter Citations that includes 358 S.E.2d 502, the cite to *Moorman.* Using Part 6 of this Shepard's series, which covers 278 S.E.2d through 443 S.E.2d, we find several cases that cite *Moorman.* (See Exhibit 4–31.) Headnotes 5 and 6 from *Moorman* correspond to the ineffective-assistance-of-counsel ruling. As Shepard's informs us that this decision followed *Moorman* (denoted by the "f" preceding "388 SE2d") and the reference is to the issue discussed in headnote 5, we might be particularly interested in looking up 388 S.E.2d 220. Naturally, we continue Shepardizing *Moorman* in all subsequent issues of Shepard's.

Another way to approach this issue is with the assistance of annotated constitutions. Our research to this point has led us to the realization that the Sixth Amendment is implicated by claims of ineffective assistance of counsel. We look up the Sixth Amendment in the Constitution volumes of the **United States Code Annotated (U.S.C.A.).** (See Exhibit 4–32.) The general outline for the Notes of Decisions directs us to section XXX, **"Effective Assistance 2101–2360."** (See Exhibit 4–33.) Under the more detailed outline accompanying section XXX, we learn that **"Alcohol abuse"** is covered in note 2157. (See Exhibit 4–34.) Following are the more recent cases that appear in connection with note 2157, which are found in the

Exhibit 4–31

SOUTHEASTERN REPORTER, 2d SERIES					Vol. 358
—489—	416SE[2]176	—498—	f 388SE[5]220	s 349SE609	
DiDonato v	416SE[3]176		392SE[6]75	362SE[1]815	—515—
Wortman	j 418SE662	North Carolina	414SE358	372SE[1]313	
1987	435SE822	v Bright	420SE[2]154	376SE[1]492	Case 2
(320NC423)	442SE499	1987	Cir. 6	385SE[1]331	
s 341SE58	Cir. 4	(320NC491)	f 970F2d[2]1545	e 385SE[2]491	Abernathy v
s 361SE73	852F2d[2]776	360SE[7]690	Tenn	j 385SE493	Delaware
f 359SE[1]506	Cir. 6	395SE[5]124	823SW225	398SE457	Consolidated
359SE[2]507	737FS[1]429	d 431SE3	Wash	402SE[1]880	Freightways
f 361SE[3]916	Idaho	89A[4]468n	776P2d172	440SE[1]588	Corp.
f 365SE[1]912	820P2d1216		Wyo		1987
365SE[2]919	Pa	—502—	751P2d1299	—515—	
376SE[1]2	571A2d435		2A[4]27s		—515—
380SE416	634A2d607	North Carolina	6A[4]16s	Case 1	
d 380SE[2]418	84A[3]411s	v Moorman			Case 3
405SE[1]923		1987	—512—	Abernathy v	Armstrong v
405SE[2]923		(320NC387)		Delaware	Armstrong
405SE[3]923		s 347SE857	Faircloth	Consolidated	1987
c 405SE[2]924		368SE[2]445	v Beard	Freightways	(320NC511)
f 416SE175		368SE[3]445	1987	Corp.	s 354SE350
416SE176		374SE412	(320NC505)	1987	367SE410

Source: SHEPARD'S Southeastern Reporter, 2d, 358 SE2d 502. Reproduced by permission of Shepard's. Further reproduction of any kind is strictly prohibited.

Exhibit 4–32

AMENDMENT VI—JURY TRIAL FOR CRIMES, AND PROCEDURAL RIGHTS

In all criminal prosecutions, the accused shall enjoy the right to a speedy and public trial, by an impartial jury of the State and district wherein the crime shall have been committed, which district shall have been previously ascertained by law, and to be informed of the nature and cause of the accusation; to be confronted with the witnesses against him; to have compulsory process for obtaining witnesses in his favor, and to have the Assistance of Counsel for his defense.

Source: U.S.C.A. Constitution Amendment VI, text, © West Publishing, used with permission.

Exhibit 4–33

Amend. 6

XII. IMPARTIALITY OF JURY GENERALLY 621–690
XIII. COMPOSITION OR SELECTION OF JURY 691–800
XIV. PUBLICITY SURROUNDING TRIAL 801–830
XV. WAIVER 831–880
XVI. STATE AND DISTRICT OF TRIAL 881–940
XVII. INFORMATION RESPECTING ACCUSATION 941–1030
XVIII. CONFRONTATION WITH WITNESSES—GENERALLY 1031–1140
XIX. PRESENCE OF DEFENDANT 1141–1190
XX. DISCOVERY, DISCLOSURE, OR INSPECTION 1191–1230
XXI. ADMISSIBILITY OF PROSECUTION EVIDENCE; HEARSAY 1231–1420
XXII. CROSS-EXAMINATION BY DEFENSE 1421–1560
XXIII. WAIVER OF RIGHT 1561–1620
XXIV. COMPULSORY PROCESS 1621–1730
XXV. ASSISTANCE OF COUNSEL—GENERALLY 1731–1790
XXVI. PROCEEDINGS TO WHICH AMENDMENT APPLIES 1791–1860
XXVII. STAGES AT WHICH REQUIRED 1861–2020
XXVIII. CHOICE OF COUNSEL 2021–2060
XXIX. APPOINTMENT OF COUNSEL 2061–2100
XXX. EFFECTIVE ASSISTANCE 2101–2360
XXXI. CONFLICT OF INTEREST 2361–2460
XXXII. WAIVER 2461–2540
XXXIII. SELF-REPRESENTATION 2541–2581

For Detailed Alphabetical Note Index, see the Various Subdivisions.

Source: U.S.C.A. Constitution Amendment VI, outline, © West Publishing, used with permission.

pamphlet that supplements the bound volume of the Sixth Amendment in U.S.C.A. (See Exhibit 4–35.)

We use an analogous strategy to examine the North Carolina Constitution's right-to-counsel provision and interpretive case law, to determine if the state constitution has been construed to demand a higher level of attorney performance than is required under the federal test announced in *Strickland v. Washington.* We find no authority to that effect. Although we are referred to several additional North Carolina cases in which ineffective-assistance-of-counsel claims have been considered, none appears to involve intoxicated defense lawyers or related facts.

Finally, we research this problem by looking up logical terms in the Descriptive Word Index to the North Carolina Digest 2d. "**Attorney and Client**" has a subheading, "**Criminal prosecution, adequacy of representation**," which refers us to "**Crim. Law 641.3**" (see Exhibit 4–36), and above that citation, "**Counsel for accused**" refers us to "**Criminal Law**" elsewhere in the Index. When we examine

Exhibit 4–34

XXX. ASSISTANCE OF COUNSEL—
EFFECTIVE ASSISTANCE

Subdivisions Index

Generally 2101
Abandonment of
    Appeal 2227
    Representation 2137
Abdication of duty to client, standards or tests of
    effectiveness 2110
Acquittal, motion for 2222
Admissibility of evidence 2264
Advantage of prosecution, interference with attor-
    ney-client relationship 2186
Advice, pleas
    Charges 2287

Conditional pleas 2288
Defenses 2289
Parole eligibility 2290
Potential punishment 2292
Probation eligibility 2291
Advice respecting right to appeal 2225
Advisory role of counsel 2140
Advocate role of counsel 2141
Alcohol abuse 2157 ◀
Alibi defense 2248
Alibi witnesses, calling or presenting of 2330
Alternative sentence 2303
Appeal or review
    Generally 2223
    Abandonment of appeal 2227
    Advice respecting right to appeal 2225

Source: U.S.C.A. Constitution Amendment VI, detailed outline, © West Publishing, used with permission.

"Criminal Law" in the Descriptive Word Index, we find the subheading, "**Assistance of Counsel**. Crim. Law 641.12(3)," and another subheading, "**Indigents. . . .—Adequacy of representation. Crim. Law 641.13(3)**." (See Exhibit 4–37.)

To pin down the optimal key number, we resort to the outline of the criminal law topic in the North Carolina Digest 2d. Key number 641.13 pertains to adequacy of representation, with several subdivisions that may be of interest, including 641.13(1), (2), and (3). (See Exhibit 4–38.) Now we have only to consult the Digest under these key numbers to review case squibs that may prove to be of interest. We naturally check both the bound volume and the pocket supplement. (See Exhibit 4–39.)

Exhibit 4–35

2157. Drug or alcohol abuse

Habeas corpus petitioner convicted of capital murder did not receive ineffective assistance of counsel, on grounds that his attorney failed to present evidence of his substance abuse and naval service; attorney had made tactical decision that substance abuse history could be more harmful than helpful, and the same was true of military service, as it had ended with discharge due to fraudulent enlistment. Sawyers v. Collins, C.A.5 (Tex.) 1993, 986 F.2d 1493, rehearing denied 992 F.2d 326, certiorari denied 113 S.Ct. 2405, 508 U.S. 933, 124 L.Ed. 2d 300.

Defendant was not deprived of effective assistance of trial counsel based upon his allegations that he could smell alcohol on his attorney's breath and that after trial, counsel entered facility for treatment of alcohol abuse; defendant pointed to no specific instances where counsel's performance during trial was deficient because of alcohol abuse nor were any such instances apparent from record at trial. Burnett v. Collins, C.A.5 (Tex.) 1993, 982 F.2d 922.

Defense counsel's alleged use of alcohol during trial was not ineffective assistance in capital murder prosecution, absent showing by defendant that alcohol use, if true, deprived him of fair trial. Russell v. Lynaugh, C.A.5 (Tex.) 1989, 892 F.2d 1205, certiorari denied 111 S.Ct. 2909, 501 U.S. 1259, 115 L.Ed. 2d 1073, rehearing denied 112 S.Ct. 27, 501 U.S. 1277, 115 L.Ed. 2d 1109.

Defense counsel's alcoholism, or even alcohol or drug use during trial, does not necessarily constitute per se violation of Sixth Amendment absent some identifiable deficient performance resulting from intoxication; even where counsel is under significant psychological duress at trial, defendant claiming ineffective assistance must present actual errors and omissions at trial that prejudiced his case. U.S. v. Jackson, N.D.Ill. 1996, 930 F.Supp. 1228.

Evidence that defense counsel was treated at hospitals at various times during state criminal trial or that counsel used prescription drugs during that period is insufficient to state claim of ineffective assistance of counsel; instead, defendant must show specific incidents of deficient performance resulting from drug use and prejudice stemming from that performance. McDougall v. Rice, W.D.N.C. 1988, 685 F.Supp. 532.

Fact that defendant's attorney was a drug user did not establish ineffective assistance of counsel per se. State v. Green, N.J.Super.A.D. 1994, 643 A.2d 18, 274 N.J.Super. 15, certification denied 645 A.2d 141, 137 N.J. 312.

Defense counsel's alleged alcoholism would not be found to have deprived defendant of any state or federal constitutional right to effective assistance, although defense counsel had been hospitalized for alcohol rehabilitation immediately prior to defendant's trial, where defendant offered no proof that hospitalization prevented counsel from preparing defendant's case, and record showed no incapacity on counsel's part as result of his allegedly being alcoholic or any concern by court for presence of intoxicated counsel. People v. Huggins, N.Y.Sup. 1989, 541 N.Y.S.2d 1016, 144 Misc.2d 49.

Source: U.S.C.A. Constitution Amendment VI, squibs, © West Publishing, used with permission.

Exhibit 4–36

40 N C D 2d—148

References are to Digest Topics and Key Numbers

ATTORNEY AND CLIENT—Cont'd
CONTINGENT fees—
  Attorney fees, see this index Attorney Fees
CONTINUANCE—
  Sickness of counsel, review of decisions. Crim
    Law 1151
  Want of time for preparation for trial, review of
    lower court's discretion in ruling on motion
    for continuance. Crim Law 1151
CONTINUING legal education. Atty & C 32(1)
CONTRACTS between, assignability. Assign 19
CONTRACTS in general. Atty & C 122–124
  Attorney fees, see this index Attorney Fees
  Behalf of client. Atty & C 81
  Liens as affected by contracts. Atty & C 176
  Partial invalidity. Contracts 137(1)
  Tortious interference with contract relation.
    Torts 12
CONVEYANCES, dealings between attorney and
  client. Atty & C 123(2)
CONVICTS—
  Recaptured escapee seeking injunction and
    damages, appointment of counsel. Atty & C
    23
CORPORATIONS—
  Practice of law by. Corp 377 1/2
  Purposes of incorporation. Corp 14(1)
  Representation by. Corp 508, 669
CORPORATIONS for law practice. Corp 14(1)
COSTS—
  Action for negligence or malpractice. Atty & C
    129(4)
  Agreement to reimburse client as void for want
    of consideration. Contracts 47

Disciplinary proceedings. Atty & C 59
COSTS, fees of attorney, item of cost—
  County taxpayer's action. Counties 196(9)
  Foreclosure of drainage assessment liens.
    Drains 90
  Municipal taxpayer's action. Mun Corp
    1000(7)
  Persons liable for costs of appeal. Costs 240
  Trust, action to establish or enforce trust.
    Trusts 377
COUNSEL for accused, see this index Criminal
  Law
COUNTIES, see this index Counties
COUNTY commissioners, attorney for board,
  remedy for wrongful removal. Counties 129
COURTESY, dignity, decorum, requirement. Atty
  & C 32(8)
COURTS—
  Assignment of counsel by. Atty & C 23
  Criticism of. Atty & C 32(8)
  District Court of United States, generally, post
    Fees. Atty & C 132
  Jurisdiction. Jurisdiction, generally, post
COURTS, criticism of. Atty & C 32(8)
COURTS-MARTIAL. Armed S 47(4)
CRIMINAL offenses. Offenses, generally, post
CRIMINAL prosecution, adequacy of representa-
  tion. Crim Law 641.13
CRITICISM of courts. Atty & C 32(8)
DAMAGES, negligence or wrongful acts. Atty &
  C 129(4)
DEALINGS between attorney and client. Atty &
  C 122–124

Source: North Carolina Digest 2d, DWI, "Attorney & Client," © West Publishing, used with permission.

Exhibit 4–37

CRIMINAL LAW—Cont'd

COUNSEL for accused. Crim Law 641–641.13

    Absence. Crim Law 641.12(2), 918(9), 936(3)

        Adjournments. Crim Law 649(3)

        Confession of crime. Crim Law 517.2

        Declarations by accused. Crim Law 412.2(4)

        Identification in court, unrepresented defendant. Crim Law 339

        Lineup, unrepresented defendant. Crim Law 339

    Accusatory stage of proceedings—

        Declarations by accused. Crim Law 412.2(2)

    Adequacy of representation. Crim Law 641.13

        Confession of crime. Crim Law 517.2

        Due process of law. Const Law 268.1(6)

        Grounds for habeas corpus. Hab Corp 486 (1–5)

        Weight of evidence. Hab Corp 721(2)

    Admissions. Crim Law 410

    Admissions without counsel, court's own motion excluding. Crim Law 406(3)

    Traffic violations. Crim Law 641.2

    Approval of defendant. Crim Law 641.10(1)

    Argument and conduct of counsel. Crim Law 701, 1154, 1171

    Assignment by court. Atty & C 23

    Assistance of counsel. Crim Law 641.12(3)

        Due process of law. Const Law 268.1

        Federal courts, state compensation for services. Atty & C 132

        Misdemeanor cases. Crim Law 641.2, 641.7(1)

        Time. Crim Law 264

    Attorney fees—

Atty & C 132

Costs 295

Counties 139

Crim Law 641.12(3)

Incompetency. Crim Law 920

    Affidavits. Crim Law 956(7)

    Due process of law. Const Law 268.1(6)

    Withdrawal of guilty plea. Crim Law 274(7)

Incompetents. Const Law 268.2(2)

In-custody interrogation, right to counsel, retrospective operation. Courts 100(1)

Indigents. Crim Law 641.6(3), 641.13(3)

    Acceptance of counsel appointed by court. Crim Law 641.10(1)

    Adequacy of representation. Crim Law 641.13(3)

    Advising of right. Crim Law 393(1), 412.2(3)

    Appointment to counsel—

        Misdemeanors. Crim Law 641.2, 641.7(1)

        Preliminary examination. Crim Law 232

        Preparation for trial—

        Predetermination, time needed. Crim Law 577

        Time. Crim Law 264

    Arguments and conduct of counsel. Crim Law 701

    Assignment of counsel—

        Atty & C 23

        Crim Law 641

    Attorney fees—

        Atty & C 132

Source: NC Digest 2d, DWI, "Criminal Law," © West Publishing, used with permission.

Exhibit 4–38

---

12 NC D 2d—43   **CRIMINAL LAW**

XII. TRIAL.—Continued
  (B) COURSE AND CONDUCT OF TRIAL IN GENERAL.—Continued.
    ⚷ 641.3 —Stage of proceedings as affecting right.—Continued.
      (9). —Photographic identification.
      (10). —Lineup or showup.
      (11). Mental examination.
    641.4 —Waiver of right to counsel.
      (1). In general; right to appear pro se.
      (2). Capacity and requisites in general.

    641.11. —Public defenders.
    641.12. —Deprivation or allowance of counsel.
      (1). Acts or omissions constituting, in general.
      (2). Presence of counsel and consultation.
      (3). Assistance and compensation of counsel.
      (4). Effect of representation or deprivation of rights.
    641.13. —Adequacy of representation.
      (1). In general.
      (2). Particular cases and problems.
      (3). —Indigent's or incompetent's counsel and public defenders.
      (4). —General qualifications of counsel.
      (5). —Pretrial proceedings; sanity hearing.
      (6). —Evidence; procurement, presentation and objections.
      (7). —Post-trial procedure and review.
      (8). —Counsel of defendant's choice or defendant pro se.

Source: NC Digest 2d, DWI, "Criminal Law—Outline," © West Publishing, used with permission.

Exhibit 4–39

N.C. 1994. Standard to be met for showing of ineffective assistance of counsel under North Carolina Constitution is identical to that under Federal Constitution. U.S.C.A. Const.Amend. 6.—State v. Bacon, 446 S.E2d 542, 337 N.C. 66, certiorari denied 115 S.Ct. 1120, 130 L.Ed.2d 1083.

N.C. 1994. When court engages in effective assistance of counsel analysis, fair assessment of attorney performance requires that every effort be made to eliminate distorting effects of hindsight, to reconstruct circumstances of counsel's challenged conduct and to evaluate conduct from counsel's perspective at the time and because of difficulties inherent in making this evaluation, court must indulge strong presumption that counsel's conduct falls within wide range of reasonable professional assistance. U.S.C.A. Const.Amend. 6.—State v. Mason, 446 S.E.2d 58, 337 N.C. 165.

N.C. 1993. Defendant bears burden of proving any constitutionally deficient performance of his trial counsel and actual prejudice. U.S.C.A. Const.Amend. 6.—State v. McHone, 435 S.E.2d 296, 334 N.C. 627, certiorari denied Van McHone v. North Carolina, 114 S.Ct. 1577, 128 L.Ed.2d 220.

N.C. 1991. Defendant is entitled to relief on grounds of ineffective assistance of counsel . . .

Defendant did not receive ineffective assistance of counsel by attorney who did not press for dismissal of claim that he had traveled in interstate commerce in furtherance of drug enterprise; there was no indication that claim would have been dismissed. U.S.C.A. Const.Amend. 6.—Id.

C.A.4 (N.C.) 1990. Defendant's claim that one member of his defense team gained admission to case by misrepresentation and lied about his own track record was not appropriate basis for claim of ineffective assistance of counsel; defendant did not show that he was prejudiced by alleged "puffing." U.S.C.A. Const.Amend. 6.—McDougall v. Dixon, 921 F.2d 518, certiorari denied 111 S.Ct. 2840, 501 U.S. 1223, 115 L.Ed.2d 1009.

Migraine headaches suffered by one member of defense team during trial in capital murder prosecution, and his treatment for depression, did not result in ineffective assistance of counsel; counsel's treating physician was present during entire trial, and counsel's capacity in courtroom was not adversely affected by hospital visits or medical attention; inasmuch as experienced criminal trial lawyer was able to conduct examination of witnesses if counsel was unable to conduct examination. U.S.C.A. Const.Amend. 6.—Id.

Closing argument presented by one member of defense team which amounted to "head-on" attack on death penalty, although unsuccessful at sentencing phase of capital murder prosecution, was not deficient, nor was defendant denied effective assistance of counsel. U.S.C.A. Const.Amend. 6.—Id.

E.D.N.C. 1991. Murder defendant's attorneys did not render ineffective assistance on basis of failing to move for change of venue, in that lack of massive pretrial publicity in the county . . .

Source: NC Digest 2d, Criminal Law 641.3, squibs, © West Publishing, used with permission.

Problem 4

Assume that the court administrator for the United States Court of Appeals for the Ninth Circuit is interested in documenting the number of reported opinions each year between 1986 and 1995 that decided claims involving ineffective assistance of counsel in criminal cases. How do we find information that would help answer this question?

When approaching a task such as this, it is a good idea to keep in mind the old adage about not reinventing the wheel. The court administrator's first step should be to determine whether anyone has already collected the sought-after information. The **American Statistics Index** provides references to federal government publications containing statistical information, and the **Statistical Reference Index** provides references to state documents and privately published reports containing statistical information. They should be checked to see if the desired information already has been compiled. In addition, the court administrator should consult either **LegalTrac** or the **ILP** for references to law review articles that may include the data being sought, or that may cite sources where the data can be found. Let us assume that the administrator has been unsuccessful in finding the information in any of these sources and therefore has reconciled herself to the fact that she is going to have to collect the data herself.

The court administrator is interested only in reported opinions from the Ninth Circuit Court of Appeals. We know from our discussion of case law research in Chapter 3 that these opinions are published in the **Federal Reporter** series. Recall that West publishes an index to federal court cases, the **Federal Practice Digest**, and that the digest provides subject indexing to cases through the West key number system. Therefore, our court administrator might attempt to compile the information by locating the appropriate key number(s) in the Federal Practice Digest and then scan the blurbs under the key number(s) for cases decided by the Ninth Circuit Court of Appeals. Of course, this process works only if she is able to locate key numbers concerning ineffective assistance of counsel in criminal cases. Because she is interested only in decisions covering 1986 through 1995, she can limit her search to the Federal Practice Digest 4th series, which covers cases decided since 1982. She can proceed to the Descriptive Word Index volumes of the Federal Practice Digest 4th and locate under the heading "**Criminal Law**," and subheading "**Counsel for accused**," another subheading "**Adequacy of representation**." (See Exhibit 4–40.) The main key number for "**Adequacy of representation**" is Crim Law ☞ 641.13. Several other key numbers are also referenced that the court administrator should investigate.

To pinpoint the key number to search more precisely, she might next proceed to the outline at the front of each topic in the digest and locate the key number's position in the key number classification system. (See Exhibit 4–41.) For Criminal Law ☞ 641.13, the court administrator would want to search the the main heading as well as subheadings, such as Criminal Law ☞ 641.13(1). The next step would be to look under all the relevant key numbers and count, for each year, the number of decisions from the Ninth Circuit that deal specifically with ineffective assistance of counsel. During this long and tedious process, the court administrator must be careful not to count the same case more than once in instances where blurbs from the same decision appear under different key numbers.

CALR Approach

The research question we are investigating is particularly well suited to **computer-assisted legal research (CALR)**. However, the researcher cannot assume

Exhibit 4–40

97 F P D 4th—392

References are to Digest Topics and Key Numbers

CRIMINAL LAW—Cont'd
CONSTITUTIONAL and statutory provisions—
Cont'd

COUNSEL for accused. Crim Law 641–641.13
  Absence. Crim Law 641.12(2), 918(9), 936(3)
    Adjournments. Crim Law 649(3)
    Confession of crime. Crim Law 517.2
    Declarations by accused. Crim Law
      412.2(4)
  Accusatory stage of proceedings—
    Declarations by accused. Crim Law
      412.2(2)

Adequacy of representation. Crim Law 641.13 ◀
  Confession of crime. Crim Law 517.2
  Disbarred attorneys. Crim Law 641.13(4)
  Due process of law. Const Law 268.1(6)
  Grounds for habeas corpus. Hab Corp
    486(1–5)
  Habeas corpus. Hab Corp 338
  Weight of evidence. Hab Corp 721(2)
Admissions. Crim Law 410
Advice by court. Crim Law 641.7(1)
  Defects as grounds for habeas corpus. Hab
    Corp 484
Advising of right to counsel—
  Confession of crime. Crim Law 517.2(3)

Source: Federal Practice Digest 4th, DWI, "Criminal Law," © West Publishing, used with permission.

that using the computer eliminates the need for critical thinking. Our court administrator should experiment with several CALR approaches and compare the results before deciding on the best research method.

WESTLAW has some advantages over LEXIS for this particular project: It has a database that includes opinions only from the Ninth Circuit Court of Appeals **(CTA9)**, and key number searching also is possible on WESTLAW. Thus, we will use WESTLAW in this example, although a search strategy could be designed for retrieving the information from LEXIS as well.

Where precise search results are required, the **Terms and Connectors** search mode should be used on WESTLAW.

First Approach. The court administrator could use a search strategy that relies on the key number or numbers that are relevant to the issue. She could have identified the appropriate key number by using the Descriptive Word Index in the printed digests or the **KEY** database (see Chapter 3 for a discussion of the **KEY** database) on WESTLAW. Let us assume that the court administrator has identified the same key numbers that we earlier depicted in Exhibit 4–40. She then might decide that the main key number, Crim Law ☞ 641.13 and its subheadings, would reference all cases, and only those cases, that deal with ineffective

Exhibit 4–41

31 F P D 4th—43   **CRIMINAL LAW**

XII. TRIAL.—Continued.
  (B) COURSE AND CONDUCT OF TRIAL IN GENERAL.—Continued.
  ⚷641.4.— Waiver of right to counsel.—Continued.
      (3). Guilty plea.
      (4). Validity and sufficiency, particular cases.
      (5). Effect of waiver or appearing pro se.

---

      (1). Acts or omissions constituting, in general.
      (2). Presence of counsel and consultation.
      (3). Assistance and compensation of counsel.
      (4). Effect of representation or deprivation of rights.
  641.13.—Adequacy of representation.
      (1). In general.
      (2). Particular cases and problems.
      (3). —Indigent's or incompetent's counsel and public defenders.
      (4). —General qualifications of counsel.
      (5). —Pretrial proceedings; sanity hearing.
      (6). —Evidence; procurement, presentation and objections.
      (7). —Post-trial procedure and review.
      (8). —Counsel of defendant's choice or defendant pro se.
  642. Appointment and services of interpreter.
  643. Appointment and services of stenographer.
  644. Attendance and conduct of officers.
  645. Right to open and close.

Source: FPD 4th, Criminal Law-outline, © West Publishing, used with permission.

assistance of counsel in a criminal context. The ensuing search strategy in the **CTA9** database looks like this: **"DA(1995) & 110k641.13!"** Remember that each key number topic is assigned a number on WESTLAW. The number for "Criminal Law" is 110. The **!** at the end of the key number picks up any finer classification of the basic key number. The court administrator then runs the key-number strategy 10 times, changing the date restriction each time so that she has a separate record for each year 1986 through 1995.

The risk incurred in choosing this approach is that relevant cases may exist that do not have the key number "Crim Law 641.13" or one of its subheadings assigned to them. Thus, the court administrator may decide to add additional key numbers, as described in the second approach.

Second Approach. The court administrator could add the key numbers listed in Exhibit 4–40, thus making the search strategy "**DA(1995) & 110k641.13! 110k517.2! 92k268.1(6)! 197k486! 197k338! 197k721(2)!**". The number **92** refers to the topic **Constitutional Law**, and the number **197** refers to the topic **Habeas Corpus**. However, if our original search strategy is too narrow in focus, this revision is too broad. The key numbers we have added are likely to refer to many issues in addition to ineffective assistance of counsel, so we are certain to retrieve references to cases having little to do with our question.

We can attempt to get around this latest problem by revising the strategy to require that certain key words appear in either the **synopsis** or the **digest** field. As we explained in Chapter 3, the **synopsis (sy)** field includes a summary of the history of a case, and the **digest (di)** field includes the headnotes from the case as well as the key numbers assigned to each headnote. Thus, the court administrator might add the following element to her search strategy: "**SY,DI(ineffective /s counsel)**". This command retrieves all cases in which "ineffective" appears in the same sentence (**/s**) as "counsel" in either the **synopsis** or **digest** field. The complete strategy would then be "**DA(1995) & 110k641.13! 110k517.2! 92k268.1(6)! 197k486! 197k338! 197k721(2)! & SY,DI(ineffective /s counsel)**". Remember that in the **Terms and Connectors** mode, the space between terms, or, in this case, key numbers, is interpreted as the Boolean operator **or**.

If our administrator is still concerned that relevant key numbers might be overlooked, she could combine a key number **topic** search with the key-word search. Recall that the **topic** field includes the topic name part of the key number. For example, for the key number Crim Law 641.13, the **topic** field would include the topic name **Criminal Law**. (See Exhibit 4–42.) The advantage of this approach is that any key number with the indicated **topics (to)** would be retrieved. This optional search strategy would look like this: "**DA(1995) & TOPIC(criminal habeas constitutional) & SY,DI(ineffective /s counsel)**".

A **topic** search does not require that the full name of the topic be used. Instead, you can just enter the word that is unique to that topic. For example, "**criminal**" will suffice for the topic **criminal law**. Similarly, "**habeas**" retrieves all references to the topic **habeas corpus**, and "**constitutional**" retrieves all references to the topic **constitutional law**. Alternatively, the researcher can include the full name of each topic in the search strategy, such as "**TO("criminal law" "habeas corpus" "constitutional law")**". As we have pointed out before, words within quotation marks are interpreted as a phrase.

Third Approach. One attractive feature of WESTLAW is that large databases are divided into smaller files according to jurisdiction and so-called "Practice Areas." At the time we worked through this example, some 40 practice areas were recognized on WESTLAW, including Criminal Justice. LEXIS also has subject files, but not one for Criminal Justice. One database in WESTLAW includes just criminal justice cases decided by the federal appellate courts. The identifier for that database is **FCJ-CTA.** This database includes all the criminal cases, so a topic

---

Exhibit 4–42

---

68 F.3d 328	R 3 OF **25**	P 3 OF 11	CTA9	**Term**

(Cite as: 68 F.3d 328)

U.S. v. Stearns

► [1]

► 110   CRIMINAL LAW

► 110XX   Trial

► 110XX(B)   Course and Conduct of Trial in General

► 110k641   Counsel for Accused

► 110k641.13   Adequacy of Representation

► 110k641.13(2)   Particular Cases and Problems

► 110k641.13(7) k. Post-trial procedure and review.

C.A.9 (Or.), 1995.

Defendant received **ineffective** assistance of **counsel** if defendant did not consent to trial **counsel's** failure to file notice of appeal, regardless of facts that defendant was convicted pursuant to guilty plea rather than at trial, that he was satisfied with his attorney at time of sentencing, and that sentencing court advised him of his right to appeal. U.S.C.A. Const.Amend. 6.

Source: CTA9, "110k641.13," © WESTLAW, used with permission.

---

and key-number search may not be necessary. To limit the results to Ninth Circuit Court of Appeals opinions, the court administrator should incorporate the following element into the search strategy: **COURT(CA9).** Recall from Chapter 3 that the **court** field actually is a subfield of the **digest** field, and it contains abbreviations for particular courts. "**CA9**" refers to the Ninth Circuit Court of Appeals. The complete search strategy is "**DA(1995) & CO(CA9) & SY,DI(ineffective /s counsel)**".

This strategy has advantages over those that use topics or key numbers because it retrieves the occasional relevant case that does not have the anticipated headnotes or key numbers assigned to it. After assessing the different search results, the court administrator finds that this last approach is the best strategy for this task: It retrieves only cases relevant to the issue, and these cases are more directly on point than cases produced by the strategies that rely on key numbers or topic restrictions. The court administrator then runs this strategy in the database 10 times, or once for each year. The screen displaying the results for 1995 is displayed in Exhibit 4–43. The notation at the top of the screen, **R 1 of 28**, indicates that we are looking at the first of 28 documents that have been retrieved.

Exhibit 4–43

Citation	Rank (R)	Page (P)	Database	Mode
95 F.3d 755	R 1 OF **28**	P 1 OF 71	FCJ-CTA	**Term**

96 Cal. Daily Op. Serv. 6744, 96 Daily Journal D.A.R. 10, 991
(Cite as: 95 F.3d 755)

Paris Hoyt CARRIGER, Petitioner-Appellant,

v.

Terry L. STEWART, Respondent-Appellee.

No. 95-99025.

United States Court of Appeals,

Ninth Circuit.

Argued (by telephone) and Submitted

Nov. 29, 1995.

Filed Nov. 29, 1995.

Opinion Withdrawn Sept. 10, 1996.

New Opinion Filed Sept. 10, 1996.

As Amended Sept. 24, 1996.

After petitioner's conviction for murder and death sentence were affirmed, ▶ 599 P.2d 788, denial of his petition for postconviction relief was also affirmed, ▶ 692 P.2d 991. Denial of petitioner's first petition for writ of habeas corpus was ultimately affirmed, ▶ 971 F.2d 329. The United States District Court for the District of Arizona, Paul G. Rosenblatt, J., denied his subsequent petition for writ of habeas corpus, and petitioner appealed. The . . .

Source: FCJ-CTA, "Ineffective counsel" result, © WESTLAW, used with permission.

The results for all the years are:

Date	Number of reported opinions
1995	28
1994	30
1993	18
1992	15
1991	17
1990	24
1989	19
1988	16
1987	13
1986	17

No search strategy is perfect. You might want to develop your own strategy for this topic to see if you can obtain better results. One possible revision would be to include "related concepts" for either "ineffective" or "counsel." Remember that WESTLAW contains a **thesaurus**, which facilitates identifying related concepts. We did not include synonymous terms, such as "attorney" for "counsel," because we assumed that the courts would use the terms "ineffective" and "counsel" in tandem, and that incorporating additional terms would only increase the likelihood of retrieving irrelevant cases. If we were researching a different issue, we might include "related concepts."

Problem 5

While serving the sixth year of a 25-year federal prison sentence, Umar Yarbrough began corresponding with Leanne Atkins, who thereafter paid Yarbrough regular visits at the prison. Following several months of visitation, the two became engaged. Yarbrough and Atkins requested permission from prison officials to formalize their marriage. This request was summarily denied. The couple then jointly filed a lawsuit alleging that the prison officials infringed their "constitutional right to marry." The prison officials have responded by asserting that a prisoner serving a 25-year sentence has no right to marry. How should this issue be resolved?

Although a constitutional issue clearly lurks in this problem, we should investigate other possible grounds for resolving the case. Courts almost always prefer to avoid deciding a constitutional question if there is an available alternative. In this instance, we should check prison regulations before we jump headfirst into constitutional research, even though we should also be armed with knowledge about what the Constitution has to say about this issue.

We begin with the **Index** to the **Code of Federal Regulations** to investigate whether any federal regulations govern this matter. Tucked away under the heading "**Prisoners**" and the subheading "**Institutional management, inmates**," we find a hodgepodge of topics, including "**Grooming, nondiscrimination, smoking, family planning . . . , 28 CFR 551.**" (See Exhibit 4–44.) The reference to "family planning" comes closer to marriage than any other subheading, so it is prudent to check title 28 of the Code of Federal Regulations, part 551, to see if there is an applicable regulation.

On locating 28 C.F.R. 551, we find that several regulations have been adopted that pertain to marriage of federal prisoners. We reproduce the topical outline and a few of the regulations in Exhibit 4–45. The regulations at parts 551.10 and 551.12 suggest that the prison officials erred when they summarily denied Yarbrough and Atkins' request to marry.

We can learn more about the constitutional dimensions of this issue by consulting other authorities. We look under "**Marriage**" in the indexes for A.L.R.2d, 3d, 4th, 5th, Federal and L. Ed. 2d to try to find a relevant A.L.R. annotation. Under "**Constitutional law** . . . —federal constitutional right to marry—Supreme Court cases," we are referred to 96 L. Ed. 2d 716. (See Exhibit 4–46.) When we consult this annotation, we find that "**Marriage of prisoner**" is specifically addressed at

Exhibit 4–44

CFR Index

Prisoners

Accident compensation for prison inmates, 28 CFR 301

Acquisition regulations, application of labor laws to Government acquisitions, Panama Canal Commission, 48 CFR 3522

Admission of inmates to institution, 28 CFR 522

Classification of inmates, 28 CFR 524

Community programs, 28 CFR 570

Computation of sentence, 28 CFR 523

Death sentences implementation in Federal cases, 28 CFR 26

Federal Acquisition Regulation, application of labor laws to Government acquisitions, 48 CFR 22

Functional literacy for State and local prisoners program, 34 CFR 489

General management and administration

Costs of incarceration fee, 28 CFR 505

General definitions, 28 CFR 500

General management policy, 28 CFR 511

Records access, 28 CFR 513

Scope of rules, 28 CFR 501

Grievance procedures standards for inmates, 28 CFR 40

Institutional management, inmates

Administrative remedy, 28 CFR 542

Contact with persons in community, 28 CFR 540

Custody, 28 CFR 552

Discipline and special housing units, 28 CFR 541

Education, 28 CFR 544

Food service, 28 CFR 547

Grooming, nondiscrimination, smoking, family planning, organizations, contributions, manuscripts, polygraph tests, and pre-trial inmates, 28 CFR 551

Inmate property, 28 CFR 553

Legal matters, 28 CFR 543

Medical services, 28 CFR 549

Prison drug program, 28 CFR 550

Religious programs, 28 CFR 548

Work and compensation, 28 CFR 545

Life skills for State and local prisoners program, 34 CFR 490

Parole, 28 CFR 572

Source: CFR, Index, "Prisoners."

§ 4[c]. (See Exhibit 4–47.) That section in turn alerts us to a Supreme Court case, *Turner v. Safley,* [482 U.S. 78], 107 S. Ct. 2254, 96 L. Ed. 2d 64 (1987), considering a prisoner's constitutional right to marry. (See Exhibit 4–48.) In fact, *Turner v. Safley* is the case that inspired the annotation we are using, and we can read the Court's opinion in this same volume of L. Ed. 2d. Shepardizing *Turner v. Safley* allows us to confirm the case's continuing validity and to find a host of decisions citing the case and applying its principles.

To find law review articles addressing this issue, we refer to 28 Index to Legal Periodicals (September 1988–August 1989), which was compiled shortly after *Turner v. Safley* was decided. We can locate helpful articles either by using the Table of Cases at the back of the volume (see Exhibit 4–49), or by looking in the

Exhibit 4–45

Bureau of Prisons, Justice

PART 551—MISCELLANEOUS

Subpart A—Grooming

Sec.
551.1 Policy.
551.2 Mustaches and beards.
551.3 Hairpieces.
551.4 Hair length.
551.5 Restrictions and exceptions.
551.6 Personal hygiene.
551.7 Bathing and clothing.

Subpart B—Marriages of Inmates

551.10 Purpose and scope.
551.11 Authority to approve a marriage.
551.12 Eligibility to marry.
551.13 Application to marry.
551.14 Special circumstances.
551.15 Furloughs.
551.16 Marriage ceremony in the institution.

Subpart C—Birth Control, Pregnancy, Child Placement, and Abortion

551.20 Purpose and scope.
551.21 Birth control.
551.22 Pregnancy.
551.23 Abortion.
551.24 Child placement.

Subpart D—Inmate Organizations

551.30 Purpose and scope.
551.31 Approval.
551.32 Dues.
551.33 Meetings.
551.34 Fund-raising projects.
551.35 Special activities.
551.36 Accountability for funds.

Subpart E—Inmate Contributions

551.50 Policy.

Subpart F—Volunteer Community Service Projects

551.60 Volunteer community service projects.

Subpart G—Administering of Polygraph Test

551.70 Purpose and scope.
551.71 Procedures.

Subpart H—Inmate Manuscripts

551.80 Definition.
551.81 Manuscript preparation.
551.82 Mailing inmate manuscripts.
551.83 Limitations on an inmate's accumulation of manuscript material.

Subpart I—Non-Discrimination Toward Inmates

551.90 Policy.

Subpart J—Pretrial Inmates

551.100 Purpose and scope.
551.101 Definitions.
551.102 Commitment prior to arraignment.
551.103 Procedure for admission.
551.104 Housing.
551.105 Custody.
551.106 Institutional employment.
551.107 Pretrial inmate reviews.
551.108 Performance pay.
551.109 Community activities.
551.110 Religious programs.
551.111 Marriage.
551.112 Education.
551.113 Counseling.
551.114 Medical, psychiatric and psychological.
551.115 Recreation.
551.116 Discipline.
551.117 Access to legal resources.
551.118 Property.
551.119 Release of funds and property of pretrial inmates.
551.120 Visiting.

continues

Exhibit 4–45   continued

Subparts K-L—(Reserved)

Subpart M—Victim and/or Witness Notification

551.150 Purpose and scope.
551.151 Definitions.
551.152 Procedures.
551.153 Cancelling the notification request.

Subpart N—Smoking/No Smoking Areas

551.160 Purpose and scope.
551.161 Definitions.
551.162 Designated no smoking areas.
551.163 Designated smoking areas.
551.164 Notice of smoking areas.

AUTHORITY: 5 U.S.C. 301; 18 U.S.C. 1512, 3621, 3622, 3624, 4001, 4005, 4042, 4081, 4082 (Repealed in part as to offenses committed on or after November 1, 1987), 4161–4166 (Repealed as to offenses committed on or after November 1, 1987), 5006–5024 (Repealed October 12, 1984 as to offenses committed after that date), 5039; 28 U.S.C. 509, 510; Pub. L. 99–500 (sec. 209); 28 CFR 0.95–0.99; Attorney General's August 6, 1991 Guidelines for Victim and Witness Assistance.
SOURCE: 44 FR 38252, June 29, 1979, unless otherwise noted.

Subpart B—Marriages of Inmates

SOURCE: 49 FR, 18385, Apr. 30, 1984, unless otherwise noted.

§ 551.10 Purpose and scope.

Source: CFR, 28 CFR 551.

The Warden shall approve an inmate's request to marry except where a legal restriction to the marriage exists, or where the proposed marriage presents a threat to the security or good order of the institution, or to the protection of the public.

The Warden may approve the use of institution facilities for an inmate's marriage ceremony. If a marriage ceremony poses a threat to the security or good order of the institution, the Warden may disapprove a marriage ceremony in the institution.

§ 551.11 Authority to approve a marriage.

(a) The Warden may approve the marriage of a federal inmate confined in a federal institution. This authority may not be delegated below the level of Acting Warden.

(b) The appropriate Community Corrections Manager may approve the request to marry of a federal inmate who is not confined in a federal institution (for example, a federal inmate who is in a community corrections center, in home confinement, in state custody, or in a local detention facility).

[49 FR 18385, Apr. 30, 1984, as amended at 58 FR 58248, Oct. 29, 1993]

§ 551.12 Eligibility to marry.

An inmate's request to marry shall be approved provided:

(a) The inmate is legally eligible to marry;
(b) The inmate is mentally competent;
(c) The intended spouse has verified, ordinarily in writing, an intention to marry the inmate; and
(d) The marriage poses no threat to institution security or good order, or to the protection of the public.

Exhibit 4–46

ALR INDEX

MARRIAGE—Cont'd

Cohabitation—Cont'd

– inducement, liability of one putative spouse to other for wrongfully inducing entry into or cohabitation under illegal, void, or nonexistent marriage, 72 ALR2d 956

College residency requirement, validity and application of provisions governing determination of residency for purpose of fixing fee differential for out-of-state students in public college, 56 ALR3d 641, §§ 5, 8, 9

Common-Law Marriage (this index)

Concealment of or misrepresentation as to prior marital status as ground for annulment of marriage, 15 ALR3d 759

Conditional Gift (this index)

Conflict of laws

generally, 82 ALR3d 1240

–Husband and Wife (this index)

–marriage license, solemnized marriage, validity as affected by absence of license required by statute, 61 ALR2d 856

–miscegenation law of state as affecting recognition of foreign marriage, 3 ALR2d 240

–nonage, conflict of laws as to validity of marriage attacked because of, 71 ALR2d 687

Consideration (this index)

Consortium (this index)

Constitutional law

–divorce, right to jury trial in state court divorce proceedings, 56 ALR4th 955

–federal constitutional right to marry—Supreme Court cases, 96 L Ed 2d 716

Constructive trust, property rights arising from relationship of couple cohabiting without marriage, 3 ALR4th 13

Source: ALR INDEX, "Constitutional Law" p. 812. Permission has been granted by the current copyright holder, West Group. Further reproduction of any kind is strictly prohibited. For additional information, please contact West Group Customer Services representative at 1-800-328-4880.

Exhibit 4–47

TURNER v SAFLEY
Reported p 64, supra
ANNOTATION

FEDERAL CONSTITUTIONAL RIGHT TO MARRY—SUPREME COURT CASES

by

John E. Theuman, J.D.

§ 1. Introduction
  [a] Scope
  [b] Related matters
§ 2. Summary and comment
§ 3. Existence and source of constitutional right to marry

§ 4. Right to marry under particular circumstances
  [a] Interracial marriage
  [b] Marriage of parent having child-support obligations
  [c] Marriage of prisoner

Source: Lawyers Edition Annotation, 96 L.Ed.2d 716. Reprinted with the permission of LEXIS-NEXIS, a division of Reed Elsevier Inc. LEXIS and NEXIS are registered trademarks of Reed Elsevier Properties Inc. FREESTYLE, KWIC, SuperKWIC and MEGA are trademarks of Reed Elsevier Properties Inc. SHEPARD'S and SHEPARDIZE are registered trademarks of Shepard's Company, a Partnership.

Exhibit 4–48

[c] **Marriage of prisoner**

Ruling that the right to marry, which is protected by the Federal Constitution, is retained by prison inmates, the United States Supreme Court held in the following case that a state prison regulation which almost completely barred prisoners from getting married was unconstitutional because it was not reasonably related to legitimate penological concerns.

A prison regulation which permitted inmates to marry only with the permission of the superintendent of the prison, and which stated that such permission should be granted only where there are "compelling reasons"—such as the birth of a child—to do so, was held to be facially infirm as infringing on the fundamental constitutional right to marry in Turner v Safley (1987, US) 96 L Ed 2d 64, 107 S Ct 2254, where the Supreme Court ruled (1) that that right does apply in the context of prison inmates and (2) that the regulation in question was not reasonably related to legitimate penological objectives. It is settled, the Supreme Court noted, that prison inmates retain those constitutional rights that are not inconsistent with their status as prisoners or with the legitimate penological objectives of the corrections system. While conceding that the right to marry is . . .

Source: Annotation, 96 L.Ed.2d 716, text. Reprinted with the permission of LEXIS-NEXIS, a division of Reed Elsevier Inc. LEXIS and NEXIS are registered trademarks of Reed Elsevier Properties Inc. FREESTYLE, KWIC, SuperKWIC and MEGA are trademarks of Reed Elsevier Properties Inc. SHEPARD'S and SHEPARDIZE are registered trademarks of Shepard's Company, a Partnership.

Subject and Author index under "**Prisons and Prisoners**." (See Exhibit 4–50.) To peruse the legal periodical literature more efficiently without having to go through several individual volumes of the ILP, we could make use of LegalTrac. As you will recall from Chapter 2, LegalTrac collects on a computer disk citations to law review and legal newspaper articles from 1980 to the present, which can be searched by subject, author, or case name.

We can find additional case law concerning a prisoner's right to marry by Shepardizing *Turner v. Safley* and by combing the footnotes that we discover in law review articles. We also can find cases by using the topic and key numbers that can be extracted from the headnotes in *Turner v. Safley* and working with the Decennial and General Digests. Annotated constitutions also help us identify case law. We illustrate this latter strategy by referring to the Fourteenth Amendment (due process of law) in West's U.S.C.A.

The Fourteenth Amendment to the U.S. Constitution includes so many important provisions, and the due process clause is so expansive in its own right, that an entire volume of U.S.C.A. is devoted to due process. (See Exhibit 4–51.) The Notes of Decisions outline accompanying the amendment includes "**Prisons and Prisoners—Generally**," at notes 3221–3450. (See Exhibit 4–52.) When we refer to that section, we find that note 3368 relates to marriage. (See Exhibit 4–53.) The more recent cases bearing on prisoners' right to marry, including *Turner v. Safley,* are annotated in the pamphlet supplement to the Fourteenth Amendment due process volume in U.S.C.A. (See Exhibit 4–54.)

Exhibit 4–49

TABLE OF CASES

Trimarchi; R. v., 63 O.R.2d 515
　30 *Crim L.Q.* 492-510 S '88
Trinidad & Tobago, In re Request for Assistance from Ministry of Legal Affairs of, 848 F.2d 1151
　82 *Am. J. Int'l L.* 824-8 O '88
Trombetta; California v., 104 S. Ct. 2528
　10 *T. Marshall L. Rev.* 594-610 Spr '85
Truck Drivers Local 807 v. Carey Transp., 816 F.2d 82
　13. *J. Corp. L.* 941-52 Spr '88
Trucke; State v., 410 N.W.2d 242 (Iowa)
　37 *Drake L. Rev.* 163-72 '87/'88
Trustees of the Office of Hawaiian Affairs v. Yamasaki, 737 P.2d 446 (Haw.)
　10 *U. Haw. L. Rev.* 345-64 Wint '88
Tucker v. Marcus, 418 N.W.2d 818 (Wis.)
　61 *Wis. B. Bull.* 29-30+ Jl '88
Tucker; United States v., 92 S. Ct. 589
　19 *Colum. Hum. Rts. L. Rev.* 123-62 Fall '87
Tufts; Commissioner v., 103 S. Ct. 1826
　41 *Baylor L. Rev.* 231-65 Spr '89
　42 *Tax Law.* 93-120 Fall '88
Tull v. United States, 107 S. Ct. 1831
　4 J. *Min. L. & Pol'y.* 359-75 '88/'89
　43 *U. Miami L. Rev.* 361-418 N '88
　1988 *Utah L. Rev.* 435-54 '88
Tully v. State of Oklahoma, 730 P.2d 1206 (Okla.)
　41 *Okla. L. Rev.* 515-29 Fall '88
Tully; Westinghouse Elec. Corp. v., 104 S. Ct. 1856
　10 *T. Marshall L. Rev.* 270-84 Fall '84/Spr '85
Tulsa Professional Collection Servs. v. Pope Estate, 108 S. Ct. 1340
　45 *J. Mo. B.* 91-5 Mr '89
　58 *Miss. L.J.* 193-205 Spr '88

　54 *Mo. L. Rev.* 189-207 Wint '89
　34 *S.D.L. Rev.* 359-80 '89
　18 *Stetson L. Rev.* 471-92 Spr '89
　11 *U. Ark. Little Rock L.J.* 603-15 '88/'89
　43 *Wash. St. B. News* 7-10 F '89
Turkette; United States v. 101 S. Ct. 2524
　62 *Tul. L. Rev.* 1419-51 Je '88
Turnage; Traynor v., 108 S. Ct. 1372
　57 *U. Cin. L. Rev.* 1443-69 '89
Turner v. Department of Employment Sec., 96 S. Ct. 249
　63 *N.Y.U. L. Rev.* 532-610 Je '88
Turner v. District of Columbia, 532 A.2d 662 (D.C.)
　27 *J. Fam. L.* 546-51 '88/'89
Turner v. Labour Party, [1987] I.R.L.R. 101
　17 *Indus. L.J.* 122-4 Je '88
Turner v. Murray, 106 S. Ct. 1683
　73 *Cornell L. Rev.* 1016-37 Jl '88
　56 *U. Chi. L. Rev.* 153-233 Wint '89
Turner v. Safley, 107 S. Ct. 2254 ←
　26 *J. Fam. L.* 856-62 '87/'88
　22 *Loy. L.A.L. Rev.* 667-715 Ja '89
　19 *Seton Hall L. Rev.* 429-55 '89
　56 *UMKC L. Rev.* 589-602 Spr '88
　33 *Vill. L. Rev.* 393-436 Ap '88
　27 *Washburn L.J.* 654-70 Spr '88
Tusch Enters. v. Coffin, 740 P.2d 1022 (Idaho)
　66 *Wash. U.L.Q.* 163-72 '88
Tuttle; City of Oklahoma City v. 105 S. Ct. 2427
　35 *UCLA L. Rev.* 1187-266 Ag '88
Tveten; Norwest Bank Nebraska v. 848 F.2d 871
　15 *Wm. Mitchell L. Rev.* 643-85 '89
Two Juveniles; Commonwealth v., 491 N.E.2d 234 (Mass.)
　21 *Suffolk U.L. Rev.* 1222-9 Wint '87

Source: Index to Legal Periodicals, 1988/89, Table of Cases, p. 749, "Turner," © H.W. Wilson Company, used with permission.

Exhibit 4–50

INDEX TO LEGAL PERIODICALS

Prisons and prisoners

  See also

  False imprisonment

  Pardon

  Parole

  Recidivism

  Rehabilitation of offenders

AIDS behind bars: prison responses and judicial deference. 62 *Temp. L. Rev.* 327-54 Spr '89

AIDS in prisons: one correctional administrator's recommended policies and procedures. T. A. Coughlin, III. 27 *Judicature* 63-6+ Je/Jl '88

Annals of the prisoners' rights movement: the contributions of Judge Merhige. R. J. Bacigal. 24 *Crim L. Bull.* 521-9 N/D '88

Censorship of inmate mail and the first amendment: the way of the circuits. 19 *Tex. Tech. L. Rev.* 1057-90 Spr '88

A comparison of prison use in England, Canada, West Germany, and the United States: a limited test of the punitive hypotheses. J. P. Lynch. 79 *J. Crim. L. & Criminology* 180-217 Spr '88

A comprehensive mental health care system for prison inmates: retrospective look at New York's ten year experience. R. T. Greene. 11 *Int'l J.L. & Psychiatry* 381-9 '88

Confinement and non-conformity: an overview from Czarist to Soviet Russia. 14 *New Eng. J. on Crim. & Civ. Confinement* 301-30 Summ '88

Constitutional law: "newly minted" standard of review for prison regulations has bittersweet impact on prisoners' rights [Turner v. Safley, 107 S. Ct. 2254]. 27 *Washburn L.J.* 654-70 Spr '88

Constitutional law: Turner v. Safley [107 S. Ct. 2254]. Prisoners' first amendment and marriage rights in conflict with prison regulations. 56 *UMKC L. Rev.* 589-602 Spr '88

Constitutional law—a prison regulation that allows inmate to inmate correspondence only if prison officials approve of such correspondence is reasonably related to legitimate prison security concerns. A regulation allowing prisoners to marry only if the prison superintendent determines there are compelling reasons for marriage is unconstitutional. Turner v. Safley, 107 S. Ct. 2254. 26 *J. Fam. L.* 856-62 '87/'88

Constitutional law—prisoners' rights—prison regulations constitutionally valid if reasonably related to legitimate penological interests—Turner v. Safley, 107 S. Ct. 2254. 19 *Seton Hall L. Rev.* 429-55 '89

Constitutional law—right to privacy—there is no reasonable expectation of privacy in a prison cell where a prison official intentionally deprives prisoner of noncontraband property when a state postdeprivation remedy is available. Hudson v. Palmer, 104 S. Ct. 3194. 10 *T. Marshall L. Rev.* 693-710 Spr '85

Source: ILP, 1988/89, "Prisons and Prisoners" p. 512, © H.W. Wilson Company, used with permission.

Exhibit 4–51

Amend. 14, § 1   CONSTITUTION

Section 1. Due Process of Law

* * * nor shall any State deprive any person of life, liberty, or property, without due process of law; * * *

Source: U.S.C.A., Constitution Amendment 14 sec. 1, © West Publishing, used with permission.

Exhibit 4–52

NOTES OF DECISIONS

Subdivisions I to XLIV appear in this volume.

I. GENERALLY 1–60
II. STANDARDS OR TESTS 61–90
III. RIGHTS PROTECTED 91–160
IV. STATE ACTION 161–310
V. METHODS OF DEPRIVATION—GENERALLY 311–350
VI. ADMINISTRATIVE RULES AND REGULATIONS 351–410
VII. LEGISLATIVE ENACTMENTS GENERALLY 411–500
VIII. OVERBROAD LEGISLATION 501–570
IX. VAGUE AND INDEFINITE LEGISLATION 571–870
X. PROCEDURAL DUE PROCESS GENERALLY 871–910
XI. ADMINISTRATIVE PROCEDURE GENERALLY 911–950
XII. CIVIL PROCEDURE—GENERALLY 951–1070
XIII. JURISDICTION GENERALLY 1071–1100
XIV. IN PERSONAM JURISDICTION 1101–1300
XV. SERVICE OF PROCESS 1301–1390
XVI. STATUTES OF LIMITATIONS 1391–1440
XVII. VERDICT, JUDGMENT, OR EXECUTION 1441–1490
XVIII. REMEDIES OR RELIEF 1491–1570
XIX. CRIMINAL PROCEDURE—GENERALLY 1571–1700
XX. POLICE MISCONDUCT 1701–1740
XXI. PROSECUTORIAL MISCONDUCT 1741–1790
XXII. JUDICIAL MISCONDUCT 1791–1840
XXIII. PRESENCE OR PHYSICAL APPEARANCE OF DEFENDANT 1841–1890
XXIV. COMPETENCY OF DEFENDANT 1891–1940
XXV. GRAND JURY 1941–2000
XXVI. INDICTMENT 2001–2070
XXVII. INFORMATION 2071–2110
XXVIII. PLEAS 2111–2180
XXIX. DEFENSES 2181–2230
XXX. EVIDENCE GENERALLY 2231–2320
XXXI. ADMISSIBILITY OF EVIDENCE GENERALLY 2321–2420
XXXII. ADMISSIBILITY OF CONFESSIONS 2421–2480
XXXIII. IDENTIFICATION OF ACCUSED 2481–2580
XXXIV. SUPPRESSION OF EVIDENCE BY PROSECUTION 2581–2710
XXXV. WITNESSES 2711–2790
XXXVI. CHARGES OR INSTRUCTIONS TO JURY 2791–2870
XXXVII. SENTENCE AND PUNISHMENT 2871–3090
XXXVIII. PROBATION 3091–3150
XXXIX. APPEAL OR REVIEW 3151–3220
XL. PRISONS AND PRISONERS—GENERALLY 3221–3450
XLI. PAROLE 3451–3530
XLII. CONTEMPT PROCEEDINGS 3531–3600

Source: U.S.C.A., Constitution Amendment 14 sec. 1, Notes-outline, © West Publishing, used with permission.

Exhibit 4–53

DUE PROCESS

XL. PRISONS AND PRISONERS—GENER-
ALLY

Subdivision Index

Generally 3221
Abortions 3370
Access to courts
   Generally 3299
   Administrative, judicial, or legislative forums
   3300
   Administrative or disciplinary segregation
   3310

Literature, newspapers, or publications
   Generally 3367
   Pretrial detainees 3243
Maid service, pretrial detainees 3244
Mail, access to courts 3304
Marriage 3368
Maximum or minimum security
   Transfers or classifications within institution
   3265
   Transfers to other institutions 3290
Medical care
   Generally 3369
   Abortions 3370

Source: U.S.C.A., Amendment 14 sec. 1, Notes outline "Prisons," © West Publishing, used with permission.

Exhibit 4–54

3368. Marriage

Inmate marriage regulation, which prohibited inmates from marrying other inmates or civilians unless prison superintendent approved marriage after finding that there were compelling reasons for doing so, was not reasonably related to any legitimate penological objective, so as to be facially invalid as denial of inmates' constitutional rights. Turner v. Safley, U.S.Mo.1987, 107 S.Ct. 2254, 482 U.S. 78, 96 L.Ed.2d 64.

County jail regulation forbidding employees from becoming socially involved with inmates in or out of jail did not violate Fourteenth Amendment due process to marry, even though inmate whom guard had married had been transferred to a different facility; rather, regulation was justified by fear that if guard became romantically involved with inmate after transfer to another facility, she might become a facilitator of unlawful communication between him and others and a potential pro-

vider of favored treatment for him, inmates would have enhanced incentive to "romance" guards, and prisoners not married or engaged to guards would attribute any differences in treatment between themselves and prisoners to the relationship harming morale. Keeney v. Heath, C.A.7 (Ind.) 1995, 57 F.3d 579.

Wisconsin correctional regulations requiring prisoner seeking to marry nonprisoner to take part in six premarital counseling sessions bore rational relationship to legitimate penological interests and, thus, did not violate prisoners' due process rights; regulations were designed to protect nonprisoners from victimization or exploitation and to contribute to prisoner rehabilitation, they provided alternative to counseling by prison chaplain, permitting marriage on demand would severely burden prison resources, and there was no obvious alternative that would . . .

Source: U.S.C.A., Amend. 14 sec. 1, Notes, Prison-marriage, © West Publishing, used with permission.

Problem 6

Assume you want to shed light on whether state prisoners' habeas corpus petitions, filed under federal statute 28 U.S.C. § 2254, are clogging the federal district courts. You want some indication of filing trends over time, so you focus on the years 1974, 1984, and 1994. One strategy would be to try to identify the number of reported federal district court decisions each of those years that involve a prisoner's petition for writ of habeas corpus. This approach seems not only cumbersome; it would also be unreliable because it would miss a large number of habeas corpus filings that are resolved through unreported decisions. Accordingly, to get a more accurate picture of filing trends, we may wish to consult other sources that provide descriptive information about the business of the federal courts.

What sources are likely to contain the information we are seeking? Books and journal articles written on the subject likely contain statistical information. We could search the holdings of various libraries for relevant books and find references to journal articles by using either ILP or LegalTrac. However, we would be fortunate indeed to find a single book or journal article containing data for all the years we are seeking. A more promising line of inquiry may be to consult references that specialize in providing statistical information. The **Sourcebook of Criminal Justice Statistics**, discussed briefly in Chapter 2, is an annual compendium of statistical information describing crime and the criminal justice system in the United States. Unfortunately, this book does not contain the statistical information necessary to answer this question.

Specialized indexes exist for documents containing statistical information. In Chapter 2, we mentioned three statistical reference series published by the Congressional Information Service (CIS). Each series is published 10 times a year, with annual cumulations. The **American Statistics Index (ASI)** contains references to U.S. government publications that include statistical information; the **Statistical Reference Index** provides references to state government and privately published documents that provide statistical information; and the **Index to International Statistics** includes references to documents with statistical information of an international nature. CIS also produces a CD-ROM product, **Statistical Masterfile**, which contains the information found in all three of the series. Since we are looking for information regarding the U.S. federal court system, the American Statistics Index seems like the logical place to begin our search. We will be using the printed set, although this same strategy could be employed using the CD-ROM Statistical Masterfile.

The ASI is divided into two parts: an Index volume and an Abstract volume. Each of the 10 issues published during the year, as well as the cumulative annual volumes, come in this two-part format. The researcher optimally uses the Index volume to gain entry to the accompanying Abstract volume, where more detailed information about possible source documents can be found. The Index volume contains several useful indexes, including a Subject and Name Index. The Abstract volume contains bibliographic information about each document, as well as a brief description of the document's contents.

Because 1994 is the most recent year for which we are seeking information, we begin with the volumes for 1994. If we look in the Subject and Name Index under the heading *"Habeas corpus,"* we find a reference that appears to be directly on point: "Prisoners petitions filed in Federal courts of appeals and district courts, by type of petition, circuit, and district, as of Mar 1994, annual rpt, 18204-11." (See Exhibit 4–55.) Although the document listed includes information only up to March 1994, the citation indicates that it is an annual report, so once we find the title for this annual series, we should be able to get the 1994 data by looking in the annual report for the following year. If we are very fortunate, the series may be comprehensive enough to provide data for 1974 and 1984 as well. Although the reference we have found looks like an excellent source, note that the first document listed under **"Habeas corpus"** also looks promising.

Notice that each reference is followed by a number. For example, the documents described as "Prisoners petitions filed . . ." is followed by "18204-11." This accession number allows the researcher to locate the bibliographic information and description of the contents of the document in the companion Abstract volume. The abstracts are arranged in numerical order by the cited accession number. The information in the Abstract volume for the entry referring us to 18204-11 is displayed in Exhibit 4–56.

The title and date of the document, **Federal Judicial Caseload Statistics**, Mar. 31, 1994, followed by "Annual. [1994]" indicate that this document is an annual publication. Note the information at the bottom of the record, **"JU10.21:994,"** which is the Superintendent of Documents (SUDOC) classification number. SUDOC numbers are assigned to most federal documents and used as a shelving scheme by most libraries. If we have access to a library with a federal government document collection, we should be able to use this number to find the document on the shelves.

Exhibit 4–55

Index by Subjects and Names

Haas, Ellen
"Food and Consumer Services: Agenda for the Future", 1504-9.1

Habeas corpus
Court civil and criminal caseloads for Federal district, appeals, and special courts, by offense, circuit, and district, 1993, annual rpt, 18204-8

Deportation and exclusion cases, writs of habeas corpus, judicial review, and declaratory judgments, FY87-93, annual rpt, 6264-2.5

Prisoners petitions filed in Federal courts of appeals and district courts, by type of petition, circuit, and district, as of Mar 1994, annual rpt, 18204-11

Habuchi, Tomonori
"Oncogene Amplification in Urothelial Cancers with p53 Gene Mutation or . . .

Source: American Statistics Index, 1994 Annual-Index "Habeas Corpus," © CIS, used with permission.

Exhibit 4–56

18204–11   **FEDERAL JUDICIAL CASELOAD STATISTICS, Mar. 31, 1994**
Annual. [1994.]
iii + 7 + App 108 p.
•Item 0729-D. † ASI/Mf/4
˙JU10.21:994. LC 79-643081.
Annual report on civil and criminal workloads in U.S. courts of appeals and district and special courts, as of Mar. 31, 1994. Also includes data on Pretrial Services Act defendants and persons under the supervision of the Federal Probation System, and juries.
Contents:
a. Narrative summary, with 1 chart and 4 tables. (p. 1–7)

b. Appeals, civil, and criminal case workload. 16 tables showing cases pending at beginning and end of year, filed, and terminated by method; time interval from filing of case to disposition; and criminal defendants involved and disposed of by type of disposition; variously by source of appeal, nature of suit and proceeding, and offense or offense type. (p. A1–A75)

c. Federal Probation System. 2 tables showing persons received for, removed from, and currently under supervision, by originator (judge, magistrate, parole, supervised and mandatory release, military and special parole, and transfer). (p. A76–A84)

Source: ASI, 1994 Annual Abstracts, p. 883, © CIS, used with permission.

The CIS also provides microfiche copies of the documents indexed in the ASI, which are arranged first by the year in which the entry appears in the ASI, and then by the accession number. Annual series receive the same accession number for each year, so we continue to use the same accession number, 18204–11, if we are looking in the microfiche collection for Federal Judicial Caseload Statistics for years other than 1994. Not all federal documents are distributed in paper format, and not all have Superintendent of Documents classification numbers assigned to them. ASI provides references to documents and the microfiche of the actual documents for many items that are not routinely distributed to libraries.

Let us return to our sample record at 18204–11 (see Exhibit 4–56). A description of the contents of the document appears under the bibliographic information. Paragraph **b.**, under "**Contents**," seems to cover the type of information we are seeking. Note the page numbers, p. A1-A75, where the information can be found, appears at the bottom of the paragraph.

We next examine the information in the Abstract volume for the other entry we found in the Index volume, accession number "**18204-8**." The beginning of the abstract for this document is displayed in Exhibit 4–57. Notice the periodicity information, "**annual**," and the publication date, "**1994**," directly beneath the lengthy title. Also, note that this document has a Superintendent of Documents classification number, **JU10.1/2:993.** The description of this document indicates that it is actually a compilation of three reports: Reports of the Proceedings of the

Exhibit 4–57

18204-8 1993 U.S. COURTS: SELECTED REPORTS. REPORTS OF THE PROCEEDINGS OF THE JUDICIAL CONFERENCE OF THE U.S.; ACTIVITIES OF THE ADMINISTRATIVE OFFICE OF THE U.S. COURTS; JUDICIAL BUSINESS OF THE U.S. COURTS
Annual. 1994.
517 p. var. paging.
•Item 0728.
GPO, price not given.
ASI/MF/8 •JU10.1/2:993.

Compilation of the following reports:
Reports of the Proceedings of the Judicial Conference of the U.S., Mar. 16 and Sept. 20, 1993. *Reports of the Proceedings* are also published separately (see ASI 18202-2). (78 p.)
Activities of the Administrative Office of the U.S. Courts, 1993. Includes 1 chart and 2 tables showing Federal judicial funding for FY93-94 and FY94 budget request, by judicial dept and selected judicial service; and Article III judgeship vacancies, nominations, and contributions, selected years FY81-93. (47 p.)
Judicial Business of the U.S Courts, 1993, Annual Report of the Director. (389 p.)
Activities of the Administrative Office of the U.S. Courts was formerly published separately. No reports for 1993 or 1992 were issued. Previous re-port, for 1991, is described in ASI 1991 Annual, 18204-2.2

Judicial Business presents data on activities of the Administrative Office of the U.S. Courts, U.S. courts of appeals, and district courts, including workloads, judges and administrative personnel, appropriations, and operating costs, for year ended Sept. 30, 1993, with trends from 1970s. Report is prepared for the fall session of the Judicial Conference of the U.S.

Judicial Business contains:
a. Main report, with text statistics, 3 charts, and 28 summary tables, listed below. (p. 1–36)
b. 10 supplemental tables, also listed below. (p. 38-57)
c. 90 detailed appendix tables, also listed below. (p. AI.1-AI.313)
d. Summary of Criminal Justice Act (CJA) program activities, with 3 tables showing FY92 expenditures for defender organizations, panel attorneys, transcripts, and administration; percent change in caseloads, by major offense, June 30, 1991-92; and number of excess compensation cases for attorneys, and largest payments, by circuit, FY92. (p. AII.1-AII.9)

Previous report, for 1992, was titled *Reports of the Proceedings of the Judicial Conference of the U.S.; Annual Report of the Director of the Administrative Office of the U.S. Courts* (see ASI 1993 Annual under this number).

ASI coverage began with report FY73 (see ASI First Annual Supplement under this number).

Source: ASI, 1994 Annual Abstracts, p. 881, © CIS, used with permission.

Judicial Conference of the U.S., Activities of the Administrative Office of the U.S. Courts, and Judicial Business of the U.S. Courts. Judging by the description, Judicial Business of the U.S. Courts seems most likely to contain the data that are of interest to us. However, as we continue to read the description, we learn that this compilation is new and that the previous report, entitled Reports of the Proceedings of the Judicial Conference of the U.S.; Annual Report of the Director of the

Administrative Office of the U.S. Courts, may provide the data we seek from 1974 and 1984.

Our next step is to find the cited documents either on the library shelves or in the ASI microfiche collection. We used the Superintendent of Documents classification number to find the annual series, Federal Judicial Caseload Statistics, in the government documents collection in our library. However, the annual for 1995 is not available, so we consult the 1995 annual for our second source, **Judicial Business of the United States Courts**. (See Exhibit 4–58.) We consult the volume's table of contents in order to locate information about state prisoners' habeas corpus petitions.(See Exhibit 4–59.) It is important to remember that state prisoners' habeas corpus petitions are considered part of the *civil* caseload of the U.S. district courts, especially because no specific reference is made in the table of contents to "prisoner petitions." There is, however, a reference under Appendix I to "U.S. District Courts—Civil," and there we locate the table containing the data we want.

Exhibit 4–58

**Reports of the Proceedings
of the Judicial Conference of the United States**

•

**Activities of the Administrative Office
of the United States Courts**

•

Judicial Business of the United States Courts

Administrative Office of the United States Courts
Thurgood Marshall Federal Judiciary Building
Washington, D.C. 20544

Source: Reports of the Proceedings of the Judicial Conference of the United States, Activities of the Administrative Office of the United States Courts, Judicial Business of the United States Courts, Administrative Office of the United States Courts.

Exhibit 4–59

Source: Reports of the Proceedings of the Judicial Conference of the United States, Activities of the Administrative Office of the United States Courts, Judicial Business of the United States Courts, Administrative Office of the United States Courts.

(See Exhibit 4–60.) This table reports that 11,918 habeas corpus petitions were filed by state prisoners during 1994.

Our next step is to locate the corresponding information for 1974 and 1984. We refer to a slightly different format of the same series for the years 1975 and 1985, Reports of the Proceedings of the Judicial Conference of the United States; Annual Report of the Director of the Administrative Office of the United States. In this instance, the table of contents provides a specific reference to data on prisoner petitions for these earlier years. (See Exhibit 4–61.) Both volumes contain the information we wanted in tabular format. (See Exhibits 4–62 and 4–63.)

Our figures for the three years are: 1974, 7,626 petitions; 1984, 8,349 petitions; and 1994, 11,913 petitions. With this information in hand, we are in a position to analyze whether state prisoners' habeas corpus petitions truly are clogging the federal district courts. We can consult either books or journal articles to gain additional perspective on the issue. Remember our earlier admonition about the possible misuse or misinterpretation of statistical information. In this instance, a quick look at the data may suggest a dramatic increase in habeas corpus petitions and that it has become too easy for prisoners to file frivolous suits. However, such a conclusion may not withstand analysis.

We check the various periodical indexes such as LegalTrac, the ILP, and **Public Affairs Information Service (PAIS)** for recent articles and books bearing on this

Exhibit 4-60

Circuit and District	Total Private Civil Cases	Contract	Real Property	FELA	Marine Personal Injury	Motor Vehicle Personal Injury	Other Personal Injury	Other Tort Actions	Antitrust	Civil Rights	Prisoner Petitions Habeas Corpus	Civil Rights	Mandamus and Other	Copyright Patent Trademark	Labor Suits	All Other
Total	190,981	26,439	2,517	1,967	2,245	4,754	32,478	3,016	658	29,636	11,918	37,925	397	6,872	14,919	15,240
DC	1,586	232	14	6	1	47	148	29	18	355	43	256	5	40	205	187
1ST	5,283	1,065	197	41	138	161	739	115	18	901	153	351	7	297	407	693
ME	499	83	6	9	18	15	111	7	–	89	18	75	–	17	17	34
MA	2,587	529	35	25	91	55	290	67	13	428	68	95	3	200	269	419
NH	588	95	7	4	5	22	94	13	–	128	27	63	1	29	28	72
RI	596	116	8	3	7	25	88	10	1	96	27	74	3	23	51	64
PR	1,013	242	141	–	17	44	156	18	4	160	13	44	–	28	42	104
2ND	18,282	3,428	266	308	161	508	2,346	377	89	2,720	731	2,203	44	1,095	2,137	1,869
CT	2,010	417	45	23	7	37	159	121	6	459	30	280	10	87	134	195
NY.N	1,315	82	4	43	2	37	126	13	4	255	103	355	4	48	111	128
NY.E	4,894	569	68	147	35	216	1,127	75	27	644	237	228	21	183	800	517
NY.S	8,603	2,213	127	84	116	179	764	154	43	1,033	272	1,018	7	715	1,007	871
NY.W	1,161	91	8	9	1	25	133	9	6	278	77	281	2	43	77	121
VT	299	56	14	2	–	14	37	5	3	51	12	41	–	19	8	37
3RD	17,781	3,221	205	531	145	775	2,148	263	92	2,771	890	3,050	23	569	1,508	1,590
DE	582	63	4	–	3	22	35	9	1	61	50	174	1	58	26	75
NJ	5,237	1,116	48	59	39	188	611	71	26	639	262	771	–	277	535	595
PA.E	7,858	1,511	89	374	73	431	1,153	101	38	1,428	308	1,137	14	145	556	500
PA.M	1,635	184	15	7	–	93	122	22	10	211	118	618	5	26	86	118
PA.W	2,161	308	15	91	25	39	164	55	17	413	141	349	1	59	297	187
VI	308	39	34	–	5	2	63	5	–	19	11	1	2	4	8	115
4TH	14,521	1,882	139	81	85	617	2,329	231	33	1,934	881	3,816	51	361	1,148	933
MD	3,564	448	38	12	17	132	754	90	5	469	205	698	25	80	334	257
NC.E	1,077	116	5	9	12	22	86	11	2	149	51	485	1	32	38	58
NC.M	532	60	2	4	2	13	45	6	2	109	35	117	4	32	37	64
NC.W	614	97	1	3	–	14	71	7	1	118	17	105	1	37	40	102
SC	2,492	414	43	10	15	201	444	38	5	408	174	454	8	39	131	108
VA.E	3,452	451	22	10	34	116	452	47	11	409	239	1,069	6	108	299	179
VA.W	1,620	98	9	12	1	58	263	18	4	139	118	697	1	21	87	94
WV.N	365	80	2	–	1	23	63	1	–	43	14	69	–	1	34	34
WV.S	805	118	17	21	3	38	151	13	3	90	28	122	5	11	148	37

Source: Reports of the Proceedings of the Judicial Conference of the United States, Activities of the Administrative Office of the United States Courts, Judicial Business of the United States Courts, Administrative Office of the United States Courts.

> **Exhibit 4–61**

Source: Reports of the Proceedings of the Judicial Conference of the United States, Activities of the Administrative Office of the United States Courts, Judicial Business of the United States Courts, Administrative Office of the United States Courts.

issue and consult library catalogs for additional books on the general topic. Among the several recent references we find is a 1994 book published by the National Center for State Courts, **Habeas Corpus in State and Federal Courts**, a book that provides a substantial analysis of state prisoners' federal habeas corpus filings, including the table displayed in Exhibit 4–64.

Although this table does not cover all the years that are relevant to our question, it provides important related information. It notes that while the absolute number of habeas corpus filings by state prisoners has increased, the proportion of prisoners filing habeas petitions has declined over time. This finding suggests that the growth in the number of habeas corpus petitions filed may be explained by increases in state prison populations, rather than by other factors. We have added this additional step not to arrive at any conclusions regarding the filing of habeas corpus petitions by state prisoners, but rather to demonstrate that limited knowledge can be a dangerous thing. A thorough research effort should produce not only data but also related materials necessary to put the data into perspective.

Exhibit 4–62

Table 24
Prisoner Petitions Filed in the U.S. District Courts
Fiscal Years 1966 to 1975

Type of Petition	Fiscal years										Percent change	
	1966	1967	1968	1969	1970	1971	1972	1973	1974	1975	1975 over 1966	1975 over 1974
Total all petitions	8,540	10,443	11,152	12,924	15,997	16,266	16,267	17,218	18,410	19,307	126.1	4.9
Petitions by Federal prisoners	2,292	2,639	2,851	3,612	4,185	4,121	4,179	4,535	4,987	5,047	120.2	1.2
U.S. Parole Board reviews	64	104	131	150	232	202	268	466	371	662	934.4	78.4
Motions to vacate sentence	863	958	1,099	1,444	1,729	1,335	1,591	1,722	1,822	1,690	95.8	-7.2
Habeas corpus	1,017	1,045	1,045	1,373	1,600	1,671	1,368	1,294	1,718	1,682	65.4	-2.1
Other prisoner petitions	348	532	576	645	624	913	952	1,053	1,076	1,013	191.1	-5.9
Mandamus, etc	333	474	516	564	488	699	700	639	631	535	60.7	-15.2
Civil rights	15	58	60	81	136	214	252	414	445	478	(¹)	7.4
Petitions by state prisoners	6,248	7,804	8,301	9,312	11,812	12,145	12,088	12,683	13,423	14,260	128.2	6.2
Habeas corpus	5,339	6,201	6,488	7,359	9,063	8,372	7,949	7,784	7,626	7,843	46.9	2.8
Other prisoner petitions	909	1,603	1,813	1,953	2,749	3,773	4,139	4,899	5,797	6,417	605.9	10.7
Mandamus, etc	691	725	741	684	719	858	791	725	561	289	-58.2	-48.5
Civil rights	218	878	1,072	1,269	2,030	2,915	3,348	4,174	5,236	6,128	2,711.0	17.0

¹Percent not computed where there are less than 25 cases.

Source: Reports of the Proceedings of the Judicial Conference of the United States, Activities of the Administrative Office of the United States Courts, Judicial Business of the United States Courts, Administrative Office of the United States Courts.

Exhibit 4-63

Table 24
U.S. District Courts
Prisoner Petitions Filed
During the Twelve Month Periods Ended June 30, 1975 through 1984

Type of Petition	1975	1976	1977	1978	1979	1980	1981	1982	1983	1984	% Change 1984/1983
Total	19,307	19,809	19,537	21,924	23,001	23,287	27,711	29,303	30,775	31,107	1.1
Petitions by Federal Prisoners	5,047	4,780	4,691	4,955	4,499	3,713	4,104	4,328	4,354	4,526	4.0
Motions to Vacate Sentence	1,690	1,693	1,921	1,924	1,907	1,322	1,248	1,186	1,311	1,427	8.8
Habeas Corpus	2,344	1,959	1,745	1,851	1,664	1,465	1,680	1,927	1,914	1,905	−0.5
Mandamus, etc.	535	626	542	544	340	323	342	381	339	372	9.7
Civil Rights	478	502	483	636	588	603	834	834	790	822	4.1
Petitions by State Prisoners	14,260	15,029	14,846	16,969	18,502	19,574	23,607	24,975	26,421	26,581	0.6
Habeas Corpus	7,843	7,833	6,866	7,033	7,123	7,031	7,790	8,059	8,532	8,349	−2.1
Mandamus, etc.	289	238	228	206	184	146	178	175	202	198	−2.0
Civil Rights	6,128	6,958	7,752	9,730	11,195	12,397	15,639	16,741	17,687	18,034	2.0

Source: Reports of the Proceedings of the Judicial Conference of the United States, Activities of the Administrative Office of the United States Courts, Judicial Business of the United States Courts, Administrative Office of the United States Courts.

Exhibit 4–64

14 *Habeas Corpus in State and Federal Courts*

These figures show that even though a smaller portion of prisoners overall are filing habeas corpus petitions in federal court, the increase in raw number of prisoners accounts for the increase in numbers of habeas petitions filed.

Table 1
State Prisoner Habeas Filings in U.S. District Courts
as Percentage of State Prisoner Population

Year	Number of State Prisoners	Habeas Corpus Filings	Percent
1961	196,453	1,020	0.52%
1962	194,886	1,408	0.72%
1963	194,155	2,106	1.08%
1964	192,627	3,694	1.92%
1965	189,855	4,845	2.55%
1966	180,409	5,839	3.24%
1967	175,317	6,201	3.54%
1968	168,211	6,488	3.86%
1969	176,384	7,359	4.17%
1970	176,391	9,063	5.14%
1971	177,113	8,372	4.73%
1972	174,379	7,949	4.56%
1973	181,396	7,784	4.29%
1974	196,105	7,626	3.89%
1975	216,462	7,843	3.62%
1976	235,853	7,833	3.32%
1977	267,936	6,866	2.56%
1978	277,473	7,033	2.53%
1979	288,086	7,123	2.47%
1980	305,458	7,031	2.30%
1981	341,255	7,790	2.28%
1982	384,689	8,059	2.09%
1983	405,312	8,492	2.10%
1984	427,739	8,309	1.94%
1985	462,284	8,482	1.83%
1986	500,725	9,012	1.80%
1987	536,784	9,480	1.77%
1988	577,672	9,852	1.71%
1989	653,392	10,543	1.61%
1990	635,974	10,809	1.70%
1991	752,525	10,323	1.00%

Source: Habeas Corpus in the Federal Courts, 1994, p. 14, © National Center for State Courts, used with permission.

Problem 7

Penny Donaldson shot and killed her husband, Ralph, following an argument in their New Jersey home. She has been charged with murder. Although Penny admits killing Ralph, she claims that Ralph was assaulting her and that she acted in self-defense. To support her defense, she plans to testify that Ralph repeatedly beat her during their marriage. Penny wants to introduce expert testimony about the battered-woman syndrome. Will such expert testimony be admissible at her trial on the issue of self-defense?

How you begin your research will be influenced by how much you already know about this issue. If you already have some understanding of the battered-woman syndrome and its relationship to a claim of self-defense, you may wish to proceed immediately to New Jersey case law that has ruled on the admissibility of expert testimony in this context. If you lack familiarity with either the battered-woman syndrome or the law of self-defense, it makes sense to do some background research before you jump into case law. Let us assume we are in the latter posture. We first turn to the secondary authorities for assistance. We begin with West's legal encyclopedia, **Corpus Juris Secundum (C.J.S.)**. Using the General Index, we find "**Self-Defense**," and the subtopic, "**Battered spouse syndrome, Crim L § 53.**" (See Exhibit 4–65.) When we look up § 53 in the volume of C.J.S. that includes criminal law, we discover a general discussion of the use of force in defense of self, another, or property, a portion of which deals specifically with the battered-woman syndrome. (See Exhibit 4–66.) We also check the pocket supplement for later cases and additional textual discussion of the issues.

Exhibit 4–65

SELF-DEFENSE
    Generally, see Title Index to Assault and Battery
Affray, Affray § 13
    Evidence, Affray § 13
    Instructions to jury, Affray § 20
Ambassadors and consuls, assault on foreign diplomat, Ambass § 20
Animals, injury to or killing trespassing animals, criminal liability, Anim § 322
Armed services,
    Conscientious objectors, conscription or draft, exemptions, Armed S § 57

Insurance beneficiaries, homicide, right to proceeds, Armed S § 228
Arrest,
    Arresting officer, use of force in making arrest Arrest § 49
    Assault and battery, resisting, Asslt&B § 19
    Misdemeanor arrests, use of force, Arrest § 49
Assault and battery, see Title Index to Assault and Battery
Attorney and client, professional ethics, Atty&C § 45
Battered spouse syndrome, **Crim L § 53**

Source: C.J.S. General Index "Battered spouse syndrome" p. 714, © West Publishing, used with permission.

Exhibit 4–66

§ 53. — Defense of Self, Another, or Property

A person is justified in the use of force against an aggressor when and to the extent it appears to him and he reasonably believes that such conduct is necessary to defend himself or another against such aggressor's imminent use of unlawful force.

Battered woman syndrome.

A battered woman syndrome is not a defense in itself, [26] and that defendant is the victim of a battering relationship is not alone sufficient to raise the issue of self-defense.[27] Under the self-defense test that accused must reasonably believe his conduct is necessary to defend himself, in a battered wife case, the objective test is how a reasonably prudent battered wife would perceive her husband's behavior.[28]

Thus, the issue of self-defense is not raised, when the victim is defendant's estranged husband, and has physically abused defendant in the past, where defendant suffers fear alone, without more, for fear alone is not sufficient to establish the appearance of imminent danger, where there is no showing that the victim has made any threatening or aggressive comments or gestures towards defendant.[29] On the other hand, self-defense is raised in such a battered wife situation where the evidence reveals that the husband threatened to kill defendant[30] or made a movement to produce a weapon,[31] for then the appearance of imminent danger is sufficiently established.

26. Wash.—State v. Walker, 700 P.2d 1168, 40 Wash.App. 658, reconsideration denied, review denied.

27. Wash.—State v. Walker, 700 P.2d 1168, 40 Wash.App. 658, reconsideration denied, review denied.

28. Kan.—State v. Hundley, 693 P.2d 475, 236 Kan. 461.

29. Wash.—State v. Walker, 700 P.2d 1168, 40 Wash.App. 658, reconsideration denied, review denied.

30. Wash.—State v. Allery, 682 P.2d 312, 101 Wash.2d 591.

31. Wash.—State v. McCullum, 656 P.2d 1064, 98 Wash.2d 484.

Source: C.J.S. Criminal Law sec. 53, © West Publishing, used with permission.

Legal encyclopedias are most useful at the earliest stages of research to provide an overview of a topic and citations to case law. We expect other authorities to discuss self-defense and the battered-woman syndrome in much greater depth. The library shelves contain several books about this issue—for example, Sara Lee Johann & Frank Osaka, **Representing . . . Battered Women Who Kill** (Charles C Thomas, 1989). We can consult many additional books as well as A.L.R. annotations. The Index to A.L.R. 2d, 3d, 4th, 5th, Federal, and L. Ed. 2d (under the heading "**Self-Defense**," in combination with different subtopics) refers us to an annotation, 18 A.L.R.4th 1153, which covers the admissibility of expert or opinion testimony on the battered-woman syndrome. (See Exhibit 4–67.) We use this annotation and its pocket part to find a New Jersey case addressing this issue, *State v. Kelly*, 97 N.J. 178, 478 A.2d 364 (1984). (See Exhibit 4–68.)

We search for recent law review analyses of the battered-woman syndrome as well. By consulting the List of Subject Headings in the front of a recent volume of

Exhibit 4–67

ALR INDEX

SELF-DEFENSE—Cont'd

Accident insurance, death or injury intentionally inflicted by another as due to accident or accidental means, 49 ALR3d 673

Adultery, relationship with assailant's wife as provocation depriving defendant of right of self-defense, 9 ALR3d 933

Arrest, right to resist excessive force used in accomplishing lawful arrest, 77 ALR3d 281

Assault and battery

–battered woman syndrome, admissibility of expert or opinion testimony on battered wife or battered woman syndrome, 18 ALR4th 1153

–character or reputation, other's character or reputation for turbulence on question of self-defense by one charged with assault or homicide, 1 ALR3d 571

Source: ALR Index, p. 1029 "Battered." Permission has been granted by the current copyright holder, West Group. Further reproduction of any kind is strictly prohibited. For additional information, please contact West Group Customer Services representative at 1-800-328-4880.

Exhibit 4–68

ANNOTATION
ADMISSIBILITY OF EXPERT OR OPINION TESTIMONY ON
BATTERED WIFE OR BATTERED WOMAN SYNDROME
by
James O. Pearson, Jr., J.D.

. . . reacted as reasonable person would have acted in similar circumstances would have been made clearer by expert testimony going to battered women's syndrome. Lentz v State (1992, Miss) 604 So 2d 343.

In prosecution for murder based on defendant's stabbing of husband, wherein defendant asserted self-defense, although record revealed that battered woman's syndrome had sufficient scientific basis to produce uniform and reasonably reliable results, conclusive ruling that expert's proffered testimony about battered woman's syndrome would satisfy state's standard of acceptability for scientific evidence would not be made, where state was not given full opportunity in trial court to

question expert's methodology in studying battered women or implicit assertion that battered woman's syndrome had been accepted by relevant scientific community. State v Kelly (1984) 97 NJ 178, 478 A2d 364.

In prosecution for murder in which defendant asserted "battered woman's syndrome" as defense and presented expert testimony that she was in fear of imminent bodily harm to either herself or her child at time she shot her husband, trial court did not violate defendant's constitutional rights against self-incrimination by ordering her to submit to psychiatric examination by state psychiatrist and . . .

Source: ALR annotation, 18 ALR 4th 1153. Permission has been granted by the current copyright holder, West Group. Further reproduction of any kind is strictly prohibited. For additional information, please contact West Group Customer Services representative at 1-800-328-4880.

the ILP (September 1994–August 1995), we find that "**Battered woman syndrome**" and "**Battered women**" both serve as headings for citations to law review articles. (See Exhibit 4–69.) Several articles are referenced under these headings. (See Exhibit 4–70.) Remember that you can expect to find a wealth of authorities cited in the footnotes of law review articles in addition to the critical discussion provided in the text. Also recall that LegalTrac sometimes is more convenient than the ILP for collecting helpful citations to legal periodicals.

The New Jersey case, *State v. Kelly,* cited in the A.L.R. annotation, provides a lead to a judicial decision that we will want to pursue. When we look up *Kelly* at 478 A.2d 364, we encounter a lengthy opinion, with 36 headnotes, that appears to be a groundbreaking decision in New Jersey on the battered-woman syndrome and its use in self-defense cases. (See Exhibit 4–71.) When we Shepardize *Kelly,* using all the appropriate volumes of Shepard's for the Atlantic Reporter 2d, our impression is confirmed that this is an important case regarding the admissibility of expert testimony about the battered-woman syndrome. We reproduce the oldest Shepard's that includes 478 A.2d 364, which reveals numerous citations to *Kelly* by courts from several jurisdictions. (See Exhibit 4–72.)

Finally, we illustrate how we might have uncovered relevant case law had we chosen to bypass secondary authorities and had instead used the New Jersey case digests and Descriptive Word Index. We find promising topics and key numbers by using the New Jersey Digest 2d Descriptive Word Index, including the supplement, under headings such as "**Battered Woman Syndrome**," "**Battered Women**" (see Exhibit 4–73), "**Expert Testimony**— . . . **Battered or abused women or spouses**" (see Exhibit 4–74), and "**Self-Defense**— . . . **homicide**" (see Exhibit 4–75). The subject of expert testimony regarding the battered-woman syndrome appears to be associated with Criminal Law ⌐ 474.4(3). We thus turn to the New Jersey Digest 2d and its supplement (see Exhibit 4–76) under that topic and key number, where we are referred to recent cases. We would be able to use additional digests, including the **Decennial Digest** and the **General Digests**, to find cases from other jurisdictions addressing this topic.

Exhibit 4–69

LIST OF SUBJECT HEADINGS

Bar associations
   See also Integrated bar
Bar examinations
Bar examiners *See* Bar examinations
Barristers
   See also Attorneys; Solicitors

Basis rules *See* Income tax/Basis rules
Battered child syndrome
Battered woman syndrome
Battered women
   See also Battered woman syndrome; Husband
   and wife

Source: Index to Legal Periodicals, 1994/95, List of Subject Headings, © H.W. Wilson Company, used with permission.

Exhibit 4–70

Washington (State)

When no one hears their cries: battered child syndrome as a defense, State v. Janes [850 P.2d 495 (Wash.1993)] R. M. Parker, student author. 19 *T. Marshall L. Rev.* 431-49 Spr '94

Battered woman syndrome

Battered woman syndrome, expert testimony, and the distinction between justification and excuse. R. F. Schopp, B. J. Sturgis, M. Sullivan. 1994 *U. Ill. L. Rev.* 45-113 '94

Battered woman syndrome testimony: Dunn v. Roberts [963 F.2d 308 (1992)], justice is done by the expansion of the battered woman syndrome. K. B. Kuhn, student author. 25 *U. Tol. L. Rev.* 1039-65 '94

Battered women and the law. M. Dowd. 30 *Trial* 62-3+ Jl '94

Beyond self-defense: the use of battered woman syndrome in duress defenses. S.D. Appel, student author. 1994 *U. Ill. L. Rev.* 955-80 '94

The dynamics of domestic violence: understanding the response from battered women. M. A. Dutton. 68 *Fla B.J.* 24-8 O '94

The legal victimization of battered women. M. E. Kampmann, student author. 15 *Women's Rts. L. Rep.* 101-13 Fall '93

Posttraumatic stress disorder among battered women: analysis of legal implications. M. A. Dutton, L. A. Goodman. 12 *Behavioral Sci. & L.* 215-34 Summ '94

Understanding battered woman syndrome. L. E. A. Walker. 31 *Trial* 30-4+ F '95

Understanding women's responses to domestic violence: a redefinition of battered woman syndrome. M. A. Dutton. 21 *Hofstra L. Rev.* 1191-242 Summ '93

Will the "real" battered woman please stand up? In search of a realistic legal definition of battered woman syndrome. A. R. Callahan, student author. 3 *Am. U. J. Gender & L.* 117-52 Fall '94

California

Deaf justice? Battered women unjustly imprisoned prior to the enactment of Evidence Code Section 1107. S. G. Baker, student author. 24 *Golden Gate U. L. Rev.* 99-130 Spr '94

Canada

Conjugal homicide and legal violence: a comparative analysis. A. Young. 31 *Osgoode Hall L.J.* 761-808 Wint '93

Florida

Clemency and the battered woman. M. R. Schlakman. 68 *Fla. B.J.* 72-4 O '94

Great Britain

Battered woman syndrome and defenses to homicide: where now? C. Wells. 14 *Legal Stud.* 266-76 Jl '94

Conjugal homicide and legal violence: a comparative analysis. A. Young. 31 *Osgoode Hall L.J.* 761-808 Wint '93

Michigan

The battered woman as criminal defendant. G. D. Rodwan, J. A. Dagher-Margosian. 73 *Mich. B.J.* 912-16 S '94

Battered women

   See also

   Battered woman syndrome

   Husband and wife

The battered woman and homelessness. G. P. Mullins. 3 *J.L. & Pol'y* 237-55 '94

Behavioral and affective correlates of Borderline Personality Organization in wife assaulters. D. G. Dutton. 17 *Int'l J.L. & Psychiatry* 265-77 Summ '94

Broken promises: a call for witness tampering sanctions in cases of child and domestic abuse. V I. Vieth. 18 *Hamline L. Rev.* 181-99 Wint '94

Caught in a web: immigrant women and domestic violence. D. L. Jang. 28 special issue *Clearinghouse Rev.* 397-405 '94

Do battered women have a right to bear arms? S. Blodgett-Ford, student author. 11 *Yale L. & Pol'y Rev.* 509-60 '93

Domestic violence: a history of arrest policies and a survey of modern laws. N. James, student author. 28 *Fam. L.Q.* 509-20 Fall '94

The federal Violence Against Women Act: an urban perspective. W. D. Haynes, L. D. Bernard. 73 *Mich. B.J.* 949-50 S '94

Source: Index to Legal Periodicals, 1994/95, "Battered Woman syndrome" p. 72, © H.W. Wilson Company, used with permission.

Exhibit 4–71

364 N.J. **478 ATLANTIC REPORTER, 2d SERIES**

97 N.J. 178
STATE of New Jersey,
Plaintiff-Respondent,
v.
Gladys KELLY, Defendant-Appellant.
Supreme Court of New Jersey.
Argued May 10, 1983.
Decided July 24, 1984.

Defendant was indicted for murder and convicted in the Superior Court, Essex County, and she appealed. The Superior Court, Appellate Division, affirmed. The Supreme Court granted certification and Wilentz, C.J., held that: (1) battered-woman's syndrome was relevant to honesty and reasonableness of defendant's belief that she was in imminent danger of death or serious injury and was appropriate subject for expert testimony; (2) defendant's expert's conclusions were sufficiently reliable under New Jersey's standards for scientific testimony, but State was entitled to further opportunity to question expert's methodology; (3) defendant's expert was sufficiently qualified; (4) exclusion of battered woman's testimony required reversal and remand for new trial; (5) defendant's 17-year-old daughter's testimony as to victim's subjecting her to physical and sexual abuse should have been admitted; (6) testimony of defendant's prior conviction of conspiracy to commit robbery was admissible; (7) prosecutor's conduct and arguments did not deny defendant fair trial; (8) instructions on provocation were deficient; and (9) five-year sentence in state prison was not excessive.

Reversed and remanded for new trial.

Handler, J., concurred in part and dissented in part with opinion.

1. Homicide ☞ **116(2)**
Self-defense exonerates person who kills in reasonable belief that such action was necessary to prevent his or her death or serious injury, even though this belief was later proven mistaken, and law accordingly requires only reasonable, not necessarily correct, judgment. N.J.S.A. 2C:3-4, subds. a, b(2).

2. Homicide ☞ **116(3)**
While it is not imperative that actual necessity exist, valid plea of self-defense will not lie absent actual, i.e., honest belief on part of defendant in necessity of using force and existence of necessity and necessity to act does not give rise to meritorious plea of self-defense where defendant was unaware of that necessity. N.J.S.A. 2D:3-4, subds. a, b(2).

3. Homicide ☞ **276**
Whether defendant actually believed in necessity of acting with deadly force to prevent imminent, grave attack is question for jury in prosecution for homicide.

4. Homicide ☞ **116(4)**
Defendant claiming privilege of self-defense must establish, in addition to honest belief, that belief in necessity to use force was reasonable, and if belief was unreasonable under circumstances, such belief does not constitute complete justification for homicide. N.J.S.A. 2C:3-4, subds. a, b(2).

5. Homicide ☞ **300(7)**
For defendant to prevail on theory of self-defense in homicide prosecution, jury need not find beyond reasonable doubt that defendant's belief was honest and reasonable, but rather, if any evidence raising issue of self-defense is adduced, either in state's or defendant's case, then jury must be instructed that state is required to prove beyond reasonable doubt self-defense claim does not accord with facts, and acquittal is required if there remains reasonable doubt whether defendant acted in self-defense. N.J.S.A. 2C:3-4, subds. a, b(2).

6. Criminal Law ☞ **469**
In prosecution for murder which resulted in conviction of reckless . . .

Source: Atlantic Reporter 2d, 478 A2d 364, © West Publishing, used with permission.

Exhibit 4–72

ATLANTIC REPORTER, 2d SERIES					Vol. 478
—364—	Cir. 2	R I	470So2d781	N Y	Ore
New Jersey	j 985F2d1170	f 612A2d731	547So2d1278	497NE51	695P2d989
v Kelly	Cir. 6	Vt	616So2d1098	488NYS2d359	S C
1984	738FS^{6}1133	575A2d193	616So2d1099	506NYS2d28	339SE122
(97NJ178)	Cir. 7	Ala	Kan	506NYS2d463	417SE90
s 453A2d859	957F2d422	603So2d414	716P2d567	520NYS2d918	Tex
486A2d^{13}843	Conn	Calif	Mass	587NYS2d470	683SW589
489A2d^{19}730	629A2d1111	251CaR829	514NE1344	N C	711SW651
f 495A2d^{11}130	Pa	264CaR180	574NE344	366SE591	W V
	502A2d^{8}257	13CaR2d337	Minn	Ohio	359SE564
	534A2d780	Colo	441NW798	551NE972	359SE565
628A2d^{1}356	j 547A2d362	833P2d760	Miss	551NE977	73A^{4}1002n
628A2d^{5}356	598A2d^{13}968	Fla	460So2d785	Okla	
		467So2D763		840P2d7	

Source: Shepard's-Atlantic Reporter 2nd, 478 A2d 364. Reproduced by permission of Shepard's. Further reproduction of any kind is strictly prohibited.

Exhibit 4–73

BANKS AND BANKING—Cont'd
DEPOSITS—
    Withdrawal. Banks 119
D'OENCH doctrine. Banks 505
FAITHLESS employee defense—
    Collections
        Forged endorsements. Banks 174
FALSE statements to insured bank. Banks 509.20
FEDERAL preemption—
    State laws or regulations. States 18.19
FIDUCIARY relationship. Banks 100
FUNDS, insured bank, misapplication. Banks 509.15
MARGIN requirements, see this index Securities Regulation
PRIVACY—
    Right to Financial Privacy Act. Banks 151
PRODUCTS liability—
    Mismanagement of custodian management securities account—
    Whether action lay in tort or contract. Action 27(5)
RACKETEERING—
    Enterprise. RICO 43
STATUTORY provisions—
    Branch banking. Banks 33
WIRE transfers—
    Stop orders. Banks 188^{1}/$_{2}$

BARRATRY
CLASS suits—
    Notice. Fed Civ Proc 177

BARS
CIVIL rights, access, see this index Civil Rights
SEX and nudity—
    Freedom of speech and press. Const Law 90.4(5)

BATTERED PERSON SYNDROME
MARITAL-LIKE relationships—
    Elements of syndrome. Assault 2

BATTERED WOMAN SYNDROME
MARITAL-LIKE relationships—
    Elements of syndrome. Assault 2
SELF-DEFENSE—
    Honesty of belief. Homic 116(3)
    Reasonableness of belief. Homic 116(4)

BATTERED WOMEN
EXPERT testimony. Crim Law 474.4(3)

BEACHES
EASEMENTS. Nav Wat 46(2)
MUNICIPAL corporations—
    Liability

Source: New Jersey Digest 2d, DWI, "Battered," © West Publishing, used with permission.

Exhibit 4–74

EXPERIMENTS
EVIDENCE—
Competency in criminal prosecutions. Crim Law
   388-388.10
EXPERT TESTIMONY
ABUSED children. Crim Law 474.4(4)
ACCIDENT reconstruction evidence, see Recon-
   struction evidence, post
AID to jury. Crim Law 469.1
ARSON. Crim Law 476.2
BALLISTICS and weapons. Crim Law 476.1
BATTERED or abused women or spouses. Crim
   Law 474.4(3)

BITEMARKS. Crim Law 475.7
BLOOD alcohol. Crim Law 475.2(4)
BODILY condition or appearance—
   Subject of expert testimony. Crim Law 473
BODILY tissue or fluids. Crim Law
   475.2(3)
CAUSE and effect—
   Subject of expert testimony. Crim
    Law 476
CHARACTER traits or profiles—
   Generally. Crim Law 474.4
   Defendant. Crim Law 474.4(5)

Source: NJ Digest 2d, DWI, "Expert Testimony," © West Publishing, used with permission.

Exhibit 4–75

SELF-DEFENSE
ASSAULT—
   Civil action. Assault 13, 30
   Criminal prosecution—
    Assault 67, 96(3)
    Homic 96
   Questions for jury. Assault 95
CONTINUANCE for evidence to show. Crim
   Law 595(6)
HOMICIDE, defense to prosecution for. Homic
   108-121
   Admissibility of evidence. Homic 186-195
   Curtilage of home, privilege of self-defense
    without retreat. Homic 118(3)
   Instructions. Homic 300
   Instructions to jury on burden of proof,
    prejudicial error. Crim Law 778(12),
    1172(2)
   Killing of husband by wife's paramour. Homic
    112(3)

Manner of repelling attack as affecting. Homic
   119
Nature and imminence of danger.
   Homic 115
Nature and purpose of attack. Homic 110
Necessity of act. Homic 96(2), 117
Presumptions and burden of proof. Homic
   151(3)
Questions for jury. Homic 276
Reasonable doubt, burden of proving falsity of
   defense, necessity for instructions. Crim
   Law 1038.2
Renewal of contest. Homic 121
Weight and sufficiency of evidence. Homic
   244
   Withdrawal after aggression. Homic 113
INSTRUCTIONS to jury—
   Crim Law 782(16)
   Homic 300

Source: NJ Digest 2d, DWI, "Self Defense," © West Publishing, used with permission.

Exhibit 4–76

🔑 **474.4(3). Battered or abused women or spouses.**

N.J.Super.A.D. 1995. Expert testimony to establish Battered Woman's Syndrome can be introduced in criminal proceeding, but application is limited to explaining victim's reactions or late reporting of events and not as evidence that the crime alleged actually occurred. N.J.S.A. 2C:12-1, subd. b(4), 2C:12-3, subd. a, 2C:14-2, subd. a.—State v. Ellis, 656 A.2d 25, 280 N.J.Super. 533.

N.J.Super.A.D. 1990. Testimony concerning battered woman syndrome was admissible to bolster victim's credibility in prosecution of defendant on charge of assault in order to explain why the victim remained with the defendant for the remainder of the day.—State v. Frost, 577 A.2d 1282, 242 N.J.Super. 601, certification denied 604 A.2d 596, 127 N.J. 321.

N.J.Super.A.D. 1990. Existence and impact of battered woman's syndrome is appropriate subject for expert testimony.—State v. Myers, 570 A.2d 1260, 239 N.J.Super. 158, certification denied 604 A.2d 598, 127 N.J. 323.

Defendant who invokes battered woman's syndrome and intends to introduce expert testimony with regard to it must submit to examination by appropriate experts selected by State, results of which will be admissible to rebut defense.—Id.

Source: NJ Digest 2d, Crim. Law 474.4(3) squibs, © West Publishing, used with permission.

Problem 8

Assume that you are familiar with Dr. Lenore Walker's published research regarding the battered-woman syndrome, including her books, **The Battered Woman Syndrome** and **Terrifying Love: Why Battered Women Kill and How Society Responds**. Further assume that Dr. Walker is scheduled to testify as a witness for the defense in a murder trial in which self-defense and the battered-woman syndrome are expected to arise as issues. Let us suppose that you are interested in identifying judicial decisions involving self-defense and the battered-woman syndrome in which Dr. Walker's research has been cited, and you also are interested in finding law review articles discussing Dr. Walker's research. How would you go about locating such references?

A series called **Shepard's Law Review Citations** provides references to court cases that have cited particular law review articles, but it covers only selected law reviews. Thus, if Dr. Walker has published an article in one of the covered law reviews, we could Shepardize the article to find cases that have cited it. Either LegalTrac or the ILP can be consulted to find law review articles authored by Dr. Walker. Using this approach, we look as far back as 1980 for references to articles written by Dr. Walker without success. Even if we had found law review articles authored by Dr. Walker and had been able to use Shepard's Law Review Citations to find citing cases, we would still miss cases that had cited Dr. Walker's books and other writings that were not published as law review articles.

This type of problem can be researched conveniently by using either LEXIS or WESTLAW. Because these computer systems allow us to look for either key words or phrases appearing in the text of documents, we can design a search strategy using Dr. Walker's name and her area of expertise to find cases and law review articles citing her work. In this example we will use the LEXIS system.

Let us assume that we are interested in both federal and state cases that make reference to Dr. Walker's work. We select the **MEGA™** library on LEXIS, and then the **MEGA** file, which contains the text of all reported federal and state judicial opinions. Remember, you can check the printed **LEXIS-NEXIS Directory of Online Services** if you need help selecting the appropriate *library* and *file* to use on the LEXIS service. We have elected to use the Boolean search mode, which, for reasons we will explain, will yield more precise results than a FREESTYLE search. We next construct our search strategy. Because we are interested in any judicial references to Dr. Walker's work, we search the **opinions** segment, which includes the text of the majority opinions as well as dissenting and concurring opinions. We might perform a very simple search by enclosing Dr. Walker's name within quotation remarks, which would look like this: **OPINION("LENORE WALKER")**. This search strategy retrieves 39 cases. (See Exhibit 4–77.) How-

Exhibit 4–77

OPINION ("LENORE WALKER")

Your search request has found 39 CASES through Level 1.
To DISPLAY these CASES press either the KWIC, FULL, CITE or SEGMTS key.
To MODIFY your search request, press the M key (for MODFY) and then the ENTER key.

For further explanations, press the H key (for HELP) and then the ENTER key.

Source: MEGA, "Lenore Walker" search. Reprinted with the permission of LEXIS-NEXIS, a division of Reed Elsevier Inc. LEXIS and NEXIS are registered trademarks of Reed Elsevier Properties Inc. FREESTYLE, KWIC, SuperKWIC and MEGA are trademarks of Reed Elsevier Properties Inc. SHEPARD'S and SHEPARDIZE are registered trademarks of Shepard's Company, a Partnership.

ever, this search would miss cases in which Dr. Walker's full name does not appear and in which her middle name or initial is included. Therefore, we might be better off to design a strategy that makes reference both to her last name and to key phrases that appear in the title of her works or describe the contents of her work.

We are fortunate that a key phrase, "**battered woman**," appears in the title of at least one of her works, and describes the subject matter of her work. Our strategy would be to **enter OPINION(WALKER w/25 "BATTERED WOM*N").** (We use the asterisk to retrieve either "**battered woman**" or "**battered women**.") The proximity requirement **w/25** is included to limit the retrieved documents to cases where the name "**Walker**" appears within 25 words of either form of the phrase. Since we know the name of another of Dr. Walker's works, **Terrifying Love**, we can include that phrase in our search strategy as well; however, because "**battered women**" appears in the subtitle of that work, any case that includes a reference to the full title would be picked up by our unmodified search strategy. Our final search strategy is **OPINION(WALKER w/25 "BATTERED WOM*N" "TERRIFYING LOVE").** We retrieve 73 cases using this new search strategy. (See Exhibit 4–78.)

We begin browsing our search results using the **.kw** (for "key word in context") option, and quickly determine that our strategy is successful in retrieving documents that cite Dr. Walker's work. (See Exhibit 4–79.)

Exhibit 4–78

OPINION (WALKER W/25 ("BATTERED WOM*N" OR "TERRIFYING LOVE"))

Your search request has found 73 CASES through Level 1.
To DISPLAY these CASES press either the KWIC, FULL, CITE or SEGMTS key.
To MODIFY your search request, press the M key (for MODFY) and then the ENTER key.

For further explanations, press the H key (for HELP) and then the ENTER key.

Source: MEGA, Walker edited. Reprinted with the permission of LEXIS-NEXIS, a division of Reed Elsevier Inc. LEXIS and NEXIS are registered trademarks of Reed Elsevier Properties Inc. FREESTYLE, KWIC, SuperKWIC and MEGA are trademarks of Reed Elsevier Properties Inc. SHEPARD'S and SHEPARDIZE are registered trademarks of Shepard's Company, a Partnership.

Exhibit 4–79

---

505 U.S. 833, *; 112 S. Ct. 2791, **;
1992 U.S. LEXIS 4751, ***1; 120 L. Ed. 2D 674

DISPOSITION: (=1) 947 F.2d 682: No. 91-902, affirmed; No. 91-744, affirmed part, reversed in part, and remanded.

OPINION:

. . . [*891] [**2828] [***108] 12.

Other studies fill in the rest of this troubling picture. Physical violence is only the most visible form of abuse. Psychological abuse, particularly social and economic isolation of women, is also common. L. [***109] Walker, The Battered [*892] Woman Syndrome 27-28 (1984). Many victims of domestic violence remain with their abusers, perhaps because they perceive no superior alternative. Herbert, Silver, & Ellard, Coping with an Abusive Relationship: I. How and Why do Women Stay?, . . .

Source: MEGA, Walker browse. Reprinted with the permission of LEXIS-NEXIS, a division of Reed Elsevier Inc. LEXIS and NEXIS are registered trademarks of Reed Elsevier Properties Inc. FREESTYLE, KWIC, SuperKWIC and MEGA are trademarks of Reed Elsevier Properties Inc. SHEPARD'S and SHEPARDIZE are registered trademarks of Shepard's Company, a Partnership.

The second part of this problem concerns locating law review articles discussing Dr. Walker's research. We might attack this issue by looking up articles on the battered-woman syndrome in either the ILP or LegalTrac and then consulting the articles to see which ones discuss Dr. Walker's research. This would be a very time-consuming way to proceed. Once again, CALR provides us with a more efficient way to approach this problem. We can use the **.cl** (for "change library") command on LEXIS to switch to the law review (**LAWREV**) library, and then enter **allrev** to proceed to the file containing the full text of law review articles. (See Exhibit 4–80.)

We use the same basic search strategy we used in the case law databases, although we must change the segment name **"opinions"** to **"text"**. This search strategy retrieves 223 documents, a considerable number to review. (See Exhibit 4–81.)

At this point, we might consider taking advantage of some of the unique features offered by the FREESTYLE search feature on LEXIS. Recall that the FREESTYLE feature is designed to retrieve documents ranked in order by relevancy. Using the FREESTYLE feature allows us to focus on the law review articles that make the most references to Dr. Walker's work, which presumably contain more than just a passing reference to her research. The default option in FREESTYLE retrieves the 25 most relevant documents.

We can use a simple search strategy incorporating the name **"Lenore Walker."** Standard citation format in more recently published law review articles requires use of an author's full name, rather than just a last name or a last name with a first

Exhibit 4–80

---

allrev

Please ENTER the NAME of the file you want to search. To see a description of a file, type its page number and press the ENTER key.

FILES—PAGE 1 Of 26 (NEXT PAGE for additional files)

NAME	PG	DESCRIP	NAME	PG	DESCRIP

— — — — — — — — — — — — — LAW REVIEW LIBRARY — — — — — — — — — — — — — —

— — — — — LAWREV GROUP FILES — — — — — — — ANNOTATIONS & INDEXES — — — —

NAME	PG	DESCRIP	NAME	PG	DESCRIP
ALLREV	1	Combined Law Review Files	ALR	1	ALR and LEd2d Annotations
LRALR	1	Combined ALLREV & ALR			
BARJNL	1	Combined Bar Journals			
WGLTXJ	1	Combined WGL Tax Journals			

— — — — — — — — — — — LAW REVIEW ARTICLES BY TOPIC — — — — — — — — — — —

NAME	PG	DESCRIP	NAME	PG	DESCRIP
ADRMLR	2	ADR and Mediation Law	FAMLR	2	Family Law
BANKLR	2	Banking Law	INSLR	2	Insurance Law
BKRTLR	2	Bankruptcy Law	IPLR	2	Intellectual Property Law
COMLR	2	Commercial Law	INTLR	2	International Law
CORPLR	2	Corporate Law	LABLR	2	Labor & Employment Law
ENVLR	2	Environmental Law	PROPLR	3	Property Law
ESTLR	2	Estate Law	SECLR	3	Securities Law
ETHICS	2	Legal Ethics Law	TAXLR	3	Tax Law

Press NEXT PAGE (or .np) to view Law Reviews by Jurisdiction files.
Press .np3 to view Individual Law Reviews and Law Journal files.

Source: LawRev, Walker. Reprinted with the permission of LEXIS-NEXIS, a division of Reed Elsevier Inc. LEXIS and NEXIS are registered trademarks of Reed Elsevier Properties Inc. FREESTYLE, KWIC, SuperKWIC and MEGA are trademarks of Reed Elsevier Properties Inc. SHEPARD'S and SHEPARDIZE are registered trademarks of Shepard's Company, a Partnership.

---

initial. We also include a phrase describing the subject matter in which we are interested, "**battered women**." We put both elements in quotations to ensure that they are searched as phrases. Including the phrases in quotations will not prevent the retrieval of documents that refer to "Lenore E. Walker" instead of "Lenore Walker," nor will it prevent retrieval of documents including the phrase "**battered woman**" instead of "**battered women.**" The FREESTYLE feature is flexible enough to pick up plural forms and middle initials even for search phrases that appear in quotations.

The search results screen allows us the option of analyzing our search strategy by using either the **.where** or **.why** commands discussed in Chapter 3. (See Exhibit 4–82.) However, in the law review file we might want to rely on the cite (**.ci**) command to display the bibliographic information for the documents. The first few cites produced seem to indicate a successful search result. (See Exhibit 4–83.)

Exhibit 4–81

TEXT (WALKER W/25 ("BATTERED WOM*N" OR "TERRIFYING LOVE"))

Your search request has found 223 ITEMS through Level 1.
To DISPLAY these ITEMS press either the KWIC, FULL, CITE or SEGMTS key.
To MODIFY your search request, press the M key (for MODFY) and then the ENTER key.

For further explanations, press the H key (for HELP) and then the ENTER key.

Source: ALLREV, result. Reprinted with the permission of LEXIS-NEXIS, a division of Reed Elsevier Inc. LEXIS and NEXIS are registered trademarks of Reed Elsevier Properties Inc. FREESTYLE, KWIC, SuperKWIC and MEGA are trademarks of Reed Elsevier Properties Inc. SHEPARD'S and SHEPARDIZE are registered trademarks of Shepard's Company, a Partnership.

Exhibit 4–82

Your FREESTYLE search has retrieved the top 25 documents based on statistical ranking. Search terms are listed in order of importance.

"LENORE WALKER" "BATTERED WOMEN"

Press ENTER to view documents in KWIC or use Full, Cite or Segmnt keys.

(=1)   Browse documents in SuperKWIC (.SK)
(=2)   Location of search terms in documents (.where)
(=3)   Number of documents with search terms (.why)
(=4)   Change document order (.sort)

For further explanation, press the H key (for HELP) and then the ENTER key.

Source: ALLREV, Walker freestyle. Reprinted with the permission of LEXIS-NEXIS, a division of Reed Elsevier Inc. LEXIS and NEXIS are registered trademarks of Reed Elsevier Properties Inc. FREESTYLE, KWIC, SuperKWIC and MEGA are trademarks of Reed Elsevier Properties Inc. SHEPARD'S and SHEPARDIZE are registered trademarks of Shepard's Company, a Partnership.

Exhibit 4–83

LEVEL 1—25 ITEMS

1. Copyright (c) 1996 University of Puget Sound Law Review Puget Sound Law Review, Winter, 1996, 19 Puget Sound L. Rev. 385, 20764 words, NOTE: State v. Riker, Battered Women Under Duress: The Concept the Washington Supreme Court Could Not Grasp, Ann-Marie Montgomery*

2. Copyright (c) Michigan Law Review 1991. Michigan Law Review, October, 1991 90 Mich. L. Rev. 1, 53602 words, ARTICLE: LEGAL IMAGES OF BATTERED WOMEN: REDEFINING THE ISSUE OF SEPARATION., Martha R. Mahoney*

3. Naval Law Review, 1996, 43 Naval L. Rev. 111, 11820 words, ARTICLE: SPOUSE BATTERING AS AGGRAVATED ASSAULT: A Proposal to Modify the UCMJ, Lieutenant Commander Peter A. Dutton, JAGC, USN*

4. Copyright (c) 1993 Golden Gate University Golden Gate University Law Review SUMMER, 1993, 23 Golden Gate U.L. Rev 829, 13566 words, COURTROOM, CODE AND CLEMENCY: REFORM IN SELF-DEFENSE JURISPRUDENCE FOR BATTERED WOMEN, Panel 1 Discussion n1

.MORE	Next Page .NP	Cite	.CI	Exit FREESTYLE .BOOL		Print Doc .PR
.WHERE	Prev Page .PP	Kwic	.KW	New Search	.NS	Print All .PA
.WHY	Next Doc .ND	Full	.FU	Modify	.M	Cmds Off .COFF
.SORT	Prev Doc .PD	SKWIC	.SK	Chg Library	.CL	Sign Off .SO

Source: ALLREV, result. Reprinted with the permission of LEXIS-NEXIS, a division of Reed Elsevier Inc. LEXIS and NEXIS are registered trademarks of Reed Elsevier Properties Inc. FREESTYLE, KWIC, SuperKWIC and MEGA are trademarks of Reed Elsevier Properties Inc. SHEPARD'S and SHEPARDIZE are registered trademarks of Shepard's Company, a Partnership.

Problem 9

Detective Goode, of the Connecticut State Police, secured a warrant to search the home of Nathaniel Martin for cocaine and related paraphernalia. The search resulted in the seizure of nearly a kilogram of the illegal drug. Martin's attorney has filed a motion to suppress the seized cocaine on the ground that the warrant authorizing the search was not supported by probable cause. The prosecutor has conceded that probable cause was lacking when the magistrate issued the search warrant. She contends, however, that Detective Goode relied on the warrant in good faith to search Martin's home and that the cocaine should not be suppressed. Is a "good faith exception" to the exclusionary rule recognized under Connecticut law? Let us assume that we are unfamiliar with the concept of a "good faith exception" to the exclusionary rule and that we want to plan our research strategy accordingly. We start with one of the most basic forms of secondary legal authority, **Words and Phrases**. Using the pocket supplement to volume 18A of Words and Phrases, which includes "**good faith exception**," we find several helpful judicial definitions of that term, with accompanying case citations. (See Exhibit 4–84.)

Exhibit 4–84

GOOD FAITH EXCEPTION

Federal "good faith exception" provides that, in the absence of allegation that magistrate abandoned is detached and neutral role, suppression is appropriate only if officers were dishonest or reckless in preparing the affidavit or could not have harbored an objectively reasonable belief in the existence of probable cause; Texas good faith exception goes beyond requirement of federal exclusionary rule and applies only if the affidavit supporting the warrant sets forth grounds for probable cause. Lockett v. State, Tex.App.–Hous. [14 Dist.], 852 S.W.2d 636, 637.

Because exclusionary rule exists to deter willful or flagrant violations of constitutional right by police officers, "good faith exception" will work to defeat suppression of evidence only where officer had reasonable grounds to believe that warrant was properly issued. State v. Jones, 6 Dist., 595 N.E.2d 485, 488, 72 Ohio App.3d 522.

"Good faith exception" to exclusionary rule provides that evidence is not to be suppressed under exclusionary rule where that evidence was discovered by officers acting in good faith and in reasonable, though mistaken, belief that they were authorized to take those actions. U.S. v. Pichany, C.A.Ind., 687 F.2d 204, 209.

Under the "good-faith exception" to the exclusionary rule, evidence will be admitted in the prosecution's case-in-chief if it is obtained by law enforcement officials acting in objectively reasonably reliance on a search warrant issued by detached and neutral magistrate, even if the affidavit on which the warrant was based is insufficient to establish probable cause. U.S. v. Maggitt, C.A.5 (Miss.), 778 F.2d 1029, 1034.

"Good faith exception" to Fourth Amendment exclusionary rule explicated in United States v. Leon does not indiscriminately do away with requirement of probable cause for issuance of search warrant, but when applicable, recognizes that officers who have acted with objective good faith have right to rely upon issuing magistrate's determination that substantial basis existed for finding probable cause. State v. Murphy, Mo.App., 693 S.W.2d 255, 266.

Under "good-faith exception," evidence seized by officer who reasonably relied on search warrant that is later deemed to be invalid is admissible if officer acted in good faith. U.S. v. Wanless, C.A.9 (Wash.), 882 F.2d 1459, 1466.

Source: WORDS and PHRASES, 118A W&P 126, "Good Faith," © West Publishing, used with permission.

We next consult a legal encyclopedia to find out more about the good faith exception to the exclusionary rule. This time, we choose **Am. Jur. 2d.** We search for likely terms in the General Index and eventually find a promising citation under the general heading **"Searches and Seizures,"** and the subtopic **"Evidence . . . —good-faith exception to exclusionary rule generally, Evid § 603-608."** (See Exhibit 4–85.) These sections in the corresponding volume of Am. Jur. 2d present an informative discussion, including mention of the fact that several state courts have declined to recognize the good faith exception to the exclusionary rule. A Connecticut case, *State v. Marsala,* 216 Conn. 150, 579 A.2d 58 [(1990)], is listed among the decisions that have rejected the good faith exception on state constitutional grounds. (See Exhibit 4–86.)

Treatises and casebooks can also be helpful starting places for research. Of the two treatises we have selected, we begin with the index to Wayne R. LaFave's

Exhibit 4–85

GENERAL INDEX

SEARCHES AND SEIZURES—Cont.d
Envelope—Cont'd
– probable cause and exigent circumstances, Search §191
Equitable relief, Search §341
Errors. Mistakes and inaccuracies, infra
Escape and escapees
– exigent circumstances, Search §76
– hot pursuit, Search §188
– scope of protection, Search §10
Estrangement, Search §100, 102
Ether, Search §82
Eviction
– hotel or motel, Search §96
– rented premises, Search §95
Evidence
  generally, Evid §601-608
– analysis, procedural guide for, Evid §602
– civil actions for wrongful interception, Search §330
– consent, Search §258
– constitutional right in particular rules of evidence, Const L §677
– criminal actions, Search §353
– destruction of evidence, Search §76
– discovery, Depos & D §3
– disposable evidence, Search §188
– documentary evidence, search of public records, Evid §1377
– Eavesdropping (this index)
– exceptions, Evid §603-607
– federal rules, Fed R Evid §6, 9, 18, 187, 414

– foreign searches and seizures, applicability exclusionary rule to, Evid §591
– Fourth Amendment exclusionary rule, Evid §601
– good-faith exception to exclusionary rule generally, Evid §603-608
– – deficiency of warrant as to place to be searched, Evid §607
– – false affidavit, warrant issued on, Evid §607
– – neutral, magistrate's failure to remain, Evid §607
– – objective reasonableness as test of good faith, Evid §605
– – probable cause, inapplicability of exception where warrant based on affidavit lacking, Evid §607
– – retroactive application of exception, Evid §604
– – silver platter doctrine, Evid §608
– – warrantless search in reliance on statute, Evid §606
– hearsay, infra
– mere evidence, seizure of, Search §160
– presumptions and burden of proof, infra
– sentencing, consideration of illegal evidence, Crim L §528
– unlawful search, generally, Search §4
– warrants, inapplicability to of Federal Rules of Evidence, Evid §15, 18
– Witnesses (this index)
Examinations. Physical characteristics and examinations, infra

Source: Am.Jur. 2d, General Index, "Evidence." Permission has been granted by the current copyright holder, West Group. Further reproduction of any kind is strictly prohibited. For additional information, please contact West Group Customer Services representative at 1-800-328-4880.

five-volume set, **Search and Seizure: A Treatise on the Fourth Amendment** (3d ed., West 1996), where we find that the "**'Good Faith' Exception**" is discussed in several sections, most prominently in section 1.3. (See Exhibit 4–87.) As promised, section 1.3 provides an extensive discussion of the good faith exception to the exclusionary rule. Note that we also are referred to an article in

Exhibit 4–86

§603. Good-faith exception to exclusionary rule

The Supreme Court has modified the Fourth Amendment exclusionary rule so as not to bar the use, in the prosecution's case in chief, of evidence obtained by officers acting in reasonable reliance on a search warrant issued by a detached and neutral magistrate which ultimately is found to be defective.[71] Although the term "good faith" does not fully capture the objective nature of the inquiry required for application of this exception, it is nevertheless known as the "good-faith exception" as a shorthand description.[72] The terms "reasonable reliance," "objective good-faith," and "objectively reasonable" are also used to refer to the conduct required of officers to whom the exception applies.[73]

A number of state courts have declined to apply the good-faith exception to the exclusionary rule, finding in their respective criminal statutes[74] or state constitutions[75] grounds for excluding the tainted evidence. While the federal Constitution establishes certain minimum levels of protection which are equally applicable to the analogous state constitutional provisions, each state has the power to provide broader standards, and go beyond the minimum floor which is established by the federal Constitution.[76]

The principal rationale for the good-faith exception is that the suppression of evidence obtained in objectively reasonable reliance on a subsequently invalidated search warrant would produce only marginal or nonexistent benefits which cannot justify the substantial cost of exclusion.[77] Application of the exclusionary rule is properly restricted to situations in which its remedial purpose is effectively advanced.[78] Exclusion of evidence will not deter police misconduct where an officer acting with objective good faith has obtained a search warrant from a judge or magistrate, and has acted within its scope.[79] In most such cases, there is no police illegality and thus nothing to deter. In the ordinary case, an officer cannot be expected to question the magistrate's probable cause determination or judgment that the form of the warrant is technically sufficient.[80] Similarly, suppressing evidence because a judge failed to make all the necessary clerical corrections with respect to a warrant, despite the judge's assurances that such changes would be made, will not serve the deterrent function of the exclusionary rule,[81] which is not designed to punish the errors of judges and magistrates, and moreover, there is no basis for believing that exclusion of evidence seized pursuant to a warrant would have a significant deterrent effect on judges or magistrates.[82]

71. Massachusetts v Sheppard, 468 US 981, 82 L Ed 2d 737, 104 S Ct 3424, on remand 394 Mass 381, 476 NE2d 541; United States v Leon, 468 US 897, 82 L Ed 2d 677, 104 S Ct 3405, reh den 468 US 1250; 82 L Ed 2d 942, 105 S Ct 52.

Annotations: Admissibility in criminal case of evidence obtained by law enforcement officer allegedly relying reasonably and in good faith on defective warrant, 82 L Ed 2d 1054.

Practice References: Cook, Constitutional Rights of the Accused 2d § 3:63.

Hunter, Federal Trial Handbook 2d § 38.2.

72. United States v Savoca (CA6 Ohio) 761 F2d 292, cert den 474 US 852, 88 L Ed 2d 126, 106 S Ct 153; United States v Leary (CA10 Colo) 846 F2d 592.

73. United States v Savoca (CA6 Ohio) 761 F2d 292, cert den 474 US 852, 88 L Ed 2d 126, 106 S Ct 153.

74. Polk v State (Tex App Dallas) 704 SW2d 929, petition for discretionary review gr (Mar 4, 1987) and petition for discretionary review ref (Mar 4, 1987) and affd (Tex Crim) 738 SW2d 274.

75. State v Marsala, 216 Conn 150, 579 A2d 58; State v Rothman (Hawaii) 779 P2d 1; People v Bigelow, 66 NY2d 417, 497 NYS2d 630, 488 NE2d 451; Commonwealth v Edmunds, 526 Pa 374, 586 A2d 887, ALR4th 1738.

76. Commonwealth v Edmunds, 526 Pa 374, 586 A2d 887.

Law Reviews: Yagla, The Good Faith Exception To The Exclusionary Rule: The Latest Ex-

continues

Exhibit 4–86 continued

ample Of "New Federalism" In The States, 71 Marquette L Rev 166 (1987).

77. United States v Leon, 468 US 897, 82 L Ed 2d 677, 104 S Ct 3405, reh den 468 US 1250, 82 L Ed 2d 942, 105 S Ct 52.

78. Illinois v Krull, 480 US 340, 94 L Ed 2d 364, 107 S Ct 1160, on remand 126 Ill 2d 235, 128 Ill Dec 105, 534 NE2d 125.

79. United States v Leon, 468 US 897, 82 L Ed 2d 677, 104 S Ct 3405, reh den 468 US 1250, 82 L Ed 2d 942, 105 S Ct 52.

80. United States v Leon, 468 US 897, 82 L Ed 2d 677, 104 S Ct 3405, reh den 468 US 1250, 82 L Ed 2d 942, 105 S Ct 52.

81. Massachusetts v Sheppard, 468 US 981, 82 L Ed 2d 737, 104 S Ct 3424, on remand 394 Mass 381, 476 NE2d 541.

82. United States v Leon, 468 US 897, 82 L

Source: Am.Jur. 2d, Evidence sec 603. Permission has been granted by the current copyright holder, West Group. Further reproduction of any kind is strictly prohibited. For additional information, please contact West Group Customer Services representative at 1-800-328-4880.

Exhibit 4–87

INDEX

GAME VIOLATIONS
Inspection for, § 10.8(e).

GARAGE
Entry of, § 2.3(d).
Looking in, § 2.3(e).

GARBAGE
Examination of, § 2.6(c).

"GOOD FAITH" EXCEPTION
Application when no warrant, § 1.3(g).
Balancing costs and benefits, § 1.3(b).
Defense to tort action, § 1.10(a).
Legislation authorizing unconstitutional search, § 1.3(h).
Leon decision, § 1.3(a).
Meaning of "good faith," § 1.3(e).
Overstated costs, § 1.3(c).
Proposed qualification of exclusionary rule, § 1.2(d).
Scope in warrant cases, § 1.3(f).
Unconstitutional police regulations and, § 1.3(i).
Understated benefits, § 1.3(d).

GOVERNMENT BENEFITS
Inspections relating to, § 10.3

Source: Search and Seizure, edited by LaFave, 3rd edition, 1996 p. 783, © West Publishing, used with permission.

West's legal encyclopedia, **Corpus Juris Secundum** (C.J.S.), and to the topic and key number, Criminal Law ⊙— 394.4. (See Exhibit 4–88.)

Our specific interest is with state constitutional law, so we also consult a treatise prepared by Barry Latzer, **State Constitutional Criminal Law** (Clark Boardman Callaghan 1995). Under "**Exclusionary Rule**," the index refers us to section 2:10 ("Good faith exception"), and section 2:13 ("*Leon* Rule . . . repudiation of"). (See Exhibit 4–89.) As we would have learned from our other references, the Supreme Court recognized a good faith exception to the exclusionary rule in *United States v. Leon*, 468 U.S. 897, 104 S. Ct. 3405, 82 L. Ed. 2d 677 (1984). At section 2:13, we are provided with a discussion of state court decisions that have rejected the good faith exception, including *State v. Marsala*, 216 Conn. 150, 579 A.2d 58 (1990). (See Exhibit 4–90.) The footnotes offer citations to other judicial decisions and law review articles that promise to be interesting.

Exhibit 4–88

---

§ 1.3 The *Leon* "Good Faith" Exception

Analysis

Subsec.
(a) The *Leon* decision.
(b) Should "costs" and "benefits" be balanced?
(c) Overstated "costs."
(d) Understated "benefits."
(e) "Good faith": generally.
(f) Scope of *Leon* exception in warrant cases.
(g) "Good faith" in without warrant cases.
(h) "Good faith" and unconstitutional legislation.
(i) "Good faith" and unconstitutional police regulations.

———

Library References:

C.J.S. Criminal Law §778.

West's Key No. Digests, Criminal Law ⊙— 394.4.

As noted earlier,[1] in recent years there has been considerable discussion of whether a so-called "good faith" exception to the exclusionary rule should be recognized. There was some reason to believe that a majority of the Supreme Court now favored such a limitation on the exclusionary rule,[2] and thus such a step seemed imminent when the . . .

Source: Search and Seizure, edited by LaFave, 3rd edition, 1996 p. 51, © West Publishing, used with permission.

Exhibit 4–89

INDEX

EXCLUSIONARY RULE
Anti-exclusionary provisions, §2:14
Application to other proceedings, §2:20
Basis for exclusion by states, §2:4
Defendant, suppression of, §2:15
Exceptions to exclusionary rule
    federal exclusionary rule, §2:3
    good faith exception, §2:10
    impeachment exception, §2:19
Federal exclusionary rule
    exceptions, §2:3
    history, §2:2
Fruits of the poisonous tree doctrine, §2:18

Good faith exception, §2:10
Impeachment exception, §2:19
Independent rules
    implication, rules by, §2:7
    states with rules, §2:6
Leon Rule
    questioning of, §2:12
    repudiation of, §2:13
    state approval of, §2:11
Nonstate agents, application to, §2:9
Private citizens, searches by, §2:16
Purposes, §2:8
Scope, §2:8

Source: State Constitutional; Criminal Law, 1995, Latzer p. Ind-3, © 1995, Clark Boardman Callaghan, used with permission.

If we want additional secondary authorities pertaining to the good faith exception, we explore the A.L.R. collection as well as law reviews. For now, however, we discuss how we could conduct our research beginning instead with primary authorities. A logical starting point would be the Connecticut Constitution. We use the **General Index** to **Connecticut General Statutes Annotated** to find the relevant section of the constitution. If we had looked for **"search and seizure"** under **"Constitution,"** we would have been referred to **"Searches and Seizures"** in the General Index. Under "Searches and Seizures," we are immediately directed to **"Const. Art. l, § 7."** (See Exhibit 4–91.)

Article l, section 7 of the Connecticut Constitution protects against unreasonable searches and seizures. (See Exhibit 4–92.) In the Notes of Decisions section of the pocket supplement, we find **"Good faith exception 26.5."** (See Exhibit 4–93.) When we turn to **note 26.5**, we are provided with brief descriptions and citations of two cases, *State v. Marsala* and *State v. Morrissey,* that appear to have rejected the good faith exception to the exclusionary rule on state constitutional grounds. (See Exhibit 4–94.)

We look up *State v. Marsala* at 579 A.2d 58 to confirm that the case holding is consistent with what has been represented in the Notes of Decisions. Three headnotes from the case are categorized under Criminal Law ⚷ 394.4(1). Recall that Professor LaFave's treatise indicated that this topic and key number correspond to the good faith exception to the exclusionary rule. (See Exhibit 4–95.)

Exhibit 4–90

§2:13. Repudiation of *Leon*

Several state supreme courts have unequivocally rejected the good faith rationale as a matter of state constitutional law, the northeastern states being especially hostile.[97]

The reasoning of the Connecticut Supreme Court, rejecting *Leon* in State v Marsala,[98] is representative of state court rationales. *Marsala* involved an affidavit that failed to recite the indicia of reliability or the basis of knowledge of two informants. The Connecticut Supreme Court unanimously reversed a lower court decision to admit some of the evidence on good faith grounds. The high court challenged the empirical assertions made in Justice White's majority opinion in *Leon*. It found that White inflated the costs of exclusion because studies show that proportionately few prosecutions were dismissed as a result of illegal searches and seizures.[99] Moreover, as Justice Brennan, dissenting in *Leon*, noted, a proper measure of the costs of the suppression doctrine would have examined the effects of the rule in cases in which evidence was actually excluded, not in all cases.[1] The costs attributed to the exclusionary rule, the Connecticut court observed, citing former Supreme Court Justice Potter Stewart, . . .

---

97. Conn—State v Marsala (1990) 216 Conn 150, 579 A2d 58.

Idaho—State v Guzman (1992) 122 Idaho 981, 842 P2d 660, reversing State v Prestwich (1989) 116 Idaho 959, 783 P2d 298.

NJ—State v Novembrino (1987) 105 NJ 95, 519 A2d 820.

NY—People v Bigelow (1985) 66 NY2d 417, 497 NYS2d 630, 488 NE2d 451 (declining on state constitutional grounds to apply a good faith exception where a search warrant was defective whether measured by the *Aguilar-Spinelli* or the *Gates* standard. "[I]f the People are permitted to use the seized evidence, the exclusionary rule's purpose is completely frustrated, a premium is placed on the illegal police action and a positive incentive is provided to others to engage in similar lawless acts in the future." Id. at 427.).

Pa—Commonwealth v Edmunds (1991) 526 Pa 374, 586 A2d 887.

Vt—State v Oakes (1991) 157 Vt 171, 598 A2d 119.

See "The Good Faith Exception to the Exclusionary Rule: An Analysis of Kansas Law," 41 University of Kansas Law Review 95 (1993); Casenotes and Comments, "Goodbye Good Faith Doctrine: Constitutional Rights Prevail with the Rejection of the *Leon* Good Faith Exception," 30 Idaho Law Review 159 (1993-94); Layrisson, "The Exclusion of Unconstitutionally Obtained Evidence and Why the Louisiana Supreme Court Should Reject United States v *Leon* on Independent State Grounds," 51 Louisiana Law Review 861 (1991); Notes, "Privacy vs. Practicality: Should Alaska Adopt the *Leon* Good Faith Exception?," 1993 Alaska Law Review 143; Yagla, "The Good Faith Exception to the Exclusionary Rule: The Latest Example of 'New Federalism' in the States," 71 Marquette Law Review 166 (1987).

98. State v Marsala (1990) 216 Conn 150, 579 A2d 58, on remand, remanded 26 Conn App 423, 601 A2d 542.

99. Id. at 64.

1. Id. at 64-65.

Source: State Constitutional; Criminal Law, 1995, Latzer p. Ind-3, p. 2-38, © 1995, Clark Boardman Callaghan, used with permission.

Exhibit 4–91

SEARCHES AND SEIZURES
▶ Generally, Const. Art. 1, §7
Alcoholic beverage tax, nonpayment, 12-454
Alcoholic Beverages, this index
Assaulting person executing warrant, 54-33d

Banks and banking, stock, tender offers or take
over bids, 36a-186
Banned hazardous substances, 21a-340
Bedding manufactured or offered for sale, 21a-235

Source: Connecticut Statutes, Index p. 623, © West Publishing, used with permission.

Exhibit 4–92

§7. Security from searches and seizures

Sec. 7. The people shall be secure in their persons, houses, papers and possessions from unreasonable searches or seizures; and no warrant to search any place, or to seize any person or things, shall issue without describing them as nearly as may be, nor without probable cause supported by oath or affirmation.

Source: Connecticut Statutes, Art. 1, sec. 7, text, © West Publishing, used with permission.

Exhibit 4–93

Notes of Decisions

Abandonment 23.5
Emergency, warrant 10.5
Field sobriety tests 19.2
Garbage 23.6
▶ Good faith exception 26.5

Questions of fact 30
Seizure 24.5
Warrant
  Emergency 10.5
Wiretaps 19.5

Source: Connecticut Statutes, Art. 1, sec. 7, Notes, © West Publishing, used with permission.

We consult different case digests under Criminal Law ⚷ 394.4(1), and related subdivisions to find cases from other jurisdictions relevant to the good faith exception issue. We have to Shepardize *State v. Marsala* in all appropriate volumes of Shepard's before we conclude our research. (See Exhibit 4–96.)

Exhibit 4–94

26.5. Good faith exception

Good faith exception to exclusionary rule does not exist under Connecticut law; exception is incompatible with provision of State Constitution prohibiting unreasonable searches or seizures and requiring probable cause for issuance of warrant. State v. Marsala (1990) 579 A.2d 58, 216 Conn. 150, on remand 601 A.2d 542, 26 Conn.App. 423.

Connecticut Constitution did not authorize good faith exception to exclusionary rule that would permit introduction of evidence at trial that resulted from search conducted, in good faith, on basis of warrant unsupported by probable cause. State v. Morrissey (1990) 577 A.2d 1060, 216 Conn. 185.

Source: Connecticut Statutes, Art. 1, sec. 7, squibs, © West Publishing, used with permission.

Exhibit 4–95

579 ATLANTIC REPORTER, 2d SERIES

216 Conn. 150
150STATE of Connecticut
v.
Michael Joseph MARSALA.
No. 13830.
Supreme Court of Connecticut.
Argued May 1, 1990.
Decided Aug. 7, 1990.

Defendant was convicted in the Superior Court, Judicial District of Fairfield, Ford, J., of selling cocaine and possession of cocaine with intent to sell, and defendant appealed. The Appellate Court, Stoughton, J., 15 Conn.App. 519, 545 A.2d 1151, found error and remanded. On remand, the Trial Court, Ford, J., determined that good-faith exception to exclusionary rule applied, and defendant appealed. The Appellate Court, Spallone, J., 19 Conn.App. 478, 563 A.3d 730, affirmed. Defendant, on granting of certification, appealed. The Supreme Court, Shea, J., held that the good faith exception to the exclusionary rule does not exist under state law.

Reversed and remanded.

1. Criminal Law ☞ 394.4(1)

Good faith exception to exclusionary rule does not exist under Connecticut law; exception is incompatible with provision of State Constitution prohibiting unreasonable searches or seizures and requiring probable cause for issuance of warrant. C.G.S.A. Const. Art. 1, §7.

2. Criminal Law ☞ 394.4(1)

Practice Book section providing that, upon motion of defendant, "the judicial authority shall suppress potential testimony or other evidence if he finds that suppression is required under the Constitution or law of the United States or the state of Connecticut" establishes no substantive standard for suppression of illegally seized evidence but, rather, depends upon applicable interpretation of either United States or State constitution to determine whether evidence should be suppressed. Practice Book 1978, §821; C.G.S.A. Const. Art. 1, §7; U.S.C.A. Const.Amend. 4.

3. Criminal Law ☞ 394.4(1)

Section of general statutes governing motions for return of unlawfully seized property and sup-

continues

Exhibit 4–95 continued

pression of evidence and Practice Book section governing return of seized property are procedural rather than substantive and, therefore, do not define extent of exclusionary rule, notwithstanding legislature's failure to adopt two proposed amendments to general statute section which would have included good faith exception. C.G.S.A. §54-33f; Practice Book 1978, §822.

4. Courts ⚷⟶ **97(1)**

Decisions of United States Supreme Court defining fundamental rights are of persuasive authority to be afforded respectful consideration, but are to be followed by Connecticut courts only when they provide no less individual protection than is guaranteed by Connecticut law.

[151]Richard Emanuel, Asst. Public Defender, with whom was G. Douglas Nash, Public Defender, for appellant (defendant).

Carolyn K. Longstreth, Asst. State's Atty., with whom, on the brief, was Donald A. Browne, State's Atty., for appellee (state).

Before PETERS, C.J., and SHEA, CALLAHAN, GLASS, COVELLO, HULL and SANTANIELLO, JJ.

SHEA, Associate Justice.

[1] The dispositive issue in this appeal is whether evidence seized by police officers in violation of our state constitution may be admitted during a criminal trial, as part of the state's case in chief, under a "good faith" exception to the exclusionary rule. The question comes to us upon certification from a decision rendered by the Appellate Court; *State v. Marsala,* 19 Conn.App. 478, 563 A.2d 730 (1989); affirming the conviction of the defendant, Michael Joseph Marsala, for two violations of the state dependency producing drug . . .

Source: Atlantic Reporter 2d, 579 A2d 58, © West Publishing, used with permission.

Exhibit 4–96

ATLANTIC REPORTER, 2d SERIES					Vol. 579
—58—	s 620A2d1293	f 610A2d[1]1231	j 620A2d756	j 631A2d321	N M
Connecticut v	577A2d[3]1062	j 612A2d1195	625A2d[4]797	f 634A2d[4]1198	819P2d1336
Marsala	579A2d[4]489	613A2d[1]228	626A2d[4]281	Pa	Utah
1990	591A2d[4]124	613A2d1309	j 627A2d928	586A2d[1]900	806P2d742
(216Ct150)	594A2d927	614A2d1234	628A2d[3]576	Vt	810P2d420
s 545A2d1151	j 594A2d932	e 614A2d[1]1236	628A2d[1]579	f 598A2d[1]121	Wyo
s 550A2d1087	594A2d[4]988	614A2d[4]1236	e 628A2d[1]584	Calif	846P2d655
s 563A2d730	f 594A2d[1]1036	j 614A2d1245	j 628A2d587	286CaR790	
s 567A2d836	606A2d5	618A2d540	628A2d1350	818P2d73	
s 601A2d542	j 607A2d370	620A2d147	630A2d[4]1323	Idaho	
s 605A2d866	607A2d443	620A2d[4]749	630A2d1351	842P2d666	

Source: Shepard's Atlantic Reporter 2d, 579 A2d 58. Reproduced by permission of Shepard's. Further reproduction of any kind is strictly prohibited.

Problem 10

The U.S. Supreme Court ruled in *United States v. Leon*, 468 U.S. 897 (1984), that the federal Constitution does not require the suppression of evidence seized by a law enforcement officer who reasonably relies on a search warrant as authority for the search, even if the warrant subsequently is determined to be invalid because it was not supported by probable cause. However, the Court also recognized that the mere issuance of a search warrant does not automatically ensure that evidence seized under the authority of the warrant will be admissible at a criminal trial. In *Leon* the Justices discuss four exceptions to the good faith exception, including (1) where a search warrant application misrepresents the truth, (2) where the magistrate or judge issuing the warrant is not impartial, and (3) where either the affidavit supporting the warrant is so lacking in probable cause or (4) the warrant itself is so facially deficient that a law enforcement officer could not reasonably rely on the validity of the warrant.

Let us assume that you are interested in ascertaining how frequently the exceptions to the good faith exception have been recognized to require the suppression of evidence seized pursuant to the issuance of an invalid search warrant in published opinions following *Leon*. How could you produce information that would help answer this question?

Throughout this text we have extolled the capabilities of CALR for the efficient retrieval of information. The primary advantage of the computer is the availability of key word and key phrase searching of the full text of documents. However, this problem presents us with a question that makes computerized key word searching of the full text of documents a poor choice. Although the issue presented in this example is complicated, we could design a computer search strategy that would retrieve documents containing all of its elements, but we would be hard-pressed to design a strategy that also limits the context in which those elements appear. In other words, we cannot limit the retrieved cases to those in which evidence was suppressed.

We encounter a similar problem if we use the West key number system, either in the printed digests or on WESTLAW, to retrieve relevant opinions. When we look at the Supreme Court Reporter version of *United States v. Leon,* 104 S. Ct. 3405, we see several pages of headnotes with West key numbers. (See Exhibits 4–97, 4–98, and 4–99.) A closer look at these headnotes and at the key numbers assigned to them indicates that several refer to the "exceptions to the good faith exception." Headnotes 8, 9, 10, 18, 20, and 21 all seem to deal with the exceptions we have identified.

We could use the key numbers assigned to these headnotes to search in either the printed digests or on WESTLAW for cases decided subsequent to *Leon*. For example, both headnote 18 and headnote 20 refer to key number **Criminal Law ☞ 394.4(6)**. We could look in the printed digests under that key number to scan blurbs dealing with cases where evidence was challenged, but we probably would still have to look up the actual cases to determine whether the evidence ultimately was suppressed. Alternatively, we could perform key number searches on WESTLAW and then examine the retrieved cases to determine if the evidence was suppressed in each case. Either way, we must eventually consult the case itself to

Exhibit 4–97

| 468 U.S. 897 | UNITED STATES v. LEON | 3405 |

UNITED STATES v. LEON
Cite as 104 S.Ct. 3405 (1984)

468 U.S. 897, 82 L.Ed.2d 677

897 **UNITED STATES, Petitioner**

v.

Alberto Antonio LEON et al.

No. 82-1771.

Argued Jan. 17, 1984.

Decided July 5, 1984.

Rehearing denied Sept. 18, 1984.

See 468 U.S. 1250, 105 S.Ct. 52.

The United States District Court for the Central District of California granted defense motions to suppress evidence. The Court of Appeals for the Ninth Circuit affirmed, 701 F.2d 187. Certiorari was granted. The Supreme Court, Justice White, held that: (1) the Fourth Amendment exclusionary rule should not be applied so as to bar the use in the prosecution's case in chief of evidence obtained by officers acting in reasonable reliance on a search warrant issued by a detached and neutral magistrate but ultimately found to be invalid; (2) standard of reasonableness is an objective one; (3) suppression is appropriate where officers have no reasonable ground for believing that the warrant was properly issued; and (4) officer's reliance on magistrate's determination of probable cause in instant case was objectively reasonable.

Judgment of Court of Appeals reversed.

Justice Blackmun filed concurring opinion.

For dissenting opinion of Justice Brennan, in which Justice Marshall joined, see 104 S.Ct. 3430.

For dissenting opinion of Justice Stevens see 104 S.Ct. 3446.

1. Searches and Seizures ⚷ 3.6(3)

Totality of the circumstances approach is the prevailing test for determining whether an informant's tip suffices to establish probable cause for issuance of a search warrant. U.S.C.A. Const.Amend. 4.

2. Federal Courts ⚷ 461

Although petition for certiorari expressly declined to seek review of determinations that search warrant was unsupported by probable cause and presented only question whether exclusionary rule should be modified in case of good-faith reliance on a search warrant, the Supreme Court had power to consider the probable cause issue and it was also within the court's authority to take the case as it came to it, accepting the Court of Appeal's conclusion that probable cause was lacking under prevailing legal standards. U.S.C.A. Const.Amend. 4; U.S.Sup.Ct.Rule 21.1(a), 28 U.S.C.A.

3. Criminal Law ⚷ 394.4(1)

Use of fruits of a past unlawful search or seizure works no new Fourth Amendment wrong. U.S.C.A. Const.Amend. 4.

4. Witnesses ⚷ 390

Evidence obtained in violation of the Fourth Amendment and inadmissible in the prosecution's case in chief may be used to impeach a defendant's direct testimony. U.S.C.A. Const.Amend. 4.

5. Witnesses ⚷ 390, 406

Evidence inadmissible in the prosecution's case in chief or otherwise as substantive evidence of guilt may be used to impeach statements made by a defendant in response to proper cross-examination reasonably suggested by defendant's direct examination.

6. Criminal Law ⚷ 394.1(3)

Perception underlying determinations that the connection between police conduct and evidence of crime may be sufficiently attenuated to permit use of that evidence at trial is a product of considerations relating to the exclusionary rule and the constitutional principles it is designed to protect. U.S.C.A. Const.Amend. 4.

7. Searches and Seizures ⚷ 3.9

Reasonable minds frequently may differ on the

continues

Exhibit 4–97   continued

question whether a particular search warrant affi-davit establishes probable cause, and preference for warrants is most appropriately effectuated by according great deference to a magistrate's deter-mination. U.S.C.A. Const.Amend. 4.

8. Searches and Seizures ⌐ **3.9**

Deference to a magistrate in search warrant matters is not boundless and deference accorded finding of probable cause does not preclude in-quiry into the knowing . . .

Source: Supreme Court Reporter, 104 S.Ct. 3405, 1st page, © West Publishing, used with permission.

Exhibit 4–98

. . . or reckless falsity of the affidavit on which that determination was based and a magistrate must purport to perform his neutral and detached func-tion and not serve merely as a rubber stamp for the police. U.S.C.A. Const.Amend. 4.

9. Searches and Seizures ⌐ **3.5**

A magistrate failing to manifest that neutrality and detachment demanded of a judicial officer when presented with a search warrant application and who acts instead as an adjunct law enforce-ment officer cannot provide valid authorization for an otherwise unconstitutional search. U.S.C.A. Const.Amend. 4.

10. Searches and Seizures ⌐ **3.9**

Reviewing courts will not defer to a search war-rant based on an affidavit that does not provide the magistrate with a substantial basis for determining

existence of probable cause. U.S.C.A. Const. Amend. 4.

---

18. Criminal Law ⌐ **394.4(5, 6)**

In applying the reasonable reliance on a search warrant exception to the Fourth Amendment ex-clusionary rule, it is necessary to consider the ob-jective reasonableness not only of the officers who eventually execute a warrant but also the officers who originally obtain it or who provide informa-tion material to the probable-cause determination and, hence, an officer cannot obtain a warrant on the basis of a "bare bones" affidavit and then rely on colleagues who are ignorant of the circum-stances under which the warrant is obtained to conduct the search. U.S.C.A. Const.Amend. 4.

Source: Supreme Court Reporter, 104 S.Ct. 3406, © West Publishing, used with permission.

determine whether the court relied on an exception to order the suppression of evidence. This promises to be a laborious process, especially in light of the sev-eral key numbers and the many years to be checked.

Do not despair. If we just take a moment to consider the problem at hand, we find that there is an easier solution. We want cases decided subsequent to *Leon* that deal with the same issues considered in *Leon.* The digests are only one source for locating subsequent cases. Another way to approach Problem 10 is to use the Shepard's series for U.S. Supreme Court decisions.

Exhibit 4–99

20. Criminal Law ☞ 394.4(6)
The reasonable reliance on a warrant exception to Fourth Amendment exclusionary rule does not apply where the issuing magistrate wholly abandoned his judicial role and, in such circumstances, no reasonably well-trained officer should rely on the warrant and an officer does not manifest objective good faith by relying on the warrant based on affidavit so lacking in indicia of probative cause as to render official belief in its existence entirely unreasonable. U.S.C.A. Const.Amend. 4.

21. Criminal Law ☞ 394.4(7)
Searches and Seizures ☞ 3.4, 3.7
Depending on the circumstances, a search warrant may be so facially deficient, i.e., in failing to particularize the place to be searched or the things to be seized, that the executing officers cannot reasonably presume it to be valid. U.S.C.A. Const.Amend. 4.

Source: Supreme Court Reporter, 104 S.Ct. 3407, © West Publishing, used with permission.

Recall that Shepard's is a citation index that provides references to documents that cite the document of interest. In Shepard's case reporter series, the citing cases can be grouped according to at least three criteria: jurisdiction, reference to a particular headnote in the original document, and **treatment code**. The treatment codes may prove most useful to us in this problem, although references to headnote numbers also may be helpful. If we can locate a treatment code that identifies relevant cases, we may have solved the Problem. Treatment codes are listed on the inside front cover of most Shepard's volumes. (See Exhibit 4–100.)

The treatment code **"d" (for "distinguished")** may be promising for Problem 10. This code signifies that "[t]he citing case is different either in law or fact, for reasons given, from the case you are Shepardizing." We could proceed through the printed Shepard's for U.S. Supreme Court cases looking for citing cases preceded by a **"d"**. We further could determine which of the "distinguished" cases makes reference to one of the headnotes that we identified earlier. For example, in Exhibit 4–101, note the reference to 634FS362, which is preceded by the **"d"** and makes reference to headnote **18**.

In Chapter 3, we discussed the advantages of Shepardizing online. We can again demonstrate some of these advantages with this example. Once we have displayed our cited case using WESTLAW, we can Shepardize it by **entering sh.** Then we can focus on citing documents with particular treatment codes and/or references to particular headnotes. We do so by using the locate (**loc**) command, followed by the letter(s) for the treatment, and the number(s) of the headnote(s). For example, the command **loc d,8,9,10,18,20,21** will retrieve only cases that receive the **distinguished** treatment and make reference to one of the identified headnotes. (See Exhibit 4–102.)

Exhibit 4–100

HISTORY AND TREATMENT ABBREVIATIONS

Abbreviations have been assigned, where applicable, to each citing case to indicate the effect the citing case had on the case you are Shepardizing. The resulting "history" (affirmed, reversed, modified, etc.) or "treatment" (followed, criticized, explained, etc.) of the case you are Shepardizing is indicated by abbreviations preceding the citing case reference. For example, the reference "f434F2d872" means that there is language on page 872 of volume 434 of the *Federal Reporter*, Second Series, that indicates the court is "following" the case you are Shepardizing. Instances in which the citing reference occurs in a dissenting opinion are indicated in the same manner. The abbreviations used to reflect both history and treatment are as follows:

History of Case

a	(affirmed)	The decision in the case you are Shepardizing was affirmed or adhered to on appeal.
cc	(connected case)	Identifies a different case from the case you are Shepardizing, but one arising out of the same subject matter or in some manner intimately connected therewith.
D	(dismissed)	An appeal from the case you are Shepardizing was dismissed.
m	(modified)	The decision in the case you are Shepardizing was changed in some way.
p	(parallel)	The citing case is substantially alike or on all fours, either in law or facts, with the case you are Shepardizing.
r	(reversed)	The decision in the case you are Shepardizing was reversed on appeal.
s	(same case)	The case you are Shepardizing involves the same litigation as the citing case, although at a different stage in the proceedings.
S	(superseded)	The citing case decision has been substituted for the decision in the case you are Shepardizing.
US	cert den	Certiorari was denied by the U.S. Supreme Court.
US	cert dis	Certiorari was dismissed by the U.S. Supreme Court.
US	cert gran	Certiorari was granted by the U.S. Supreme Court.
US	reh den	Rehearing was denied by the U.S. Supreme Court.
US	reh dis	Rehearing was dismissed by the U.S. Supreme Court.
v	(vacated)	The decision in the case you are Shepardizing has been vacated.

Treatment of Case

c	(criticized)	The citing case disagrees with the reasoning/decision for the case you are Shepardizing.
d	(distinguished)	The citing case is different either in law or fact, for reasons given, from the case you are Shepardizing.
e	(explained)	The case you are Shepardizing is interpreted in some significant way. Not merely a restatement of facts.
Ex	(Examiner's decision)	The case you are Shepardizing was cited in an Administrative Agency Examiner's Decision.
f	(followed)	The citing case refers to the case you are Shepardizing as controlling authority.

continues

Exhibit 4–100   continued

h	(harmonized)	An apparent inconsistency between the citing case and the case you are Shepardizing is explained and shown not to exist.
j	(dissenting opinion)	The case is cited in a dissenting opinion.
L	(limited)	The citing case refuses to extend the holding of the case you are Shepardizing beyond the precise issues involved.
o	(overruled)	The ruling in the case you are Shepardizing is expressly overruled.
q	(questioned)	The citing case questions the continuing validity of precedential value of the case you are Shepardizing.

Source: Shepard's Cases-treatment. Reproduced by permission of Shepard's. Further reproduction of any kind is strictly prohibited.

Exhibit 4–101

SUPREME COURT REPORTER					Vol. 104
999F2d12	625FS[20]449	635FS[20]574	d 683FS[16]288	f 771FS[161]297	830FS[20]33
d 9F3d216	634FS[16]359	639FS[23]1027	d 699FS[16]982	f 774FS[16]44	f 830FS36
d 594FS[161]238	d 634FS[17]362	652FS[161]359	704FS[131]173	f 778FS101	Cir. 2
594FS1247	d 634FS[18]362	656FS[23]276	715FS[14]32	f 795FS[161]233	745F2d[16]758
610FS[2]375	d 634FS[20]362	665FS[20]974	771FS[19]1263	805FS1042	745F2d[23]758
d 610FS[18]377	d 634FS[21]362				

Source: Shepard's Supreme Court Reporter 104 SCT 3405. Reproduced by permission of Shepard's. Further reproduction of any kind is strictly prohibited.

We can consult any of the cited cases to verify whether the evidence in fact was suppressed. (See Exhibit 4–103.) If we check these cases and determine that they meet our criteria, then we are on our way to solving this problem. You might want to fine-tune the search strategy to include other treatment codes or headnotes, or you might want to rely exclusively on the treatment code **d** without any headnote references. You should experiment to determine if you can improve on our search strategy.

Exhibit 4–102

FOR EDUCATIONAL USE ONLY   SHEPARD'S   Page 1 of 3
Citations to: 104 S.Ct. 3405
United States v Leon 1984

Coverage: > View coverage information for this result
Located: D, 8, 9, 10, 18, 20, 21

Retrieval No.	– – – Analysis – – –		– – – – Citation – – – Cir. 1	Headnote No.
1	D	Distinguished	841 F.2d 1, 4	18
2	D	Distinguished	867 F.2d 36, 44	20
3	D	Distinguished	610 F.Supp. 371, 377	18
4	D	Distinguished	634 F.Supp. 358, 362	18
5	D	Distinguished	634 F.Supp. 358, 362	20
6	D	Distinguished	634 F.Supp. 358, 362	21
			Cir. 2	
7	D	Distinguished	975 F.2d 72, 77	20

Source: Shepard's 104 SCT 3405. Reproduced by permission of Shepard's. Further reproduction of any kind is strictly prohibited.

Exhibit 4–103

(Cite as: 841 F.2d 1)

UNITED STATES of America, Appellant,

v.

Leoncio L. DIAZ, a/k/a Leonel Diaz, et al., Defendants, Appellees.

No. 87-1341.

United States Court of Appeals,

First Circuit.

Heard Oct. 9, 1987.

Decided Feb. 29, 1988.

Rehearing En Banc Denied April 28, 1988.

Defendants were charged with wire fraud and bribery, and they moved to suppress records seized pursuant to search warrant. The United States District Court for the District of Puerto Rico, > 656 F.Supp. 271, Jaime Pieras, Jr., J., granted motion. Government appealed. The Court of Appeals, Torruella, Circuit Judge, held that: (1) probable cause supported seizure of some records, but not others; (2) search warrant was overbroad; and (3) warrant was not so deficient that agent still reasonably and in good faith believed that it adequately authorized search.

Affirmed in part, reversed in part, and remanded.

(Cite as: 841 F.2d 1, *4)

to any product," was warranted as likely to produce evidence of criminal activity. The accounts payable records would also indicate the amounts purchased of different types of produce. Similarly, all cash tickets and records of cash funds are relevant because the government believed that IRSI was paying off Matos and certain employees from the cash funds it maintained on the premises, and was depositing cash proceeds of the fraud into these funds. Finally, the cash receipts journal is relevant since the affidavit states that "produce claimed as damaged are sold for cash."

> [3] A problem remains as to one of the categories listed, however. There is nothing in the affidavit to imply that any evidence of crime could be gleaned from "all bank account records." In fact, the affidavit seems to establish that the fraudulent transactions were all cash transactions, presumably to minimize the risk of discovery. [FN1]

> FN1. Since there is a complete lack of justification for the seizure of these documents in view of Special Agent Coffey's own description of the scheme, we find that the good faith exception to the exclusionary rule, set forth in the United States v Leon, > 468 U.S. 897, 104 S. Ct. 3405, 82 L.Ed.2d 677 (1984) and discussed more fully below, is inapplicable here.

Source: CTA, 841 F2d 1, © WESTLAW, used with permission.

CONCLUSION

We conclude this chapter much as we began. Remember that how we have chosen to tackle the research problems is not the only way they could have been completed, may not be the best way, and might not have been the way that you would have found most efficient. Legal research is an interactive process. It depends on the researcher, the research problem, and the references available to solve the problem. It is a sufficiently creative and serendipitous exercise that we should expect and welcome different approaches. There is no one formula for success; rather, there are multiple opportunities for it. You simply must understand what options are available, acquire some familiarity with the references and computer commands, and be willing to devote some time and energy to your research task.

The process of learning is interactive. It depends on information or substantive knowledge, the methods or tools for acquiring knowledge and building on its base, and the talents and work habits of the involved individual. We have no doubt that the ability to pose questions and understand how to go about finding answers to them is far more valuable in the long run than is possessing a stockpile of information. As the familiar adage provides: Give people fish, and they will not go hungry today; but teach them how to fish, and they will never go hungry again. Legal research skills are your ticket to a continuing education. Practice them. Enjoy their challenges. Teach them to others. They are necessary if you are to be able to find answers to what you do not know, and to keep in touch with how the law is changing. Like riding a bicycle or learning to swim, once you get the hang of legal research techniques, you are not likely to forget them. They can be an enormous asset to your studies, your career, and the depth of your understanding about law and its processes.

Index